RECURRENT NEURAL NETWORKS FOR PREDICTION

WILEY SERIES IN ADAPTIVE AND LEARNING SYSTEMS FOR SIGNAL PROCESSING, COMMUNICATIONS, AND CONTROL

Editor: Simon Haykin

Beckerman/ADAPTIVE COOPERATIVE SYSTEMS

Chen and Gu/CONTROL-ORIENTED SYSTEM IDENTIFICATION: An H Approach

Cherkassky and Mulier/LEARNING FROM DATA: Concepts, Theory and Methods

Diamantaras and Kung/PRINCIPAL COMPONENT NEURAL NETWORKS: Theory and Applications

Haykin and Puthusserypady/CHAOTIC DYNAMICS OF SEA CLUTTER

Haykin/NONLINEAR DYNAMICAL SYSTEMS: Feedforward Neural Network Perspectives

Haykin/UNSUPERVISED ADAPTIVE FILTERING, VOLUME I: Blind Source Separation

Haykin/UNSUPERVISED ADAPTIVE FILTERING, VOLUME II: Blind Deconvolution

Hines/FUZZY AND NEURAL APPROACHES IN ENGINEERING

Hrycej/NEUROCONTROL: Towards an Industrial Control Methodology

Krstic, Kanellakopoulos, and Kokotovic/NONLINEAR AND ADAPTIVE CONTROL DESIGN

Mann/INTELLIGENT IMAGE PROCESSING

Nikias and Shao/SIGNAL PROCESSING WITH ALPHA-STABLE DISTRIBUTIONS AND APPLICATIONS

Passino and Burgess/STABILITY ANALYSIS OF DISCRETE EVENT SYSTEMS

Sanchez-Peña and Sznaier/ROBUST SYSTEMS THEORY AND APPLICATIONS

Tao and Kokotovic/ADAPTIVE CONTROL OF SYSTEMS WITH ACTUATOR AND SENSOR NONLINEARITIES

Van Hulle/FAITHFUL REPRESENTATIONS AND TOPOGRAPHIC MAPS: From Distortion- to Information-Based Self-Organization

Vapnik/STATISTICAL LEARNING THEORY

Werbos/THE ROOTS OF BACKPROPAGATION: From Ordered Derivatives to Neural Networks and Political Forecasting

Yee and Haykin/REGULARIZED RADIAL-BASIS FUNCTION NETWORKS: Theory and Applications

RECURRENT NEURAL NETWORKS FOR PREDICTION

LEARNING ALGORITHMS, ARCHITECTURES AND STABILITY

Danilo P. Mandic
School of Information Systems,
University of East Anglia, UK

Jonathon A. Chambers
Department of Electronic and Electrical Engineering,
University of Bath, UK

JOHN WILEY & SONS, LTD
Chichester • New York • Weinheim • Brisbane • Singapore • Toronto

Baffins Lane, Chichester,
West Sussex, PO19 1UD, England

National 01243 779777
International (+44) 1243 779777

e-mail (for orders and customer service enquiries): cs-books@wiley.co.uk

Visit our Home Page on http://www.wiley.co.uk or http://www.wiley.com

Other Wiley Editorial Offices

John Wiley & Sons, Inc., New York, USA

WILEY-VCH Verlag GmbH, Weinheim, Germany

John Wiley & Sons Australia, Ltd, Queensland

John Wiley & Sons (Canada) Ltd, Ontario

John Wiley & Sons (Asia) Pte Ltd, Singapore

Library of Congress Cataloging-in-Publication Data

Mandic, Danilo P.
Recurrent neural networks for prediction : learning algorithms, architectures, and stability / Danilo P. Mandic, Jonathon A. Chambers.
p. cm -- (Wiley series in adaptive and learning systems for signal processing, communications, and control)
Includes bibliographical references and index.
ISBN 0-471-49517-4 (alk. paper)
1. Machine learning. 2. Neural networks (Computer science) I. Chambers, Jonathon A. II. Title. III. Adaptive and learning systems for signal processing, communications, and control.

Q325.5 .M36 2001
006.3′2--dc21 2001033418

British Library Cataloguing in Publication Data

A catalogue record for this book is available from the British Library

ISBN 0-471-49517-4

Produced from LaTeX files supplied by the author, typeset by T&T Productions Ltd, London.
Printed and bound by CPI Antony Rowe, Eastbourne

This book is printed on acid-free paper responsibly manufactured from sustainable forestry, in which at least two trees are planted for each one used for paper production.

To our students and families

Contents

Preface

New technologies in engineering, physics and biomedicine are creating problems in which nonstationarity, nonlinearity, uncertainty and complexity play a major role. Solutions to many of these problems require the use of nonlinear processors, among which neural networks are one of the most powerful. Neural networks are appealing because they learn by example and are strongly supported by statistical and optimisation theories. They not only complement conventional signal processing techniques, but also emerge as a convenient alternative to expand signal processing horizons.

The use of recurrent neural networks as identifiers and predictors in nonlinear dynamical systems has increased significantly. They can exhibit a wide range of dynamics, due to feedback, and are also tractable nonlinear maps.

In our work, neural network models are considered as massively interconnected nonlinear adaptive filters. The emphasis is on dynamics, stability and spatio-temporal behaviour of recurrent architectures and algorithms for prediction. However, wherever possible the material has been presented starting from feedforward networks and building up to the recurrent case.

Our objective is to offer an accessible self-contained research monograph which can also be used as a graduate text. The material presented in the book is of interest to a wide population of researchers working in engineering, computing, science, finance and biosciences. So that the topics are self-contained, we assume familiarity with the basic concepts of analysis and linear algebra. The material presented in Chapters 1–6 can serve as an advanced text for courses on neural adaptive systems. The book encompasses traditional and advanced learning algorithms and architectures for recurrent neural networks. Although we emphasise the problem of time series prediction, the results are applicable to a wide range of problems, including other signal processing configurations such as system identification, noise cancellation and inverse system modelling. We harmonise the concepts of learning algorithms, embedded systems, representation of memory, neural network architectures and causal–noncausal dealing with time. A special emphasis is given to stability of algorithms – a key issue in real-time applications of adaptive systems.

This book has emerged from the research that D. Mandic has undertaken while at Imperial College of Science, Technology and Medicine, London, UK. The work was continued within the vibrant culture of the University of East Anglia, Norwich, UK.

Acknowledgements

Danilo Mandic acknowledges Dr M. Razaz for providing a home from home in the Bioinformatics Laboratory, School of Information Systems, University of East Anglia. Many thanks to the people from the lab for creating a congenial atmosphere at work. The Dean of the School of Information Systems, Professor V. Rayward-Smith and his predecessor Dr J. Glauert, deserve thanks for their encouragement and support. Dr M. Bozic has done a tremendous job on proofreading the mathematics. Dr W. Sherliker has contributed greatly to Chapter 10. Dr D. I. Kim has proofread the mathematically involved chapters. I thank Dr G. Cawley, Dr M. Dzamonja, Dr A. James and Dr G. Smith for proofreading the manuscript in its various phases. Dr R. Harvey has been of great help throughout. Special thanks to my research associates I. Krcmar and Dr R. Foxall for their help with some of the experimental results. H. Graham has always been at hand with regard to computing problems. Many of the results presented here have been achieved while I was at Imperial College, where I greatly benefited from the unique research atmosphere in the Signal Processing Section of the Department of Electrical and Electronic Engineering.

Jonathon Chambers acknowledges the outstanding PhD researchers with whom he has had the opportunity to interact, they have helped so much towards his orientation in adaptive signal processing. He also acknowledges Professor P. Watson, Head of the Department of Electronic and Electrical Engineering, University of Bath, who has provided the opportunity to work on the book during its later stages. Finally, he thanks Mr D. M. Brookes and Dr P. A. Naylor, his former colleagues, for their collaboration in research projects.

Danilo Mandic
Jonathon Chambers

List of Abbreviations

ACF	Autocorrelation function
AIC	Akaike Information Criterion
ANN	Artificial Neural Network
AR	Autoregressive
ARIMA	Autoregressive Integrated Moving Average
ARMA	Autoregressive Moving Average
ART	Adaptive Resonance Theory
AS	Asymptotic Stability
ATM	Asynchronous Transfer Mode
BIC	Bayesian Information Criterion
BC	Before Christ
BIBO	Bounded Input Bounded Output
BP	Backpropagation
BPTT	Backpropagation Through Time
CM	Contraction Mapping
CMT	Contraction Mapping Theorem
CNN	Cellular Neural Network
DC	Direct Current
DR	Data Reusing
DSP	Digital Signal Processing
DVS	Deterministic Versus Stochastic
ECG	Electrocardiagram
EKF	Extended Kalman Filter
ERLS	Extended Recursive Least Squares
ES	Exponential Stability
FCRNN	Fully Connected Recurrent Neural Network
FFNN	Feedforward Neural Network
FIR	Finite Impulse Response
FPI	Fixed Point Iteration
GAS	Global Asymptotic Stability
GD	Gradient Descent
HOS	Higher-Order Statistics
HRV	Heart Rate Variability
i.i.d.	Independent Identically Distributed
IIR	Infinite Impulse Response
IVT	Intermediate Value Theorem
KF	Kalman Filter

LMS	Least Mean Square
LPC	Linear Predictive Coding
LRGF	Locally Recurrent Globally Feedforward
LUT	Look-Up Table
MA	Moving Average
MLP	Multi-Layer Perceptron
MMSE	Minimum Mean Square Error
MPEG	Moving Pictures Experts Group
MSLC	Multiple Sidelobe Cancellation
MSE	Mean Squared Error
MVT	Mean Value Theorem
NARMA	Nonlinear Autoregressive Moving Average
NARMAX	Nonlinear Autoregressive Moving Average with eXogenous input
NARX	Nonlinear Autoregressive with eXogenous input
NGD	Nonlinear Gradient Descent
NNGD	Normalised Nonlinear Gradient Descent
NLMS	Normalised Least Mean Square
NMA	Nonlinear Moving Average
NN	Neural Network
NO_2	Nitrogen dioxide
NRTRL	Normalised RTRL algorithm
pdf	probability density function
PG	Prediction Gain
PRNN	Pipelined Recurrent Neural Network
PSD	Power Spectral Density
RAM	Random Access Memory
RBF	Radial Basis Function
RBP	Recurrent Backpropagation
RLS	Recursive Least Squares
RNN	Recurrent Neural Network
ROM	Read-Only Memory
R–R	Distance between two consecutive R waves in ECG
RTRL	Real-Time Recurrent Learning
SG	Stochastic Gradient
SP	Signal Processing
VLSI	Very Large Scale Integration
WGN	White Gaussian Noise
WWW	World Wide Web

Mathematical Notation

z'	First derivative of variable z
$\{\cdot\}$	Set of elements
α	Momentum constant
β	Slope of the nonlinear activation function Φ
γ	Contraction constant, gain of the activation function
Γ	Modular group of compositions of Möbius transformations
$\boldsymbol{\Gamma}(k)$	Adaptation gain vector
δ_i	Gradient at ith neuron
δ_{ij}	Kronecker delta function
$\epsilon(k)$	Additive noise
η	Learning rate
$\nabla_x Y$	Gradient of Y with respect to x
$\lvert\cdot\rvert$	Modulus operator
$\lVert\cdot\rVert_p$	Vector or matrix p-norm operator
$*$	Convolution operator
$\propto$	Proportional to
$\mathbf{0}$	Null vector
λ	Forgetting factor
λ_i	Eigenvalues of a matrix
μ	Step size in the LMS algorithm
μ	Mean of a random variable
$\nu(k)$	Additive noise
Ω_k	Set of nearest neighbours vectors
$\lvert\Omega_k\rvert$	Number of elements in set Ω_k
$\pi^j_{n,l}$	Sensitivity of the jth neuron to the change in $w_{n,l}$
Φ	Nonlinear activation function of a neuron
Φ'_x	First derivative of Φ with respect to x
$\boldsymbol{\Pi}$	Matrix of gradients at the output neuron of an RNN
∂	Partial derivative operator
$\sum$	Summation operator
σ	Variance
$\sigma(x)$	General sigmoid function
$\boldsymbol{\Sigma}(k)$	Sample input matrix
τ	Delay operator
$\boldsymbol{\Theta}$	General nonlinear function
$\boldsymbol{\Theta}(k)$	Parameter vector

$(\cdot)^{\mathrm{T}}$	Vector or matrix transpose
$\mathcal{C}_{\mathrm{L}}(\cdot)$	Computational load
B	Backshift operator
$\mathbb{C}$	Set of complex numbers
$C^n(a,b)$	The class of n-times continuously differentiable functions on an open interval (a,b)
$d(k)$	Desired response
$\deg(\cdot)$	Degree of a polynomial
$\mathrm{diag}[\cdot]$	Diagonal elements of a matrix
$E(k)$	Cost function
$E[\cdot]$	Expectation operator
$E[y \mid x]$	Conditional expectation operator
$e(k)$	Instantaneous prediction error
$\bar{e}(k)$	*A posteriori* prediction error
$F(\cdot,\cdot)$	Nonlinear approximation function
$G(\cdot)$	Basis function
$\boldsymbol{I}$	Identity matrix
$\inf$	Infimum
$J(k)$	Cost function
$\mathcal{J}$	Non-negative error measure
$\boldsymbol{J}$	Jacobian
k	Discrete time index
$\boldsymbol{H}$	Hessian matrix
$H(z)$	Transfer function in z-domain
H^{∞}	The infinity norm quadratic optimisation
L	Lipschitz constant
$\mathcal{L}_p$	$\mathcal{L}_p$ norm
$\lim$	Limit
$\max$	Maximum
$\min$	Minimum
$\mathbb{N}$	Set of natural numbers
$\mathcal{N}(\mu,\sigma^2)$	Normal random process with mean μ and variance σ^2
$\mathcal{O}(\cdot)$	Order of computational complexity
$P(f)$	Power spectral density
$q(k)$	Measurement noise
$\mathbb{R}$	Set of real numbers
$\mathbb{R}^+$	$\{x \in \mathbb{R} \mid x > 0\}$
$\mathbb{R}^n$	The n-dimensional Euclidean space
R_{p}	Prediction gain
$\boldsymbol{R}_{x,y}$, $\boldsymbol{R}_{\boldsymbol{x},\boldsymbol{y}}$	Correlation matrix between vectors $\boldsymbol{x}$ and $\boldsymbol{y}$
S	Class of sigmoidal functions
$\mathrm{sgn}(\cdot)$	Signum function
$\mathrm{span}(\cdot)$	Span of a vector
$\sup$	Supremum
$\boldsymbol{T}$	Contraction operator
$\mathrm{tr}\{\cdot\}$	Trace operator (sum of diagonal elements of a matrix)

$\boldsymbol{u}(k)$	Input vector to an RNN
v	Internal activation potential of a neuron
$v(k)$	System noise
$\boldsymbol{w}(k)$	Weight vector
$\Delta\boldsymbol{w}(k)$	Correction to the weight vector
$\tilde{\boldsymbol{w}}(k)$	Optimal weight vector
$\breve{\boldsymbol{w}}(k)$	Weight error vector
$\hat{\boldsymbol{w}}(k)$	Weight vector estimate
$w_{i,j}$	Weight connecting neuron j to neuron i
$\boldsymbol{W}(k)$	Weight matrix of an NN
$\Delta\boldsymbol{W}(k)$	Correction to the weight matrix
$\boldsymbol{x}(k), \boldsymbol{X}(k)$	External input vector to an NN
$x_l(k) \in \boldsymbol{X}(k)$	The lth element of vector $\boldsymbol{X}(k)$
$\hat{x}$	Estimated value of x
x^*	Fixed point of the sequence $\{x\}$
y	Output of an NN
$y_{i,j}$	Output of the jth neuron of the ith module of the PRNN
z^{-k}	The kth-order time delay
$\mathcal{Z}$	The $\mathcal{Z}$ transform
$\mathcal{Z}^{-1}$	The inverse $\mathcal{Z}$ transform

1

Introduction

Artificial neural network (ANN) models have been extensively studied with the aim of achieving human-like performance, especially in the field of pattern recognition. These networks are composed of a number of nonlinear computational elements which operate in parallel and are arranged in a manner reminiscent of biological neural interconnections. ANNs are known by many names such as connectionist models, parallel distributed processing models and neuromorphic systems (Lippmann 1987). The origin of connectionist ideas can be traced back to the Greek philosopher, Aristotle, and his ideas of mental associations. He proposed some of the basic concepts such as that memory is composed of simple elements connected to each other via a number of different mechanisms (Medler 1998).

While early work in ANNs used anthropomorphic arguments to introduce the methods and models used, today neural networks used in engineering are related to algorithms and computation and do not question how brains might work (Hunt *et al.* 1992). For instance, recurrent neural networks have been attractive to physicists due to their isomorphism to spin glass systems (Ermentrout 1998). The following properties of neural networks make them important in signal processing (Hunt *et al.* 1992): they are nonlinear systems; they enable parallel distributed processing; they can be implemented in VLSI technology; they provide learning, adaptation and data fusion of both qualitative (symbolic data from artificial intelligence) and quantitative (from engineering) data; they realise multivariable systems.

The area of neural networks is nowadays considered from two main perspectives. The first perspective is cognitive science, which is an interdisciplinary study of the mind. The second perspective is connectionism, which is a theory of information processing (Medler 1998). The neural networks in this work are approached from an engineering perspective, i.e. to make networks efficient in terms of topology, learning algorithms, ability to approximate functions and capture dynamics of time-varying systems. From the perspective of connection patterns, neural networks can be grouped into two categories: feedforward networks, in which graphs have no loops, and recurrent networks, where loops occur because of feedback connections. Feedforward networks are static, that is, a given input can produce only one set of outputs, and hence carry no memory. In contrast, recurrent network architectures enable the information to be temporally memorised in the networks (Kung and Hwang 1998). Based on training by example, with strong support of statistical and optimisation theories

(Cichocki and Unbehauen 1993; Zhang and Constantinides 1992), neural networks are becoming one of the most powerful and appealing nonlinear signal processors for a variety of signal processing applications. As such, neural networks expand signal processing horizons (Chen 1997; Haykin 1996b), and can be considered as massively interconnected nonlinear adaptive filters. Our emphasis will be on dynamics of recurrent architectures and algorithms for prediction.

1.1 Some Important Dates in the History of Connectionism

In the early 1940s the pioneers of the field, McCulloch and Pitts, studied the potential of the interconnection of a model of a neuron. They proposed a computational model based on a simple neuron-like element (McCulloch and Pitts 1943). Others, like Hebb were concerned with the adaptation laws involved in neural systems. In 1949 Donald Hebb devised a learning rule for adapting the connections within artificial neurons (Hebb 1949). A period of early activity extends up to the 1960s with the work of Rosenblatt (1962) and Widrow and Hoff (1960). In 1958, Rosenblatt coined the name 'perceptron'. Based upon the perceptron (Rosenblatt 1958), he developed the theory of statistical separability. The next major development is the new formulation of learning rules by Widrow and Hoff in their Adaline (Widrow and Hoff 1960). In 1969, Minsky and Papert (1969) provided a rigorous analysis of the perceptron. The work of Grossberg in 1976 was based on biological and psychological evidence. He proposed several new architectures of nonlinear dynamical systems (Grossberg 1974) and introduced adaptive resonance theory (ART), which is a real-time ANN that performs supervised and unsupervised learning of categories, pattern classification and prediction. In 1982 Hopfield pointed out that neural networks with certain symmetries are analogues to spin glasses.

A seminal book on ANNs is by Rumelhart *et al.* (1986). Fukushima explored competitive learning in his biologically inspired Cognitron and Neocognitron (Fukushima 1975; Widrow and Lehr 1990). In 1971 Werbos developed a backpropagation learning algorithm which he published in his doctoral thesis (Werbos 1974). Rumelhart *et al.* rediscovered this technique in 1986 (Rumelhart *et al.* 1986). Kohonen (1982), introduced self-organised maps for pattern recognition (Burr 1993).

1.2 The Structure of Neural Networks

In neural networks, computational models or nodes are connected through weights that are adapted during use to improve performance. The main idea is to achieve good performance via dense interconnection of simple computational elements. The simplest node provides a linear combination of N weights $w_1, \dots, w_N$ and N inputs $x_1, \dots, x_N$, and passes the result through a nonlinearity Φ, as shown in Figure 1.1.

Models of neural networks are specified by the net topology, node characteristics and training or learning rules. From the perspective of connection patterns, neural networks can be grouped into two categories: feedforward networks, in which graphs have no loops, and recurrent networks, where loops occur because of feedback connections. Neural networks are specified by (Tsoi and Back 1997)

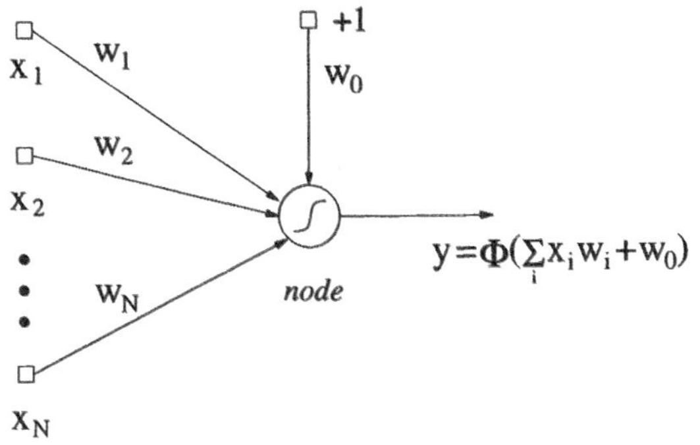

Figure 1.1 Connections within a node

- Node: typically a sigmoid function;
- Layer: a set of nodes at the same hierarchical level;
- Connection: constant weights or weights as a linear dynamical system, feedforward or recurrent;
- Architecture: an arrangement of interconnected neurons;
- Mode of operation: analogue or digital.

Massively interconnected neural nets provide a greater degree of robustness or fault tolerance than sequential machines. By robustness we mean that small perturbations in parameters will also result in small deviations of the values of the signals from their nominal values.

In our work, hence, the term *neuron* will refer to an operator which performs the mapping

$$\text{Neuron:} \quad \mathbb{R}^{N+1} \rightarrow \mathbb{R} \tag{1.1}$$

as shown in Figure 1.1. The equation

$$y = \Phi\left(\sum_{i=1}^{N} w_i x_i + w_0\right) \tag{1.2}$$

represents a mathematical description of a neuron. The input vector is given by $\boldsymbol{x} = [x_1, \dots, x_N, 1]^{\mathrm{T}}$, whereas $\boldsymbol{w} = [w_1, \dots, w_N, w_0]^{\mathrm{T}}$ is referred to as the weight vector of a neuron. The weight w_0 is the weight which corresponds to the bias input, which is typically set to unity. The function $\Phi : \mathbb{R} \rightarrow (0, 1)$ is monotone and continuous, most commonly of a sigmoid shape. A set of interconnected neurons is a neural network (NN). If there are N input elements to an NN and M output elements of an NN, then an NN defines a continuous mapping

$$\text{NN:} \quad \mathbb{R}^{N} \rightarrow \mathbb{R}^{M}. \tag{1.3}$$

1.3 Perspective

Before the 1920s, prediction was undertaken by simply extrapolating the time series through a global fit procedure. The beginning of modern time series prediction was in 1927 when Yule introduced the autoregressive model in order to predict the annual number of sunspots. For the next half century the models considered were linear, typically driven by white noise. In the 1980s, the state-space representation and machine learning, typically by neural networks, emerged as new potential models for prediction of highly complex, nonlinear and nonstationary phenomena. This was the shift from rule-based models to data-driven methods (Gershenfeld and Weigend 1993).

Time series prediction has traditionally been performed by the use of linear parametric autoregressive (AR), moving-average (MA) or autoregressive moving-average (ARMA) models (Box and Jenkins 1976; Ljung and Soderstrom 1983; Makhoul 1975), the parameters of which are estimated either in a block or a sequential manner with the least mean square (LMS) or recursive least-squares (RLS) algorithms (Haykin 1994). An obvious problem is that these processors are linear and are not able to cope with certain nonstationary signals, and signals whose mathematical model is not linear. On the other hand, neural networks are powerful when applied to problems whose solutions require knowledge which is difficult to specify, but for which there is an abundance of examples (Dillon and Manikopoulos 1991; Gent and Sheppard 1992; Townshend 1991). As time series prediction is conventionally performed entirely by inference of future behaviour from examples of past behaviour, it is a suitable application for a neural network predictor. The neural network approach to time series prediction is non-parametric in the sense that it does not need to know any information regarding the process that generates the signal. For instance, the order and parameters of an AR or ARMA process are not needed in order to carry out the prediction. This task is carried out by a process of learning from examples presented to the network and changing network weights in response to the output error.

Li (1992) has shown that the recurrent neural network (RNN) with a sufficiently large number of neurons is a realisation of the nonlinear ARMA (NARMA) process. RNNs performing NARMA prediction have traditionally been trained by the real-time recurrent learning (RTRL) algorithm (Williams and Zipser 1989a) which provides the training process of the RNN 'on the run'. However, for a complex physical process, some difficulties encountered by RNNs such as the high degree of approximation involved in the RTRL algorithm for a high-order MA part of the underlying NARMA process, high computational complexity of $\mathcal{O}(N^4)$, with N being the number of neurons in the RNN, insufficient degree of nonlinearity involved, and relatively low robustness, induced a search for some other, more suitable schemes for RNN-based predictors.

In addition, in time series prediction of nonlinear and nonstationary signals, there is a need to learn long-time temporal dependencies. This is rather difficult with conventional RNNs because of the problem of vanishing gradient (Bengio *et al.* 1994). A solution to that problem might be NARMA models and nonlinear autoregressive moving average models with exogenous inputs (NARMAX) (Siegelmann *et al.* 1997) realised by recurrent neural networks. However, the quality of performance is highly dependent on the order of the AR and MA parts in the NARMAX model.

The main reasons for using neural networks for prediction rather than classical time series analysis are (Wu 1995)

- they are computationally at least as fast, if not faster, than most available statistical techniques;
- they are self-monitoring (i.e. they learn how to make accurate predictions);
- they are as accurate if not more accurate than most of the available statistical techniques;
- they provide iterative forecasts;
- they are able to cope with nonlinearity and nonstationarity of input processes;
- they offer both parametric and nonparametric prediction.

1.4 Neural Networks for Prediction: Perspective

Many signals are generated from an inherently nonlinear physical mechanism and have statistically non-stationary properties, a classic example of which is speech. Linear structure adaptive filters are suitable for the nonstationary characteristics of such signals, but they do not account for nonlinearity and associated higher-order statistics (Shynk 1989). Adaptive techniques which recognise the nonlinear nature of the signal should therefore outperform traditional linear adaptive filtering techniques (Haykin 1996a; Kay 1993). The classic approach to time series prediction is to undertake an analysis of the time series data, which includes modelling, identification of the model and model parameter estimation phases (Makhoul 1975). The design may be iterated by measuring the closeness of the model to the real data. This can be a long process, often involving the derivation, implementation and refinement of a number of models before one with appropriate characteristics is found.

In particular, the most difficult systems to predict are

- those with non-stationary dynamics, where the underlying behaviour varies with time, a typical example of which is speech production;
- those which deal with physical data which are subject to noise and experimentation error, such as biomedical signals;
- those which deal with short time series, providing few data points on which to conduct the analysis, such as heart rate signals, chaotic signals and meteorological signals.

In all these situations, traditional techniques are severely limited and alternative techniques must be found (Bengio 1995; Haykin and Li 1995; Li and Haykin 1993; Niranjan and Kadirkamanathan 1991).

On the other hand, neural networks are powerful when applied to problems whose solutions require knowledge which is difficult to specify, but for which there is an abundance of examples (Dillon and Manikopoulos 1991; Gent and Sheppard 1992; Townshend 1991). From a system theoretic point of view, neural networks can be considered as a conveniently parametrised class of nonlinear maps (Narendra 1996).

There has been a recent resurgence in the field of ANNs caused by new net topologies, VLSI computational algorithms and the introduction of massive parallelism into neural networks. As such, they are both universal function approximators (Cybenko 1989; Hornik *et al.* 1989) and arbitrary pattern classifiers. From the Weierstrass Theorem, it is known that polynomials, and many other approximation schemes, can approximate arbitrarily well a continuous function. Kolmogorov's theorem (a negative solution of Hilbert's 13th problem (Lorentz 1976)) states that any continuous function can be approximated using only linear summations and nonlinear but continuously increasing functions of only one variable. This makes neural networks suitable for universal approximation, and hence prediction. Although sometimes computationally demanding (Williams and Zipser 1995), neural networks have found their place in the area of nonlinear autoregressive moving average (NARMA) (Bailer-Jones *et al.* 1998; Connor *et al.* 1992; Lin *et al.* 1996) prediction applications. Comprehensive survey papers on the use and role of ANNs can be found in Widrow and Lehr (1990), Lippmann (1987), Medler (1998), Ermentrout (1998), Hunt *et al.* (1992) and Billings (1980).

Only recently, neural networks have been considered for prediction. A recent competition by the Santa Fe Institute for Studies in the Science of Complexity (1991–1993) (Weigend and Gershenfeld 1994) showed that neural networks can outperform conventional linear predictors in a number of applications (Waibel *et al.* 1989). In journals, there has been an ever increasing interest in applying neural networks. A most comprehensive issue on recurrent neural networks is the issue of the *IEEE Transactions of Neural Networks*, vol. 5, no. 2, March 1994. In the signal processing community, there has been a recent special issue 'Neural Networks for Signal Processing' of the *IEEE Transactions on Signal Processing*, vol. 45, no. 11, November 1997, and also the issue 'Intelligent Signal Processing' of the *Proceedings of IEEE*, vol. 86, no. 11, November 1998, both dedicated to the use of neural networks in signal processing applications.

Figure 1.2 shows the frequency of the appearance of articles on recurrent neural networks in common citation index databases. Figure 1.2(a) shows number of journal and conference articles on recurrent neural networks in IEE/IEEE publications between 1988 and 1999. The data were gathered using the IEL Online service, and these publications are mainly periodicals and conferences in electronics engineering. Figure 1.2(b) shows the frequency of appearance for BIDS/ATHENS database, between 1988 and 2000,[1] which also includes non-engineering publications. From Figure 1.2, there is a clear growing trend in the frequency of appearance of articles on recurrent neural networks. Therefore, we felt that there was a need for a research monograph that would cover a part of the area with up to date ideas and results.

1.5 Structure of the Book

The book is divided into 12 chapters and 10 appendices. An introduction to connectionism and the notion of neural networks for prediction is included in Chapter 1. The fundamentals of adaptive signal processing and learning theory are detailed in Chapter 2. An initial overview of network architectures for prediction is given in Chapter 3.

[1] At the time of writing, only the months up to September 2000 were covered.

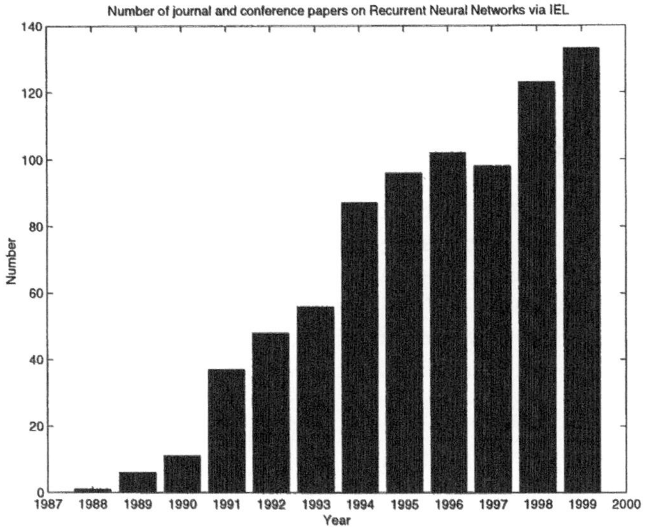

(a) Appearance of articles on Recurrent Neural Networks in IEE/IEEE publications in period 1988–1999

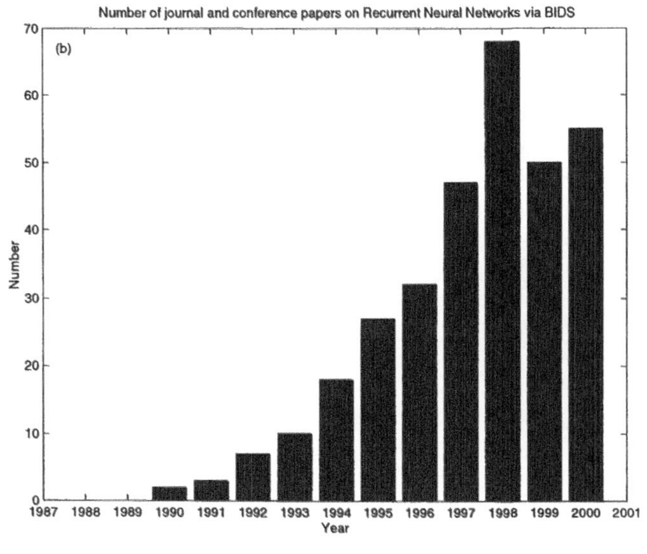

(b) Appearance of articles on Recurrent Neural Networks in BIDS database in period 1988–2000

Figure 1.2 Appearance of articles on RNNs in major citation databases. (a) Appearance of articles on recurrent neural networks in IEE/IEEE publications in period 1988–1999. (b) Appearance of articles on recurrent neural networks in BIDS database in period 1988–2000.

Chapter 4 contains a detailed discussion of activation functions and new insights are provided by the consideration of neural networks within the framework of modular groups from number theory. The material in Chapter 5 builds upon that within Chapter 3 and provides more comprehensive coverage of recurrent neural network architectures together with concepts from nonlinear system modelling. In Chapter 6, neural networks are considered as nonlinear adaptive filters whereby the necessary learning strategies for recurrent neural networks are developed. The stability issues for certain recurrent neural network architectures are considered in Chapter 7 through the exploitation of fixed point theory and bounds for global asymptotic stability are derived. *A posteriori* adaptive learning algorithms are introduced in Chapter 8 and the synergy with data-reusing algorithms is highlighted. In Chapter 9, a new class of normalised algorithms for online training of recurrent neural networks is derived. The convergence of online learning algorithms for neural networks is addressed in Chapter 10. Experimental results for the prediction of nonlinear and nonstationary signals with recurrent neural networks are presented in Chapter 11. In Chapter 12, the exploitation of inherent relationships between parameters within recurrent neural networks is described. Appendices A to J provide background to the main chapters and cover key concepts from linear algebra, approximation theory, complex sigmoid activation functions, a precedent learning algorithm for recurrent neural networks, terminology in neural networks, *a posteriori* techniques in science and engineering, contraction mapping theory, linear relaxation and stability, stability of general nonlinear systems and deseasonalising of time series. The book concludes with a comprehensive bibliography.

1.6 Readership

This book is targeted at graduate students and research engineers active in the areas of communications, neural networks, nonlinear control, signal processing and time series analysis. It will also be useful for engineers and scientists working in diverse application areas, such as artificial intelligence, biomedicine, earth sciences, finance and physics.

2

Fundamentals

2.1 Perspective

Adaptive systems are at the very core of modern digital signal processing. There are many reasons for this, foremost amongst these is that adaptive filtering, prediction or identification do not require explicit *a priori* statistical knowledge of the input data. Adaptive systems are employed in numerous areas such as biomedicine, communications, control, radar, sonar and video processing (Haykin 1996a).

2.1.1 Chapter Summary

In this chapter the fundamentals of adaptive systems are introduced. Emphasis is first placed upon the various structures available for adaptive signal processing, and includes the predictor structure which is the focus of this book. Basic learning algorithms and concepts are next detailed in the context of linear and nonlinear structure filters and networks. Finally, the issue of modularity is discussed.

2.2 Adaptive Systems

Adaptability, in essence, is the ability to react in sympathy with disturbances to the environment. A system that exhibits adaptability is said to be adaptive. Biological systems are adaptive systems; animals, for example, can adapt to changes in their environment through a learning process (Haykin 1999a).

A generic adaptive system employed in engineering is shown in Figure 2.1. It consists of

- a set of adjustable parameters (weights) within some filter structure;
- an error calculation block (the difference between the desired response and the output of the filter structure);
- a control (learning) algorithm for the adaptation of the weights.

The type of learning represented in Figure 2.1 is so-called supervised learning, since the learning is directed by the desired response of the system. Here, the goal

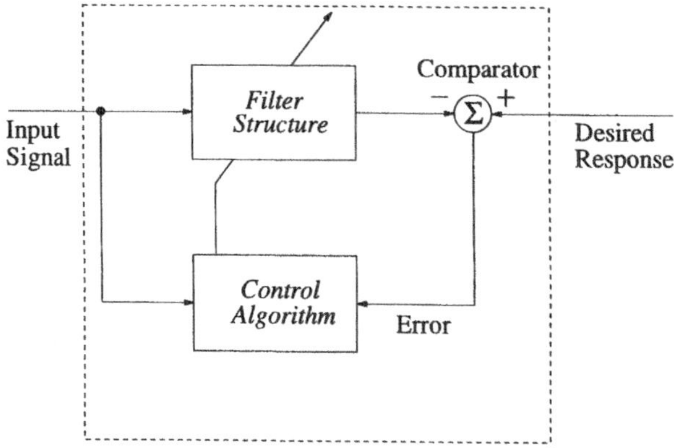

Figure 2.1 Block diagram of an adaptive system

is to adjust iteratively the free parameters (weights) of the adaptive system so as to minimise a prescribed cost function in some predetermined sense.[1] The filter structure within the adaptive system may be linear, such as a finite impulse response (FIR) or infinite impulse response (IIR) filter, or nonlinear, such as a Volterra filter or a neural network.

2.2.1 Configurations of Adaptive Systems Used in Signal Processing

Four typical configurations of adaptive systems used in engineering are shown in Figure 2.2 (Jenkins *et al.* 1996). The notions of an adaptive filter and adaptive system are used here interchangeably.

For the system identification configuration shown in Figure 2.2(a), both the adaptive filter and the unknown system are fed with the same input signal $x(k)$. The error signal $e(k)$ is formed at the output as $e(k) = d(k) - y(k)$, and the parameters of the adaptive system are adjusted using this error information. An attractive point of this configuration is that the desired response signal $d(k)$, also known as a teaching or training signal, is readily available from the unknown system (plant). Applications of this scheme are in acoustic and electrical echo cancellation, control and regulation of real-time industrial and other processes (plants). The knowledge about the system is stored in the set of converged weights of the adaptive system. If the dynamics of the plant are not time-varying, it is possible to identify the parameters (weights) of the plant to an arbitrary accuracy.

If we desire to form a system which inter-relates noise components in the input and desired response signals, the noise cancelling configuration can be implemented (Figure 2.2(b)). The only requirement is that the noise in the primary input and the reference noise are correlated. This configuration subtracts an estimate of the noise from the received signal. Applications of this configuration include noise cancellation

[1] The aim is to minimise some function of the error e. If $E[e^2]$ is minimised, we consider minimum mean squared error (MSE) adaptation, the statistical expectation operator, $E[\cdot]$, is due to the random nature of the inputs to the adaptive system.

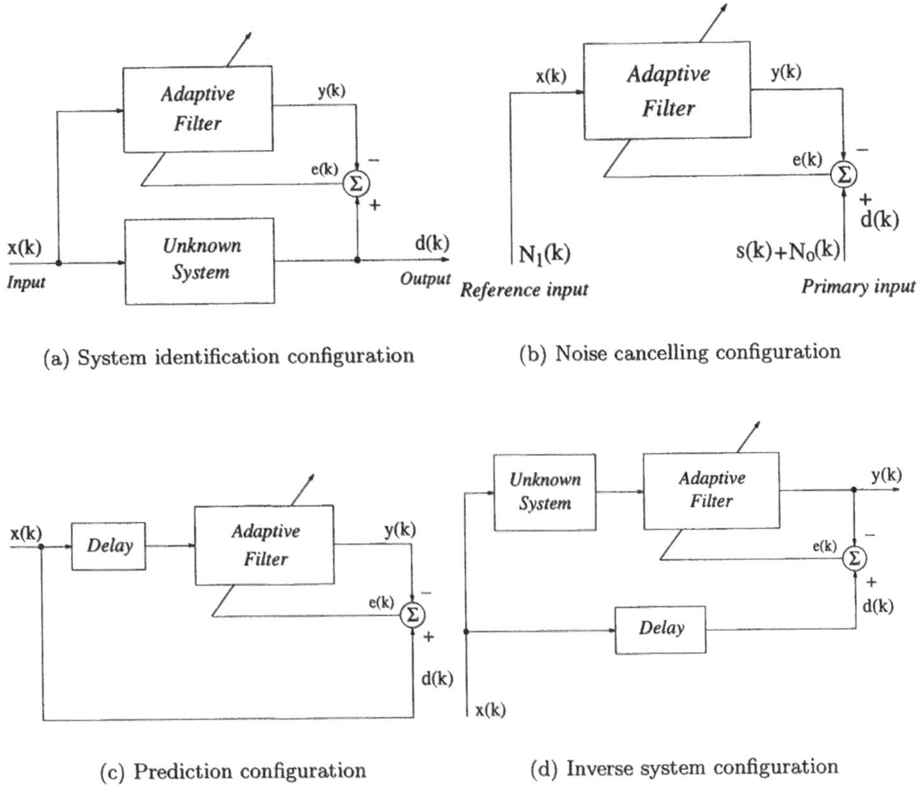

(a) System identification configuration (b) Noise cancelling configuration

(c) Prediction configuration (d) Inverse system configuration

Figure 2.2 Configurations for applications of adaptive systems

in acoustic environments and estimation of the foetal ECG from the mixture of the maternal and foetal ECG (Widrow and Stearns 1985).

In the adaptive prediction configuration, the desired signal is the input signal advanced relative to the input of the adaptive filter, as shown in Figure 2.2(c). This configuration has numerous applications in various areas of engineering, science and technology and most of the material in this book is dedicated to prediction. In fact, prediction may be considered as a basis for any adaptation process, since the adaptive filter is trying to predict the desired response.

The inverse system configuration, shown in Figure 2.2(d), has an adaptive system cascaded with the unknown system. A typical application is adaptive channel equalisation in telecommunications, whereby an adaptive system tries to compensate for the possibly time-varying communication channel, so that the transfer function from the input to the output of Figure 2.2(d) approximates a pure delay.

In most adaptive signal processing applications, parametric methods are applied which require *a priori* knowledge (or postulation) of a specific model in the form of differential or difference equations. Thus, it is necessary to determine the appropriate model order for successful operation, which will underpin data length requirements. On the other hand, nonparametric methods employ general model forms of integral

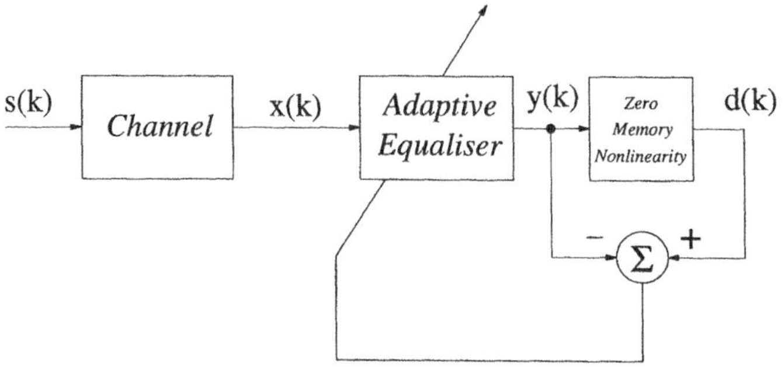

Figure 2.3 Block diagram of a blind equalisation structure

equations or functional expansions valid for a broad class of dynamic nonlinearities. The most widely used nonparametric methods are referred to as the Volterra–Wiener approach and are based on functional expansions.

2.2.2 Blind Adaptive Techniques

The presence of an explicit desired response signal, $d(k)$, in all the structures shown in Figure 2.2 implies that conventional, supervised, adaptive signal processing techniques may be applied for the purpose of learning. When no such signal is available, it may still be possible to perform learning by exploiting so-called blind, or unsupervised, methods. These methods exploit certain *a priori* statistical knowledge of the input data. For a single signal, this knowledge may be in the form of its constant modulus property, or, for multiple signals, their mutual statistical independence (Haykin 2000). In Figure 2.3 the structure of a blind equaliser is shown, notice the desired response is generated from the output of a zero-memory nonlinearity. This nonlinearity is implicitly being used to test the higher-order (i.e. greater than second-order) statistical properties of the output of the adaptive equaliser. When ideal convergence of the adaptive filter is achieved, the zero-memory nonlinearity has no effect upon the signal $y(k)$ and therefore $y(k)$ has identical statistical properties to that of the channel input $s(k)$.

2.3 Gradient-Based Learning Algorithms

We provide a brief introduction to the notion of gradient-based learning. The aim is to update iteratively the weight vector $\boldsymbol{w}$ of an adaptive system so that a nonnegative error measure $\mathcal{J}(\,\cdot\,)$ is reduced at each time step k,

$$\mathcal{J}(\boldsymbol{w} + \Delta\boldsymbol{w}) < \mathcal{J}(\boldsymbol{w}), \tag{2.1}$$

where $\Delta\boldsymbol{w}$ represents the change in $\boldsymbol{w}$ from one iteration to the next. This will generally ensure that after training, an adaptive system has captured the relevant properties of the unknown system that we are trying to model. Using a Taylor series expansion

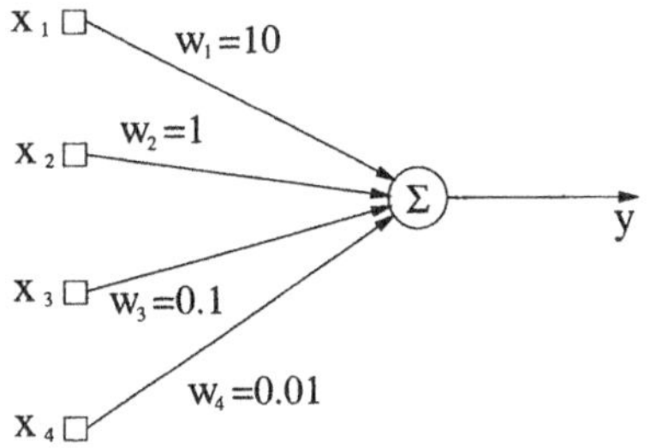

Figure 2.4 Example of a filter with widely differing weights

to approximate the error measure, we obtain[2]

$$\mathcal{J}(\boldsymbol{w}) + \Delta \boldsymbol{w} \frac{\partial \mathcal{J}(\boldsymbol{w})}{\partial \boldsymbol{w}} + \mathcal{O}(\boldsymbol{w}^2) < \mathcal{J}(\boldsymbol{w}). \tag{2.2}$$

This way, with the assumption that the higher-order terms in the left-hand side of (2.2) can be neglected, (2.1) can be rewritten as

$$\Delta \boldsymbol{w} \frac{\partial \mathcal{J}(\boldsymbol{w})}{\partial \boldsymbol{w}} < 0. \tag{2.3}$$

From (2.3), an algorithm that would continuously reduce the error measure on the run, should change the weights in the opposite direction of the gradient $\partial \mathcal{J}(\boldsymbol{w})/\partial \boldsymbol{w}$, i.e.

$$\Delta \boldsymbol{w} = -\eta \frac{\partial \mathcal{J}}{\partial \boldsymbol{w}}, \tag{2.4}$$

where η is a small positive scalar called the learning rate, step size or adaptation parameter.

Examining (2.4), if the gradient of the error measure $\mathcal{J}(\boldsymbol{w})$ is steep, large changes will be made to the weights, and conversely, if the gradient of the error measure $\mathcal{J}(\boldsymbol{w})$ is small, namely a flat error surface, a larger step size η may be used. Gradient descent algorithms cannot, however, provide a sense of importance or hierarchy to the weights (Agarwal and Mammone 1994). For example, the value of weight w_1 in Figure 2.4 is 10 times greater than w_2 and 1000 times greater than w_4. Hence, the component of the output of the filter within the adaptive system due to w_1 will, on the average, be larger than that due to the other weights. For a conventional gradient algorithm, however, the change in w_1 will not depend upon the relative sizes of the coefficients, but the relative sizes of the input data. This deficiency provides the motivation for certain partial update gradient-based algorithms (Douglas 1997).

It is important to notice that *gradient-descent-based algorithms inherently forget old data*, which leads to a problem called vanishing gradient and has particular importance for learning in filters with recursive structures. This issue is considered in more detail in Chapter 6.

---

[2] The explanation of the $\mathcal{O}$ notation can be found in Appendix A.

2.4 A General Class of Learning Algorithms

To introduce a general class of learning algorithms and explain in very crude terms relationships between them, we follow the approach from Guo and Ljung (1995). Let us start from the linear regression equation,

$$y(k) = \boldsymbol{x}^{\mathrm{T}}(k)\boldsymbol{w}(k) + \nu(k), \tag{2.5}$$

where $y(k)$ is the output signal, $\boldsymbol{x}(k)$ is a vector comprising the input signals, $\nu(k)$ is a disturbance or noise sequence, and $\boldsymbol{w}(k)$ is an unknown time-varying vector of weights (parameters) of the adaptive system. Variation of the weights at time k is denoted by $\boldsymbol{n}(k)$, and the weight change equation becomes

$$\boldsymbol{w}(k) = \boldsymbol{w}(k-1) + \boldsymbol{n}(k). \tag{2.6}$$

Adaptive algorithms can track the weights only approximately, hence for the following analysis we use the symbol $\hat{\boldsymbol{w}}$. A general expression for weight update in an adaptive algorithm is

$$\hat{\boldsymbol{w}}(k+1) = \hat{\boldsymbol{w}}(k) + \eta\boldsymbol{\Gamma}(k)(y(k) - \boldsymbol{x}^{\mathrm{T}}(k)\hat{\boldsymbol{w}}(k)), \tag{2.7}$$

where $\boldsymbol{\Gamma}(k)$ is the adaptation gain vector, and η is the step size. To assess how far an adaptive algorithm is from the optimal solution we introduce the weight error vector, $\breve{\boldsymbol{w}}(k)$, and a sample input matrix $\boldsymbol{\Sigma}(k)$ as

$$\breve{\boldsymbol{w}}(k) = \boldsymbol{w}(k) - \hat{\boldsymbol{w}}(k), \qquad \boldsymbol{\Sigma}(k) = \boldsymbol{\Gamma}(k)\boldsymbol{x}^{\mathrm{T}}(k). \tag{2.8}$$

Equations (2.5)–(2.8) yield the following weight error equation:

$$\breve{\boldsymbol{w}}(k+1) = (\boldsymbol{I} - \eta\boldsymbol{\Sigma}(k))\breve{\boldsymbol{w}}(k) - \eta\boldsymbol{\Gamma}(k)\nu(k) + \boldsymbol{n}(k+1). \tag{2.9}$$

For different gains $\boldsymbol{\Gamma}(k)$, the following three well-known algorithms can be obtained from (2.7).[3]

1. The least mean square (LMS) algorithm:

$$\boldsymbol{\Gamma}(k) = \boldsymbol{x}(k). \tag{2.10}$$

2. Recursive least-squares (RLS) algorithm:

$$\boldsymbol{\Gamma}(k) = P(k)\boldsymbol{x}(k), \tag{2.11}$$

$$P(k) = \frac{1}{1-\eta}\left[P(k-1) - \eta\frac{P(k-1)\boldsymbol{x}(k)\boldsymbol{x}^{\mathrm{T}}(k)P(k-1)}{1-\eta+\eta\boldsymbol{x}^{\mathrm{T}}(k)P(k-1)\boldsymbol{x}(k)}\right]. \tag{2.12}$$

3. Kalman filter (KF) algorithm (Guo and Ljung 1995; Kay 1993):

$$\boldsymbol{\Gamma}(k) = \frac{P(k-1)\boldsymbol{x}(k)}{R+\eta\boldsymbol{x}^{\mathrm{T}}(k)P(k-1)\boldsymbol{x}(k)}, \tag{2.13}$$

$$P(k) = P(k-1) - \frac{\eta P(k-1)\boldsymbol{x}(k)\boldsymbol{x}^{\mathrm{T}}(k)P(k-1)}{R+\eta\boldsymbol{x}^{\mathrm{T}}(k)P(k-1)\boldsymbol{x}(k)} + \eta Q. \tag{2.14}$$

[3] Notice that the role of η in the RLS and KF algorithm is different to that in the LMS algorithm. For RLS and KF we may put $\eta = 1$ and introduce a forgetting factor instead.

The KF algorithm is the optimal algorithm in this setting if the elements of $\boldsymbol{n}(k)$ and $\nu(k)$ in (2.5) and (2.6) are Gaussian noises with a covariance matrix $Q > 0$ and a scalar value $R > 0$, respectively (Kay 1993). All of these adaptive algorithms can be referred to as sequential estimators, since they refine their estimate as each new sample arrives. On the other hand, block-based estimators require all the measurements to be acquired before the estimate is formed.

Although the most important measure of quality of an adaptive algorithm is generally the covariance matrix of the weight tracking error $E[\breve{\boldsymbol{w}}(k)\breve{\boldsymbol{w}}^{\mathrm{T}}(k)]$, due to the statistical dependence between $\boldsymbol{x}(k)$, $\nu(k)$ and $\boldsymbol{n}(k)$, precise expressions for this covariance matrix are extremely difficult to obtain.

To undertake statistical analysis of an adaptive learning algorithm, the classical approach is to assume that $\boldsymbol{x}(k)$, $\nu(k)$ and $\boldsymbol{n}(k)$ are statistically independent. Another assumption is that the homogeneous part of (2.9)

$$\breve{\boldsymbol{w}}(k+1) = (\boldsymbol{I} - \eta\boldsymbol{\Sigma}(k))\breve{\boldsymbol{w}}(k) \tag{2.15}$$

and its averaged version

$$E[\breve{\boldsymbol{w}}(k+1)] = (\boldsymbol{I} - \eta E[\boldsymbol{\Sigma}(k)])E[\breve{\boldsymbol{w}}(k)] \tag{2.16}$$

are exponentially stable in stochastic and deterministic senses (Guo and Ljung 1995).

2.4.1 Quasi-Newton Learning Algorithm

The quasi-Newton learning algorithm utilises the second-order derivative of the objective function[4] to adapt the weights. If the change in the objective function between iterations in a learning algorithm is modelled with a Taylor series expansion, we have

$$\Delta E(\boldsymbol{w}) = E(\boldsymbol{w} + \Delta\boldsymbol{w}) - E(\boldsymbol{w}) \approx (\nabla_{\boldsymbol{w}} E(\boldsymbol{w}))^{\mathrm{T}}\Delta\boldsymbol{w} + \tfrac{1}{2}\Delta\boldsymbol{w}^{\mathrm{T}}\boldsymbol{H}\Delta\boldsymbol{w}. \tag{2.17}$$

After setting the differential with respect to $\Delta\boldsymbol{w}$ to zero, the weight update equation becomes

$$\Delta\boldsymbol{w} = -\boldsymbol{H}^{-1}\nabla_{\boldsymbol{w}} E(\boldsymbol{w}). \tag{2.18}$$

The Hessian $\boldsymbol{H}$ in this equation determines not only the direction but also the step size of the gradient descent.

To conclude: adaptive algorithms mainly differ in their form of adaptation gains. The gains can be roughly divided into two classes: gradient-based gains (e.g. LMS, quasi-Newton) and Riccati equation-based gains (e.g. KF and RLS).

2.5 A Step-by-Step Derivation of the Least Mean Square (LMS) Algorithm

Consider a set of input–output pairs of data described by a mapping function f:

$$d(k) = f(\boldsymbol{x}(k)), \qquad k = 1, 2, \ldots, N. \tag{2.19}$$

---

[4] The term *objective function* will be discussed in more detail later in this chapter.

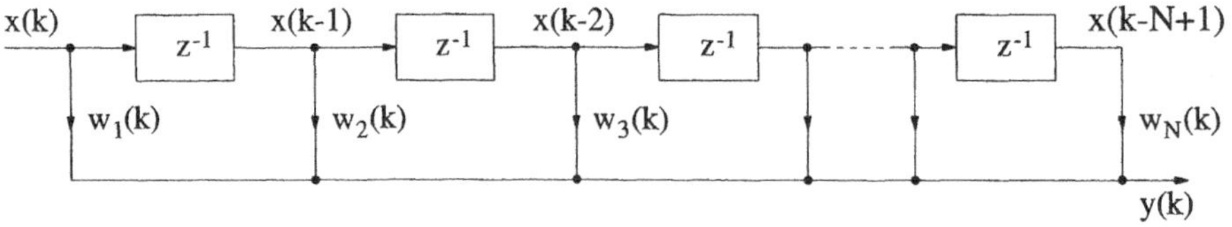

Figure 2.5 Structure of a finite impulse response filter

The function $f(\cdot)$ is assumed to be unknown. Using the concept of adaptive systems explained above, the aim is to approximate the unknown function $f(\cdot)$ by a function $F(\cdot, \boldsymbol{w})$ with adjustable parameters $\boldsymbol{w}$, in some prescribed sense. The function F is defined on a system with a known architecture or structure. It is convenient to define an instantaneous performance index,

$$J(\boldsymbol{w}(k)) = [d(k) - F(\boldsymbol{x}(k), \boldsymbol{w}(k))]^2, \tag{2.20}$$

which represents an energy measure. In that case, function F is most often just the inner product $F = \boldsymbol{x}^{\mathrm{T}}(k)\boldsymbol{w}(k)$ and corresponds to the operation of a linear FIR filter structure. As before, the goal is to find an optimisation algorithm that minimises the cost function $J(\boldsymbol{w})$. The common choice of the algorithm is motivated by the method of steepest descent, and generates a sequence of weight vectors $\boldsymbol{w}(1), \boldsymbol{w}(2), \dots,$ as

$$\boldsymbol{w}(k+1) = \boldsymbol{w}(k) - \eta \boldsymbol{g}(k), \qquad k = 0, 1, 2, \dots, \tag{2.21}$$

where $\boldsymbol{g}(k)$ is the gradient vector of the cost function $J(\boldsymbol{w})$ at the point $\boldsymbol{w}(k)$

$$\boldsymbol{g}(k) = \left.\frac{\partial J(\boldsymbol{w})}{\partial \boldsymbol{w}}\right|_{\boldsymbol{w}=\boldsymbol{w}(k)}. \tag{2.22}$$

The parameter η in (2.21) determines the behaviour of the algorithm:

- for η small, algorithm (2.21) converges towards the global minimum of the error performance surface;
- if the value of η approaches some critical value η_{c}, the trajectory of convergence on the error performance surface is either oscillatory or overdamped;
- if the value of η is greater than η_{c}, the system is unstable and does not converge.

These observations can only be visualised in two dimensions, i.e. for only two parameter values $w_1(k)$ and $w_2(k)$, and can be found in Widrow and Stearns (1985). If the approximation function F in the gradient descent algorithm (2.21) is linear we call such an adaptive system a linear adaptive system. Otherwise, we describe it as a nonlinear adaptive system. Neural networks belong to this latter class.

2.5.1 The Wiener Filter

Suppose the system shown in Figure 2.1 is modelled as a linear FIR filter (shown in Figure 2.5), we have $F(\boldsymbol{x}, \boldsymbol{w}) = \boldsymbol{x}^{\mathrm{T}}\boldsymbol{w}$, dropping the k index for convenience. Consequently, the instantaneous cost function $J(\boldsymbol{w}(k))$ is a quadratic function of the weight vector. The Wiener filter is based upon minimising the ensemble average of this instantaneous cost function, i.e.

$$J_{\text{Wiener}}(\boldsymbol{w}(k)) = E[[d(k) - \boldsymbol{x}^{\mathrm{T}}(k)\boldsymbol{w}(k)]^2] \tag{2.23}$$

and assuming $d(k)$ and $x(k)$ are zero mean and jointly wide sense stationary. To find the minimum of the cost function, we differentiate with respect to $\boldsymbol{w}$ and obtain

$$\frac{\partial J_{\text{Wiener}}}{\partial \boldsymbol{w}} = -2E[e(k)\boldsymbol{x}(k)], \tag{2.24}$$

where $e(k) = d(k) - \boldsymbol{x}^{\mathrm{T}}(k)\boldsymbol{w}(k)$.

At the Wiener solution, this gradient equals the null vector $\mathbf{0}$. Solving (2.24) for this condition yields the Wiener solution,

$$\boldsymbol{w} = \boldsymbol{R}_{\boldsymbol{x},\boldsymbol{x}}^{-1}\boldsymbol{r}_{\boldsymbol{x},d}, \tag{2.25}$$

where $\boldsymbol{R}_{\boldsymbol{x},\boldsymbol{x}} = E[\boldsymbol{x}(k)\boldsymbol{x}^{\mathrm{T}}(k)]$ is the autocorrelation matrix of the zero mean input data $\boldsymbol{x}(k)$ and $\boldsymbol{r}_{\boldsymbol{x},d} = E[\boldsymbol{x}(k)d(k)]$ is the crosscorrelation between the input vector and the desired signal $d(k)$. The Wiener formula has the same general form as the block least-squares (LS) solution, when the exact statistics are replaced by temporal averages.

The RLS algorithm, as in (2.12), with the assumption that the input and desired response signals are jointly ergodic, approximates the Wiener solution and asymptotically matches the Wiener solution.

More details about the derivation of the Wiener filter can be found in Haykin (1996a, 1999a).

2.5.2 Further Perspective on the Least Mean Square (LMS) Algorithm

To reduce the computational complexity of the Wiener solution, which is a block solution, we can use the method of steepest descent for a recursive, or sequential, computation of the weight vector $\boldsymbol{w}$. Let us derive the LMS algorithm for an adaptive FIR filter, the structure of which is shown in Figure 2.5. In view of a general adaptive system, this FIR filter becomes the filter structure within Figure 2.1. The output of this filter is

$$y(k) = \boldsymbol{x}^{\mathrm{T}}(k)\boldsymbol{w}(k). \tag{2.26}$$

Widrow and Hoff (1960) utilised this structure for adaptive processing and proposed instantaneous values of the autocorrelation and crosscorrelation matrices to calculate the gradient term within the steepest descent algorithm. The cost function they proposed was

$$J(k) = \tfrac{1}{2}e^2(k), \tag{2.27}$$

which is again based upon the instantaneous output error $e(k) = d(k) - y(k)$. In order to derive the weight update equation we start from the instantaneous gradient

$$\frac{\partial J(k)}{\partial \boldsymbol{w}(k)} = e(k)\frac{\partial e(k)}{\partial \boldsymbol{w}(k)}. \tag{2.28}$$

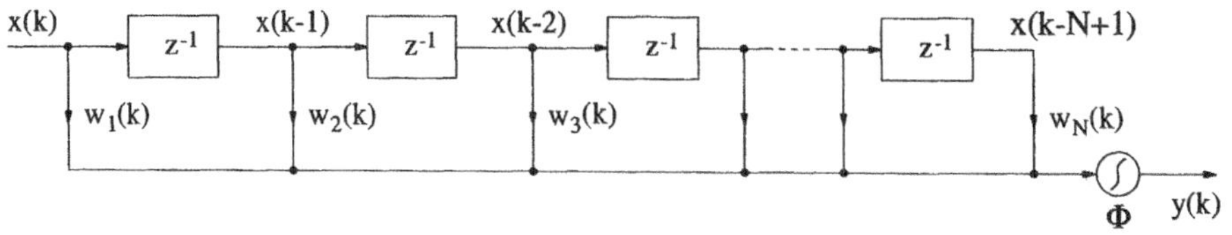

Figure 2.6 The structure of a nonlinear adaptive filter

Following the same procedure as for the general gradient descent algorithm, we obtain

$$\frac{\partial e(k)}{\partial \boldsymbol{w}(k)} = -\boldsymbol{x}(k) \tag{2.29}$$

and finally

$$\frac{\partial J(k)}{\partial \boldsymbol{w}(k)} = -e(k)\boldsymbol{x}(k). \tag{2.30}$$

The set of equations that describes the LMS algorithm is given by

$$\left.\begin{aligned} y(k) &= \sum_{i=1}^{N} x_i(k)w_i(k) = \boldsymbol{x}^{\mathrm{T}}(k)\boldsymbol{w}(k), \\ e(k) &= d(k) - y(k), \\ \boldsymbol{w}(k+1) &= \boldsymbol{w}(k) + \eta e(k)\boldsymbol{x}(k). \end{aligned}\right\} \tag{2.31}$$

The LMS algorithm is a very simple yet extremely popular algorithm for adaptive filtering. It is also optimal in the H^{∞} sense which justifies its practical utility (Hassibi *et al.* 1996).

2.6 On Gradient Descent for Nonlinear Structures

Adaptive filters and neural networks are formally equivalent, in fact, the structures of neural networks are generalisations of linear filters (Maass and Sontag 2000; Nerrand *et al.* 1991). Depending on the architecture of a neural network and whether it is used online or offline, two broad classes of learning algorithms are available:

- techniques that use a direct computation of the gradient, which is typical for linear and nonlinear adaptive filters;
- techniques that involve backpropagation, which is commonplace for most offline applications of neural networks.

Backpropagation is a computational procedure to obtain gradients necessary for adaptation of the weights of a neural network contained within its hidden layers and is not radically different from a general gradient algorithm.

As we are interested in neural networks for real-time signal processing, we will analyse online algorithms that involve direct gradient computation. In this section we introduce a learning algorithm for a nonlinear FIR filter, whereas learning algorithms for online training of recurrent neural networks will be introduced later. Let us start from a simple nonlinear FIR filter, which consists of the standard FIR filter cascaded with a memoryless nonlinearity Φ as shown in Figure 2.6. This structure can be seen as a single neuron with a dynamical FIR synapse. This FIR synapse provides memory to the neuron. The output of this filter is given by

$$y(k) = \Phi(\boldsymbol{x}^{\mathrm{T}}(k)\boldsymbol{w}(k)). \tag{2.32}$$

The nonlinearity $\Phi(\,\cdot\,)$ after the tap-delay line is typically a sigmoid. Using the ideas from the LMS algorithm, if the cost function is given by

$$J(k) = \tfrac{1}{2}e^2(k) \tag{2.33}$$

we have

$$e(k) = d(k) - \Phi(\boldsymbol{x}^{\mathrm{T}}(k)\boldsymbol{w}(k)), \tag{2.34}$$

$$\boldsymbol{w}(k+1) = \boldsymbol{w}(k) - \eta\nabla_{\boldsymbol{w}}J(k), \tag{2.35}$$

where $e(k)$ is the instantaneous error at the output neuron, $d(k)$ is some teaching (desired) signal, $\boldsymbol{w}(k) = [w_1(k), \dots, w_N(k)]^{\mathrm{T}}$ is the weight vector and $\boldsymbol{x}(k) = [x_1(k), \dots, x_N(k)]^{\mathrm{T}}$ is the input vector.

The gradient $\nabla_{\boldsymbol{w}}J(k)$ can be calculated as

$$\frac{\partial J(k)}{\partial \boldsymbol{w}(k)} = e(k)\frac{\partial e(k)}{\partial \boldsymbol{w}(k)} = -e(k)\Phi'(\boldsymbol{x}^{\mathrm{T}}(k)\boldsymbol{w}(k))\boldsymbol{x}(k), \tag{2.36}$$

where $\Phi'(\,\cdot\,)$ represents the first derivative of the nonlinearity $\Phi(\cdot)$ and the weight update Equation (2.35) can be rewritten as

$$\boldsymbol{w}(k+1) = \boldsymbol{w}(k) + \eta\Phi'(\boldsymbol{x}^{\mathrm{T}}(k)\boldsymbol{w}(k))e(k)\boldsymbol{x}(k). \tag{2.37}$$

This is the weight update equation for a direct gradient algorithm for a nonlinear FIR filter.

2.6.1 Extension to a General Neural Network

When deriving a direct gradient algorithm for a general neural network, the network architecture should be taken into account. For large networks for offline processing, classical backpropagation is the most convenient algorithm. However, for online learning, extensions of the previous algorithm should be considered.

2.7 On Some Important Notions From Learning Theory

In this section we discuss in more detail the inter-relations between the error, error function and objective function in learning theory.

2.7.1 Relationship Between the Error and the Error Function

The error at the output of an adaptive system is defined as the difference between the output value of the network and the target (desired output) value. For instance, the instantaneous error $e(k)$ is defined as

$$e(k) = d(k) - y(k). \tag{2.38}$$

The instantaneous error can be positive, negative or zero, and is therefore not a good candidate for the criterion function for training adaptive systems. Hence we look for another function, called the *error* function, that is a function of the instantaneous error, but is suitable as a criterion function for learning. Error functions are also called *loss* functions. They are defined so that an increase in the error function corresponds to a reduction in the quality of learning, and they are nonnegative. An error function can be defined as

$$E(N) = \sum_{i=0}^{N} e^2(i) \tag{2.39}$$

or as an average value

$$\bar{E}(N) = \frac{1}{N+1} \sum_{i=0}^{N} e^2(i). \tag{2.40}$$

2.7.2 The Objective Function

The *objective* function is a function that we want to minimise during training. It can be equal to an error function, but often it may include other terms to introduce constraints. For instance in generalisation, too large a network might lead to overfitting. Hence the objective function can consist of two parts, one for the error minimisation and the other which is either a penalty for a large network or a penalty term for excessive increase in the weights of the adaptive system or some other chosen function (Tikhonov *et al.* 1998). An example of such an objective function for online learning is

$$J(k) = \frac{1}{N} \sum_{i=1}^{N} (e^2(k-i+1) + G(\|w(k-i+1)\|_2^2)), \tag{2.41}$$

where G is some linear or nonlinear function. We often use symbols E and J interchangeably to denote the cost function.

2.7.3 Types of Learning with Respect to the Training Set and Objective Function

Batch learning

Batch learning is also known as epochwise, or offline learning, and is a common strategy for offline training. The idea is to adapt the weights once the whole training set has been presented to an adaptive system. It can be described by the following steps.

1. Initialise the weights
2. Repeat
    - Pass all the training data through the network

- Sum the errors after each particular pattern
- Update the weights based upon the total error
- Stop if some prescribed error performance is reached

The counterpart of batch learning is so-called incremental learning, online, or pattern learning. The procedure for this type of learning is as follows.

1. Initialise the weights
2. Repeat
   - Pass one pattern through the network
   - Update the weights based upon the instantaneous error
   - Stop if some prescribed error performance is reached

The choice of the type of learning is very much dependent upon application. Quite often, for networks that need initialisation, we perform one type of learning in the initialisation procedure, which is by its nature an offline procedure, and then use some other learning strategy while the network is running. Such is the case with recurrent neural networks for online signal processing (Mandic and Chambers 1999f).

2.7.4 Deterministic, Stochastic and Adaptive Learning

Deterministic learning is an optimisation technique based on an objective function which always produces the same result, no matter how many times we recompute it. Deterministic learning is always offline.

Stochastic learning is useful when the objective function is affected by noise and local minima. It can be employed within the context of a gradient descent learning algorithm. The idea is that the learning rate gradually decreases during training and hence the steps on the error performance surface in the beginning of training are large which speeds up training when far from the optimal solution. The learning rate is small when approaching the optimal solution, hence reducing misadjustment. This gradual reduction of the learning rate can be achieved by e.g. annealing (Kirkpatrick *et al.* 1983; Rose 1998; Szu and Hartley 1987).

The idea behind the concept of adaptive learning is to forget the past when it is no longer relevant and adapt to the changes in the environment. The terms 'adaptive learning' or 'gear-shifting' are sometimes used for gradient methods in which the learning rate is changed during training.

2.7.5 Constructive Learning

Constructive learning deals with the change of architecture or interconnections in the network during training. Neural networks for which topology can change over time are called *ontogenic* neural networks (Fiesler and Beale 1997). Two basic classes of constructive learning are network growing and network pruning. In the network growing approach, learning begins with a network with no hidden units, and if the error is too big, new hidden units are added to the network, training resumes, and so on. The most used algorithm based upon network growing is the so-called cascade-correlation algorithm (Hoehfeld and Fahlman 1992). Network pruning starts from a large network and if the error in learning is smaller than allowed, the network size is reduced until the desired ratio between accuracy and network size is reached (Reed 1993; Sum *et al.* 1999).

2.7.6 Transformation of Input Data, Learning and Dimensionality

A natural question is whether to linearly/nonlinearly transform the data before feeding them to an adaptive processor. This is particularly important for neural networks, which are nonlinear processors. If we consider each neuron as a basic component of a neural network, then we can refer to a general neural network as a system with componentwise nonlinearities. To express this formally, consider a scalar function $\sigma : \mathbb{R} \to \mathbb{R}$ and systems of the form,

$$\boldsymbol{y}(k) = \boldsymbol{\sigma}(\boldsymbol{A}\boldsymbol{x}(k)), \tag{2.42}$$

where the matrix $\boldsymbol{A}$ is an $N \times N$ matrix and $\boldsymbol{\sigma}$ is applied componentwise

$$\boldsymbol{\sigma}(x_1(k), \ldots, x_N(k)) = (\sigma(x_1(k)), \ldots, \sigma(x_N(k))). \tag{2.43}$$

Systems of this type arise in a wide variety of situations. For a linear σ, we have a linear system. If the range of σ is finite, the state vector of (2.42) takes values from a finite set, and dynamical properties can be analysed in time which is polynomial in the number of possible states. Throughout this book we are interested in functions, σ, and combination matrices, $\boldsymbol{A}$, which would guarantee a fixed point of this mapping. Neural networks are commonly of the form (2.42). In such a context we call σ the activation function. Results of Siegelmann and Sontag (1995) show that saturated linear systems (piecewise linear) can represent Turing machines, which is achieved by encoding the transition rules of the Turing machine in the matrix $\boldsymbol{A}$.

The curse of dimensionality

The curse of dimensionality (Bellman 1961) refers to the exponential growth of computation needed for a specific task as a function of the dimensionality of the input space. In neural networks, a network quite often has to deal with many irrelevant inputs which, in turn, increase the dimensionality of the input space. In such a case, the network uses much of its resources to represent and compute irrelevant information, which hampers processing of the desired information. A remedy for this problem is preprocessing of input data, such as feature extraction, and to introduce some importance function to the input samples. The curse of dimensionality is particularly prominent in unsupervised learning algorithms. Radial basis functions are also prone to this problem. Selection of a neural network model must therefore be suited for a particular task. Some *a priori* information about the data and scaling of the inputs can help to reduce the severity of the problem.

Transformations on the input data

Activation functions used in neural networks are centred around a certain value in their output space. For instance, the mean of the logistic function is 0.5, whereas the tanh function is centred around zero. Therefore, in order to perform efficient prediction, we should match the range of the input data, their mean and variance, with the range of the chosen activation function. There are several operations that we could perform on the input data, such as the following.

1. Normalisation, which in this context means dividing each element of the input vector $\boldsymbol{x}(k)$ by its squared norm, i.e. $x_i(k) \in \boldsymbol{x}(k) \to x_i(k)/\|\boldsymbol{x}(k)\|_2^2$.

2. Rescaling, which means transforming the input data in the manner that we multiply/divide them by a constant and also add/subtract a constant from the data.[5]

3. Standardisation, which is borrowed from statistics, where, for instance, a random Gaussian vector is standardised if its mean is subtracted from it, and the vector is then divided by its standard deviation. The resulting random variable is called a 'standard normal' random variable with zero mean and unity standard deviation. Some examples of data standardisation are

   - Standardisation to zero mean and unity standard deviation can be performed as

     $$\text{mean} = \frac{\sum_i X_i}{N}, \qquad \text{std} = \sqrt{\frac{\sum_i (X_i - \text{mean})^2}{N-1}}.$$

     The standardised quantity becomes $S_i = (X_i - \text{mean})/\text{std}$.

   - Standardise X to midrange 0 and range 2. This can be achieved by

     $$\text{midrange} = \tfrac{1}{2}(\max_i X_i + \min_i X_i), \qquad \text{range} = \max_i X_i - \min_i X_i,$$

     $$S_i = \frac{X_i - \text{midrange}}{\text{range}/2}.$$

4. Principal component analysis (PCA) represents the data by a set of unit norm vectors called normalised eigenvectors. The eigenvectors are positioned along the directions of greatest data variance. The eigenvectors are found from the covariance matrix $\boldsymbol{R}$ of the input dataset. An eigenvalue λ_i, $i = 1, \dots, N$, is associated with each eigenvector. Every input data vector is then represented by a linear combination of eigenvectors.

As pointed out earlier, standardising input variables has an effect on training, since steepest descent algorithms are sensitive to scaling due to the change in the weights being proportional to the value of the gradient and the input data.

---

[5] In real life a typical rescaling is transforming the temperature from Celsius into Fahrenheit scale.

Nonlinear transformations of the data

This method to transform the data can help when the dynamic range of the data is too high. In that case, for instance, we typically apply the log function to the input data. The log function is often applied in the error and objective functions for the same purposes.

2.8 Learning Strategies

To construct an optimal neural approximating model we have to determine an appropriate training set containing all the relevant information of the process and define a suitable topology that matches the complexity and performance requirements. The training set construction issue requires four entities to be considered (Alippi and Piuri 1996; Bengio 1995; Haykin and Li 1995; Shadafan and Niranjan 1993):

- the number of training data samples N_{D};
- the number of patterns N_{P} constituting a batch;
- the number of batches N_{B} to be extracted from the training set;
- the number of times the generic batch is presented to the network during learning.

The assumption is that the training set is sufficiently rich so that it contains all the relevant information necessary for learning.

The requirement coincides with the hypothesis that the training data have been generated by a fully exciting input signal, such as white noise, which is able to excite all the process dynamics. White noise is a persistently exciting input signal and is used for the driving component of moving average (MA), autoregressive (AR) and autoregressive moving average (ARMA) models.

2.9 General Framework for the Training of Recurrent Networks by Gradient-Descent-Based Algorithms

In this section we summarise some of the important concepts mentioned earlier.

2.9.1 Adaptive Versus Nonadaptive Training

The training of a network makes use of two sequences, the sequence of inputs and the sequence of corresponding desired outputs. If the network is first trained (with a training sequence of finite length) and subsequently used (with the fixed weights obtained from training), this mode of operation is referred to as *non-adaptive* (Nerrand *et al.* 1994). Conversely, the term *adaptive* refers to the mode of operation whereby the network is trained permanently throughout its application (with a training sequence of infinite length). Therefore, the adaptive network is suitable for input processes which exhibit statistically non-stationary behaviour, a situation which is normal in the fields of adaptive control and signal processing (Bengio 1995; Haykin 1996a; Haykin and

Li 1995; Khotanzad and Lu 1990; Narendra and Parthasarathy 1990; Nerrand *et al.* 1994).

2.9.2 Performance Criterion, Cost Function, Training Function

The computation of the coefficients during training aims at finding a system whose operation is optimal with respect to some performance criterion which may be either qualitative, e.g. (subjective) quality of speech reconstruction, or quantitative, e.g. maximising signal to noise ratio for spatial filtering. The goal is to define a positive *training function* which is such that a decrease of this function through modifications of the coefficients of the network leads to an improvement of the performance of the system (Bengio 1995; Haykin and Li 1995; Nerrand *et al.* 1994; Qin *et al.* 1992). In the case of non-adaptive training, the training function is defined as a function of all the data of the training set (in such a case, it is usually termed as a *cost function*). The minimum of the cost function corresponds to the optimal performance of the system. Training is an optimisation procedure, conventionally using gradient-based methods.

In the case of adaptive training, it is impossible, in most instances, to define a time-independent cost function whose minimisation leads to a system that is optimal with respect to the performance criterion. Therefore, the training function is time dependent. The modification of the coefficients is computed continually from the gradient of the training function. The latter involves the data pertaining to a time window of finite length, which shifts in time (sliding window) and the coefficients are updated at each sampling time.

2.9.3 Recursive Versus Nonrecursive Algorithms

A nonrecursive algorithm employs a cost function (i.e. a training function defined on a fixed window), whereas a recursive algorithm makes use of a training function defined on a sliding window of data. An adaptive system must be trained by a recursive algorithm, whereas a non-adaptive system may be trained either by a nonrecursive or by a recursive algorithm (Nerrand *et al.* 1994).

2.9.4 Iterative Versus Noniterative Algorithms

An iterative algorithm performs coefficient modifications several times from a set of data pertaining to a given data window, a non-iterative algorithm makes only one (Nerrand *et al.* 1994). For instance, the conventional LMS algorithm (2.31) is thus a recursive, non-iterative algorithm operating on a sliding window.

2.9.5 Supervised Versus Unsupervised Algorithms

A supervised learning algorithm performs learning by using *a teaching signal*, i.e. the desired output signal, while an unsupervised learning algorithm, as in blind signal processing, has no reference signal as a teaching input signal. An example of a supervised learning algorithm is the *delta rule*, while unsupervised learning algorithms are,

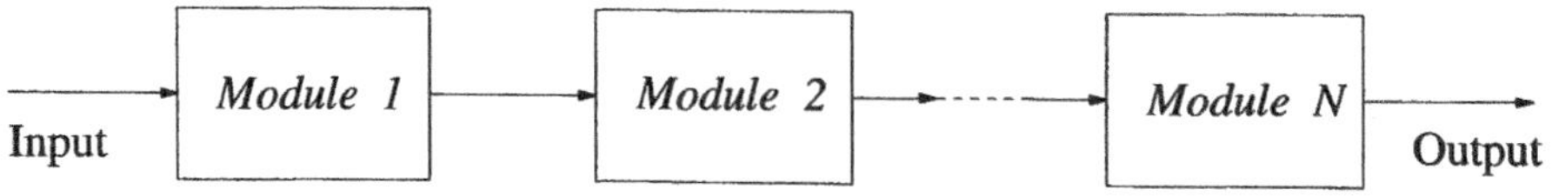

Figure 2.7 A cascaded realisation of a general system

for example, the *reinforcement learning algorithm* and the *competitive rule ('winner takes all') algorithm*, whereby there is some sense of concurrency between the elements of the network structure (Bengio 1995; Haykin and Li 1995).

2.9.6 Pattern Versus Batch Learning

Updating the network weights by *pattern learning* means that the weights of the network are updated immediately after each pattern is fed in. The other approach is to take all the data as a whole batch, and the network is not updated until the entire batch of data is processed. This approach is referred to as *batch learning* (Haykin and Li 1995; Qin *et al.* 1992).

It can be shown (Qin *et al.* 1992) that while considering feedforward networks (FFN), after one training sweep through all the data, the pattern learning is a first-order approximation of the batch learning with respect to the learning rate η. Therefore, the FFN pattern learning approximately implements the FFN batch learning after one batch interval. After multiple sweeps through the training data, the difference between the FFN pattern learning and FFN batch learning is of the order[6] $\mathcal{O}(\eta^2)$. Therefore, for small training rates, the FFN pattern learning approximately implements FFN batch learning after multiple sweeps through the training data. For recurrent networks, the weight updating slopes for pattern learning and batch learning are different[7] (Qin *et al.* 1992). However, the difference could also be controlled by the learning rate η. The difference will converge to zero as quickly as η goes to zero[8] (Qin *et al.* 1992).

2.10 Modularity Within Neural Networks

The hierarchical levels in neural network architectures are synapses, neurons, layers and neural networks, and will be discussed in Chapter 5. The next step would be combinations of neural networks. In this case we consider modular neural networks. Modular neural networks are composed of a set of smaller subnetworks (modules), each performing a subtask of the complete problem. To depict this problem, let us recourse to the case of linear adaptive filters described by a transfer function in the

[6] In fact, if the data being processed exhibit highly stationary behaviour, then the average error calculated after FFN batch learning is very close to the instantaneous error calculated after FFN pattern learning, e.g. the speech data can be considered as being stationary within an observed frame. That forms the basis for use of various real-time and recursive learning algorithms, e.g. RTRL.

[7] It can be shown (Qin *et al.* 1992) that for feedforward networks, the updated weights for both pattern learning and batch learning adapt at the same slope (derivative $dw/d\eta$) with respect to the learning rate η. For recurrent networks, this is not the case.

[8] In which case we have a very slow learning process.

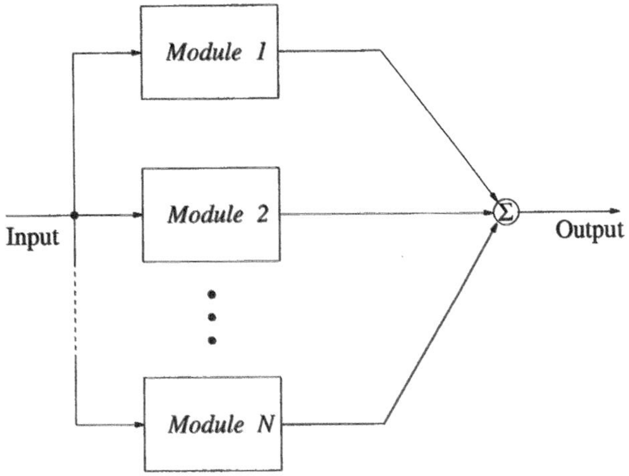

Figure 2.8 A parallel realisation of a general system

z-domain $H(z)$ as

$$H(z) = \frac{\sum_{k=0}^{M} b(k)z^{-k}}{1 + \sum_{k=1}^{N} a(k)z^{-k}}. \tag{2.44}$$

We can rearrange this function either in a cascaded manner as

$$H(z) = A \prod_{k=1}^{\max\{M,N\}} \frac{1 - \beta_k z^{-1}}{1 - \alpha_k z^{-1}}, \tag{2.45}$$

or in a parallel manner as

$$H(z) = \sum_{k=1}^{N} \frac{A_k}{1 - \alpha_k z^{-1}}, \tag{2.46}$$

where for simplicity we have assumed first-order poles and zeros of $H(z)$. A cascaded realisation of a general system is shown in Figure 2.7, whereas a parallel realisation of a general system is shown in Figure 2.8. We can also combine neural networks in these two configurations. An example of cascaded neural network is the so-called pipelined recurrent neural network, whereas an example of a parallel realisation of a neural network is the associative Gaussian mixture model, or winner takes all network. Taking into account that neural networks are nonlinear systems, we talk about nested modular architectures instead of cascaded architectures. The nested neural scheme can be written as

$$F(W, X) = \Phi\biggl(\sum_n w_n \Phi\biggl(\sum_i v_i \Phi\biggl(\cdots \Phi\biggl(\sum_j u_j X_j\biggr)\cdots\biggr)\biggr)\biggr), \tag{2.47}$$

where Φ is a sigmoidal function. It corresponds to a multilayer network of units that sum their inputs with 'weights' $W = \{w_n, v_i, u_j, \ldots\}$ and then perform a sigmoidal

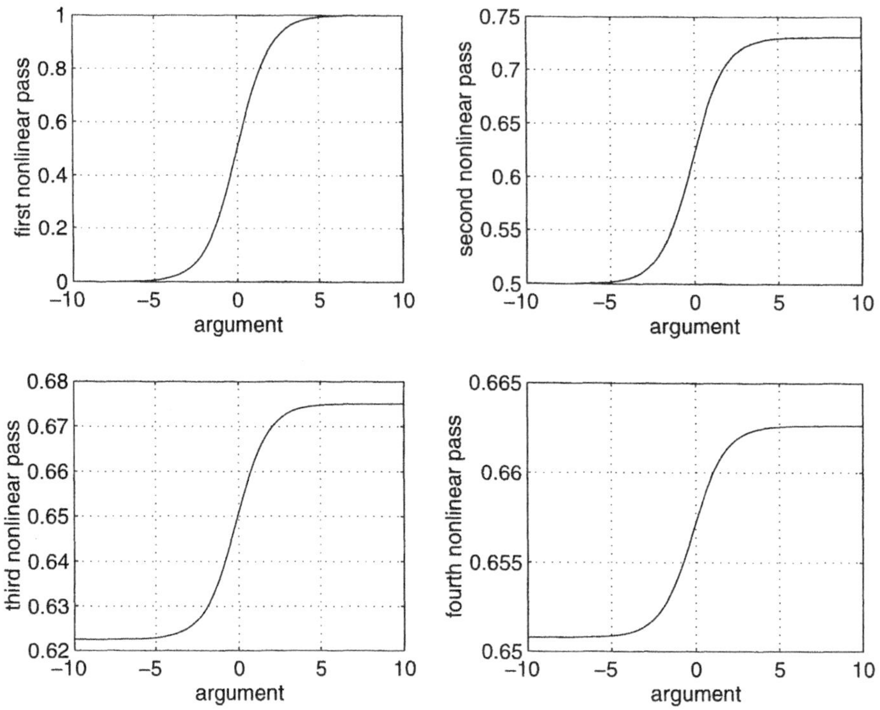

Figure 2.9 Effects of nesting sigmoid nonlinearities: first, second, third and fourth pass

transformation of this sum. Its motivation is that the function

$$F(W, X) = \Phi\Big(\sum_n w_n \Phi\Big(\sum_j u_j X_j\Big)\Big) \tag{2.48}$$

can approximate arbitrarily well any continuous multivariate function (Funahashi 1989; Poggio and Girosi 1990).

Since we use sigmoid 'squashing' activation functions, modular structures contribute to a general stability issue. The effects of a simple scheme of nested sigmoids are shown in Figure 2.9. From Figure 2.9 we see that pure nesting successively reduces the range of the output signal, bringing this composition of nonlinear functions to the fixed point of the employed nonlinearity for sufficiently many nested sigmoids.

Modular networks possess some advantages over classical networks, since the overall complex function is simplified and modules possibly do not have hidden units which speeds up training. Also, input data might be decomposable into subsets which can be fed to separate modules. Utilising modular neural networks has not only computational advantages but also development advantages, improved efficiency, improved interpretability and easier hardware implementation. Also, there are strong suggestions from biology that modular structures are exploited in cognitive mechanisms (Fiesler and Beale 1997).

2.11 Summary

Configurations of general adaptive systems have been provided, and the prediction configuration has been introduced within this framework. Gradient-descent-based learning algorithms have then been developed for these configurations, with an emphasis on the LMS algorithm. A thorough discussion of learning modes and learning parameters is given. Finally, modularity within neural networks has been addressed.

3

Network Architectures for Prediction

3.1 Perspective

The architecture, or structure, of a predictor underpins its capacity to represent the dynamic properties of a statistically nonstationary discrete time input signal and hence its ability to predict or forecast some future value. This chapter therefore provides an overview of available structures for the prediction of discrete time signals.

3.2 Introduction

The basic building blocks of all discrete time predictors are adders, delayers, multipliers and for the nonlinear case zero-memory nonlinearities. The manner in which these elements are interconnected describes the architecture of a predictor. The foundations of linear predictors for statistically stationary signals are found in the work of Yule (1927), Kolmogorov (1941) and Wiener (1949). The later studies of Box and Jenkins (1970) and Makhoul (1975) were built upon these fundamentals. Such linear structures are very well established in digital signal processing and are classified either as finite impulse response (FIR) or infinite impulse response (IIR) digital filters (Oppenheim *et al.* 1999). FIR filters are generally realised without feedback, whereas IIR filters[1] utilise feedback to limit the number of parameters necessary for their realisation. The presence of feedback implies that the consideration of stability underpins the design of IIR filters. In statistical signal modelling, FIR filters are better known as moving average (MA) structures and IIR filters are named autoregressive (AR) or autoregressive moving average (ARMA) structures. The most straightforward version of nonlinear filter structures can easily be formulated by including a nonlinear operation in the output stage of an FIR or an IIR filter. These represent simple examples of nonlinear autoregressive (NAR), nonlinear moving average (NMA) or nonlinear autoregressive moving average (NARMA) structures (Nerrand *et al.* 1993). Such filters have immediate application in the prediction of discrete time random signals that arise from some

[1] FIR filters can be represented by IIR filters, however, in practice it is not possible to represent an arbitrary IIR filter with an FIR filter of finite length.

nonlinear physical system, as for certain speech utterances. These filters, moreover, are strongly linked to single neuron neural networks.

The neuron, or node, is the basic processing element within a neural network. The structure of a neuron is composed of multipliers, termed synaptic weights, or simply weights, which scale the inputs, a linear combiner to form the activation potential, and a certain zero-memory nonlinearity to model the activation function. Different neural network architectures are formulated by the combination of multiple neurons with various interconnections, hence the term connectionist modelling (Rumelhart *et al.* 1986). Feedforward neural networks, as for FIR/MA/NMA filters, have no feedback within their structure. Recurrent neural networks, on the other hand, similarly to IIR/AR/NAR/NARMA filters, exploit feedback and hence have much more potential structural richness. Such feedback can either be local to the neurons or global to the network (Haykin 1999b; Tsoi and Back 1997). When the inputs to a neural network are delayed versions of a discrete time random input signal the correspondence between the architectures of nonlinear filters and neural networks is evident.

From a biological perspective (Marmarelis 1989), the *prototypical* neuron is composed of a cell body (soma), a tree-like element of fibres (dendrites) and a long fibre (axon) with sparse branches (collaterals). The axon is attached to the soma at the *axon hillock*, and, together with its collaterals, ends at synaptic terminals (boutons), which are employed to pass information onto their neurons through *synaptic junctions*. The soma contains the nucleus and is attached to the trunk of the dendritic tree from which it receives incoming information. The dendrites are conductors of input information to the soma, i.e. input ports, and usually exhibit a high degree of arborisation.

The possible architectures for nonlinear filters or neural networks are manifold. The state-space representation from system theory is established for linear systems (Kailath 1980; Kailath *et al.* 2000) and provides a mechanism for the representation of structural variants. An insightful canonical form for neural networks is provided by Nerrand *et al.* (1993), by the exploitation of state-space representation which facilitates a unified treatment of the architectures of neural networks.[2]

3.3 Overview

The chapter begins with an explanation of the concept of prediction of a statistically stationary discrete time random signal. The building blocks for the realisation of linear and nonlinear predictors are then discussed. These same building blocks are also shown to be the basic elements necessary for the realisation of a neuron. Emphasis is placed upon the particular zero-memory nonlinearities used in the output of nonlinear filters and activation functions of neurons.

An aim of this chapter is to highlight the correspondence between the structures in nonlinear filtering and neural networks, so as to remove the apparent boundaries between the work of practitioners in control, signal processing and neural engineering. Conventional linear filter models for discrete time random signals are introduced and,

---

[2] ARMA models also have a canonical (up to an invariant) representation.

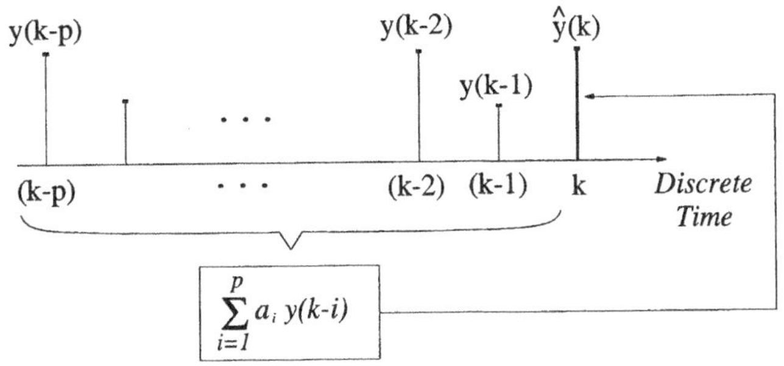

Figure 3.1 Basic concept of linear prediction

with the aid of statistical modelling, motivate the structures for linear predictors; their nonlinear counterparts are then developed.

A feedforward neural network is next introduced in which the nonlinear elements are distributed throughout the structure. To employ such a network as a predictor, it is shown that short-term memory is necessary, either at the input or integrated within the network. Recurrent networks follow naturally from feedforward neural networks by connecting the output of the network to its input. The implications of local and global feedback in neural networks are also discussed.

The role of state-space representation in architectures for neural networks is described and this leads to a canonical representation. The chapter concludes with some comments.

3.4 Prediction

A real discrete time random signal $\{y(k)\}$, where k is the discrete time index and $\{\cdot\}$ denotes the set of values, is most commonly obtained by sampling some analogue measurement. The voice of an individual, for example, is translated from pressure variation in air into a continuous time electrical signal by means of a microphone and then converted into a digital representation by an analogue-to-digital converter. Such discrete time random signals have statistics that are time-varying, but on a short-term basis, the statistics may be assumed to be time invariant.

The principle of the prediction of a discrete time signal is represented in Figure 3.1 and forms the basis of linear predictive coding (LPC) which underlies many compression techniques. The value of signal $y(k)$ is predicted on the basis of a sum of p past values, i.e. $y(k-1), y(k-2), \dots, y(k-p)$, weighted, by the coefficients a_i, $i = 1, 2, \dots, p$, to form a prediction, $\hat{y}(k)$. The prediction error, $e(k)$, thus becomes

$$e(k) = y(k) - \hat{y}(k) = y(k) - \sum_{i=1}^{p} a_i y(k-i). \tag{3.1}$$

The estimation of the parameters a_i is based upon minimising some function of the error, the most convenient form being the mean square error, $E[e^2(k)]$, where $E[\cdot]$ denotes the statistical expectation operator, and $\{y(k)\}$ is assumed to be statistically wide sense stationary,[3] with zero mean (Papoulis 1984). A fundamental advantage of the mean square error criterion is the so-called orthogonality condition, which implies that

$$E[e(k)y(k-j)] = 0, \qquad j = 1, 2, \ldots, p, \tag{3.2}$$

is satisfied only when a_i, $i = 1, 2, \ldots, p$, take on their optimal values. As a consequence of (3.2) and the linear structure of the predictor, the optimal weight parameters may be found from a set of linear equations, named the Yule–Walker equations (Box and Jenkins 1970),

$$\begin{bmatrix} r_{yy}(0) & r_{yy}(1) & \cdots & r_{yy}(p-1) \\ r_{yy}(1) & r_{yy}(0) & \cdots & r_{yy}(p-2) \\ \vdots & \vdots & \ddots & \vdots \\ r_{yy}(p-1) & r_{yy}(p-2) & \cdots & r_{yy}(0) \end{bmatrix} \begin{bmatrix} a_1 \\ a_2 \\ \vdots \\ a_p \end{bmatrix} = \begin{bmatrix} r_{yy}(1) \\ r_{yy}(2) \\ \vdots \\ r_{yy}(p) \end{bmatrix}, \tag{3.3}$$

where $r_{yy}(\tau) = E[y(k)y(k+\tau)]$ is the value of the autocorrelation function of $\{y(k)\}$ at lag τ. These equations may be equivalently written in matrix form as

$$\boldsymbol{R}_{yy}\boldsymbol{a} = \boldsymbol{r}_{yy}, \tag{3.4}$$

where $\boldsymbol{R}_{yy} \in \mathbb{R}^{p \times p}$ is the autocorrelation matrix and $\boldsymbol{a}, \boldsymbol{r}_{yy} \in \mathbb{R}^p$ are, respectively, the parameter vector of the predictor and the crosscorrelation vector. The Toeplitz symmetric structure of $\boldsymbol{R}_{yy}$ is exploited in the Levinson–Durbin algorithm (Hayes 1997) to solve for the optimal parameters in $\mathcal{O}(p^2)$ operations. The quality of the prediction is judged by the minimum mean square error (MMSE), which is calculated from $E[e^2(k)]$ when the weight parameters of the predictor take on their optimal values. The MMSE is calculated from $r_{yy}(0) - \sum_{i=1}^{p} a_i r_{yy}(i)$.

Real measurements can only be assumed to be locally wide sense stationary and therefore, in practice, the autocorrelation function values must be estimated from some finite length measurement in order to employ (3.3). A commonly used, but statistically biased and low variance (Kay 1993), autocorrelation estimator for application to a finite length N measurement, $\{y(0), y(1), \ldots, y(N-1)\}$, is given by

$$\hat{r}_{yy}(\tau) = \frac{1}{N} \sum_{k=0}^{N-\tau-1} y(k)y(k+\tau), \qquad \tau = 0, 1, 2, \ldots, p. \tag{3.5}$$

These estimates would then replace the exact values in (3.3) from which the weight parameters of the predictor are calculated. This procedure, however, needs to be repeated for each new length N measurement, and underlies the operation of a block-based predictor.

A second approach to the estimation of the weight parameters $\boldsymbol{a}(k)$ of a predictor is the sequential, adaptive or learning approach. The estimates of the weight parameters are refined at each sample number, k, on the basis of the new sample $y(k)$ and the prediction error $e(k)$. This yields an update equation of the form

$$\hat{\boldsymbol{a}}(k+1) = \hat{\boldsymbol{a}}(k) + \eta f(e(k), \boldsymbol{y}(k)), \qquad k \geqslant 0, \tag{3.6}$$

[3] Wide sense stationarity implies that the mean is constant, the autocorrelation function is only a function of the time lag and the variance is finite.

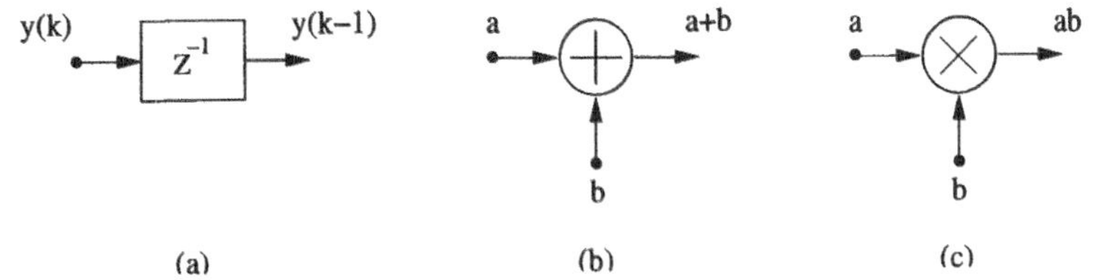

Figure 3.2 Building blocks of predictors: (a) delayer, (b) adder, (c) multiplier

where η is termed the adaptation gain, $f(\cdot)$ is some function dependent upon the particular learning algorithm, whereas $\hat{\boldsymbol{a}}(k)$ and $\boldsymbol{y}(k)$ are, respectively, the estimated weight vector and the predictor input vector. Without additional prior knowledge, zero or random values are chosen for the initial values of the weight parameters in (3.6), i.e. $\hat{a}_i(0) = 0$, or n_i, $i = 1, 2, \ldots, p$, where n_i is a random variable drawn from a suitable distribution. The sequential approach to the estimation of the weight parameters is particularly suitable for operation of predictors in statistically nonstationary environments. Both the block and sequential approach to the estimation of the weight parameters of predictors can be applied to linear and nonlinear structure predictors.

3.5 Building Blocks

In Figure 3.2 the basic building blocks of discrete time predictors are shown. A simple delayer has input $y(k)$ and output $y(k-1)$, note that the sampling period is normalised to unity. From linear discrete time system theory, the delay operation can also be conveniently represented in $\mathcal{Z}$-domain notation as the z^{-1} operator[4] (Oppenheim *et al.* 1999). An adder, or sumer, simply produces an output which is the sum of all the components at its input. A multiplier, or scaler, used in a predictor generally has two inputs and yields an output which is the product of the two inputs. The manner in which delayers, adders and multipliers are interconnected determines the architecture of linear predictors. These architectures, or structures, are shown in block diagram form in the ensuing sections.

To realise nonlinear filters and neural networks, zero-memory nonlinearities are required. Three zero-memory nonlinearities, as given in Haykin (1999b), with inputs $v(k)$ and outputs $\Phi(k)$ are described by the following operations:

$$\text{Threshold:} \qquad \Phi(v(k)) = \begin{cases} 0, & v(k) < 0, \\ 1, & v(k) \geqslant 0, \end{cases} \tag{3.7}$$

$$\text{Piecewise-linear:} \qquad \Phi(v(k)) = \begin{cases} 0, & v(k) \leqslant -\frac{1}{2}, \\ v(k), & -\frac{1}{2} < v(k) < +\frac{1}{2}, \\ 1, & v(k) \geqslant \frac{1}{2}, \end{cases} \tag{3.8}$$

$$\text{Logistic:} \qquad \Phi(v(k)) = \frac{1}{1 + \mathrm{e}^{-\beta v(k)}}, \qquad \beta \geqslant 0. \tag{3.9}$$

[4] The z^{-1} operator is a delay operator such that $\mathcal{Z}(y(k-1)) = z^{-1}\mathcal{Z}(y(k))$.

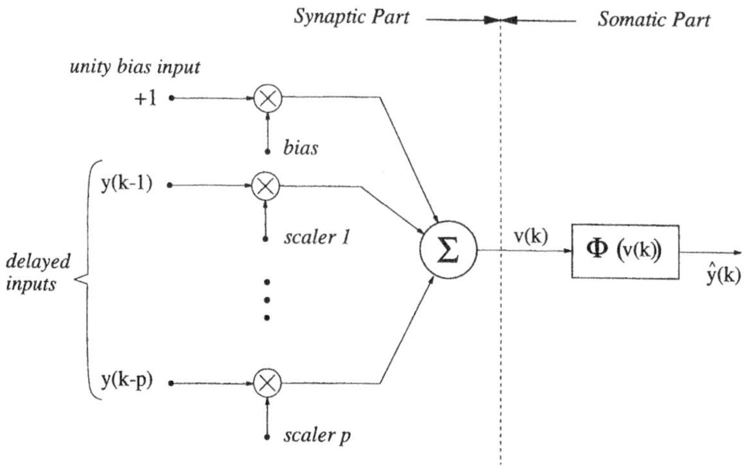

Figure 3.3 Structure of a neuron for prediction

The most commonly used nonlinearity is the logistic function since it is continuously differentiable and hence facilitates the analysis of the operation of neural networks. This property is crucial in the development of first- and second-order learning algorithms. When $\beta \to \infty$, moreover, the logistic function becomes the unipolar threshold function. The logistic function is a strictly nondecreasing function which provides for a gradual transition from linear to nonlinear operation. The inclusion of such a zero-memory nonlinearity in the output stage of the structure of a linear predictor facilitates the design of nonlinear predictors.

The threshold nonlinearity is well-established in the neural network community as it was proposed in the seminal work of McCulloch and Pitts (1943), however, it has a discontinuity at the origin. The piecewise-linear model, on the other hand, operates in a linear manner for $|v(k)| < \frac{1}{2}$ and otherwise saturates at zero or unity. Although easy to implement, neither of these zero-memory nonlinearities facilitates the analysis of the operation of nonlinear structures, because of badly behaved derivatives.

Neural networks are composed of basic processing units named neurons, or nodes, in analogy with the biological elements present within the human brain (Haykin 1999b). The basic building blocks of such artificial neurons are identical to those for nonlinear predictors. The block diagram of an artificial neuron[5] is shown in Figure 3.3. In the context of prediction, the inputs are assumed to be delayed versions of $y(k)$, i.e. $y(k-i)$, $i = 1, 2, \dots, p$. There is also a constant bias input with unity value. These inputs are then passed through $(p+1)$ multipliers for scaling. In neural network parlance, this operation in scaling the inputs corresponds to the role of the synapses in physiological neurons. A sumer then linearly combines (in fact this is an affine transformation) these scaled inputs to form an output, $v(k)$, which is termed the induced local field or activation potential of the neuron. Save for the presence of the bias input, this output is identical to the output of a linear predictor. This component of the neuron, from a biological perspective, is termed the synaptic part (Rao and Gupta 1993). Finally,

[5] The term 'artificial neuron' will be replaced by 'neuron' in the sequel.

$v(k)$ is passed through a zero-memory nonlinearity to form the output, $\hat{y}(k)$. This zero-memory nonlinearity is called the (nonlinear) activation function of a neuron and can be referred to as the somatic part (Rao and Gupta 1993). Such a neuron is a static mapping between its input and output (Hertz *et al.* 1991) and is very different from the dynamic form of a biological neuron. The synergy between nonlinear predictors and neurons is therefore evident. The structural power of neural networks in prediction results, however, from the interconnection of many such neurons to achieve the overall predictor structure in order to distribute the underlying nonlinearity.

3.6 Linear Filters

In digital signal processing and linear time series modelling, linear filters are well-established (Hayes 1997; Oppenheim *et al.* 1999) and have been exploited for the structures of predictors. Essentially, there are two families of filters: those without feedback, for which their output depends only upon current and past input values; and those with feedback, for which their output depends both upon input values and past outputs. Such filters are best described by a constant coefficient difference equation, the most general form of which is given by

$$y(k) = \sum_{i=1}^{p} a_i y(k-i) + \sum_{j=0}^{q} b_j e(k-j), \tag{3.10}$$

where $y(k)$ is the output, $e(k)$ is the input,[6] a_i, $i = 1, 2, \ldots, p$, are the (AR) feedback coefficients and b_j, $j = 0, 1, \ldots, q$, are the (MA) feedforward coefficients. In causal systems, (3.10) is satisfied for $k \geqslant 0$ and the initial conditions, $y(i)$, $i = -1, -2, \ldots, -p$, are generally assumed to be zero. The block diagram for the filter represented by (3.10) is shown in Figure 3.4. Such a filter is termed an autoregressive moving average (ARMA(p, q)) filter, where p is the order of the autoregressive, or feedback, part of the structure, and q is the order of the moving average, or feedforward, element of the structure. Due to the feedback present within this filter, the impulse response, namely the values of $y(k)$, $k \geqslant 0$, when $e(k)$ is a discrete time impulse, is infinite in duration and therefore such a filter is termed an infinite impulse response (IIR) filter within the field of digital signal processing.

The general form of (3.10) is simplified by removing the feedback terms to yield

$$y(k) = \sum_{j=0}^{q} b_j e(k-j). \tag{3.11}$$

Such a filter is termed moving average (MA(q)) and has a finite impulse response, which is identical to the parameters b_j, $j = 0, 1, \ldots, q$. In digital signal processing, therefore, such a filter is named a finite impulse response (FIR) filter. Similarly, (3.10)

[6] Notice $e(k)$ is used as the filter input, rather than $x(k)$, for consistency with later sections on prediction error filtering.

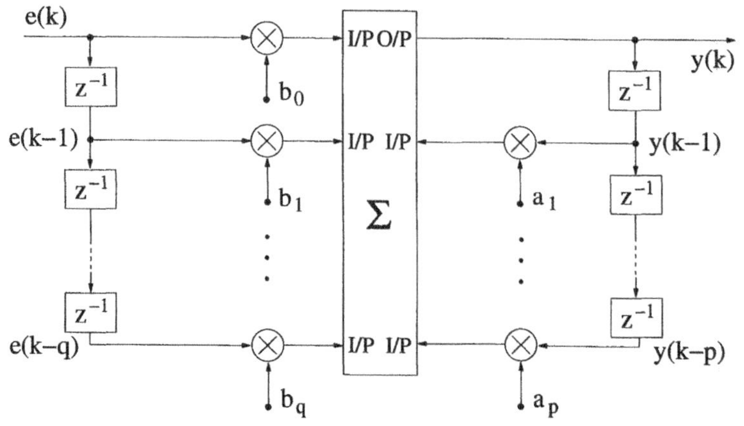

Figure 3.4 Structure of an autoregressive moving average filter (ARMA(p, q))

is simplified to yield an autoregressive (AR(p)) filter

$$y(k) = \sum_{i=1}^{p} a_i y(k-i) + e(k), \tag{3.12}$$

which is also termed an IIR filter. The filter described by (3.12) is the basis for modelling the speech production process (Makhoul 1975). The presence of feedback within the AR(p) and ARMA(p, q) filters implies that selection of the a_i, $i = 1, 2, \dots, p$, coefficients must be such that the filters are BIBO stable, i.e. a bounded output will result from a bounded input (Oppenheim *et al.* 1999).[7] The most straightforward way to test stability is to exploit the $\mathcal{Z}$-domain representation of the transfer function of the filter represented by (3.10):

$$H(z) = \frac{Y(z)}{E(z)} = \frac{b_0 + b_1 z^{-1} + \cdots + b_q z^{-q}}{1 - a_1 z^{-1} - \cdots - a_p z^{-p}} = \frac{N(z)}{D(z)}. \tag{3.13}$$

To guarantee stability, the p roots of the denominator polynomial of $H(z)$, i.e. the values of z for which $D(z) = 0$, the poles of the transfer function, must lie within the unit circle in the z-plane, $|z| < 1$. In digital signal processing, cascade, lattice, parallel and wave filters have been proposed for the realisation of the transfer function described by (3.13) (Oppenheim *et al.* 1999). For prediction applications, however, the direct form, as in Figure 3.4, and lattice structures are most commonly employed.

In signal modelling, rather than being deterministic, the input $e(k)$ to the filter in (3.10) is assumed to be an independent identically distributed (i.i.d.) discrete time random signal. This input is an integral part of a rational transfer function discrete time signal model. The filtering operations described by Equations (3.10)–(3.12),

[7] This type of stability is commonly denoted as BIBO stability in contrast to other types of stability, such as global asymptotic stability (GAS).

together with such an i.i.d. input with prescribed finite variance σ_e^2, represent respectively, ARMA(p,q), MA(q) and AR(p) signal models. The autocorrelation function of the input $e(k)$ is given by $\sigma_e^2\delta(k)$ and therefore its power spectral density (PSD) is $P_e(f) = \sigma_e^2$, for all f. The PSD of an ARMA model is therefore

$$P_y(f) = |H(f)|^2 P_e(f) = \sigma_e^2 |H(f)|^2, \qquad f \in (-\tfrac{1}{2}, \tfrac{1}{2}], \tag{3.14}$$

where f is the normalised frequency. The quantity $|H(f)|^2$ is the magnitude squared frequency domain transfer function found from (3.13) by replacing $z = e^{j2\pi f}$. The role of the filter is therefore to shape the PSD of the driving noise to match the PSD of the physical system. Such an ARMA model is well motivated by the Wold decomposition, which states that any stationary discrete time random signal can be split into the sum of uncorrelated deterministic and random components. In fact, an ARMA(∞,∞) model is sufficient to model any stationary discrete time random signal (Theiler *et al.* 1993).

3.7 Nonlinear Predictors

If a measurement is assumed to be generated by an ARMA(p,q) model, the optimal conditional mean predictor of the discrete time random signal $\{y(k)\}$

$$\hat{y}(k) = E[y(k) \mid y(k-1), y(k-2), \dots, y(0)] \tag{3.15}$$

is given by

$$\hat{y}(k) = \sum_{i=1}^{p} a_i y(k-i) + \sum_{j=1}^{q} b_j \hat{e}(k-j), \tag{3.16}$$

where the residuals $\hat{e}(k-j) = y(k-j) - \hat{y}(k-j)$, $j = 1, 2, \dots, q$. Notice the predictor described by (3.16) utilises the past values of the actual measurement, $y(k-i)$, $i = 1, 2, \dots, p$; whereas the estimates of the unobservable input signal, $e(k-j)$, $j = 1, 2, \dots, q$, are formed as the difference between the actual measurements and the past predictions. The feedback present within (3.16), which is due to the residuals $\hat{e}(k-j)$, results from the presence of the MA(q) part of the model for $y(k)$ in (3.10). No information is available about $e(k)$ and therefore it cannot form part of the prediction. On this basis, the simplest form of nonlinear autoregressive moving average NARMA(p,q) model takes the form,

$$y(k) = \Theta\biggl(\sum_{i=1}^{p} a_i y(k-i) + \sum_{j=1}^{q} b_j e(k-j)\biggr) + e(k), \tag{3.17}$$

where $\Theta(\,\cdot\,)$ is an unknown differentiable zero memory nonlinear function. Notice $e(k)$ is not included within $\Theta(\,\cdot\,)$ as it is unobservable. The term NARMA(p,q) is adopted to define (3.17), since save for the $e(k)$, the output of an ARMA(p,q) model is simply passed through the zero-memory nonlinearity $\Theta(\,\cdot\,)$.

The corresponding NARMA(p,q) predictor is given by

$$\hat{y}(k) = \Theta\biggl(\sum_{i=1}^{p} a_i y(k-i) + \sum_{j=1}^{q} b_j \hat{e}(k-j)\biggr), \tag{3.18}$$

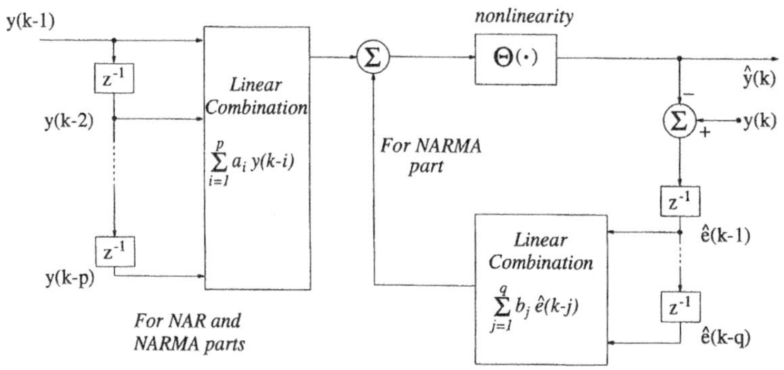

Figure 3.5 Structure of NARMA(p, q) and NAR(p) predictors

where the residuals $\hat{e}(k-j) = y(k-j) - \hat{y}(k-j)$, $j = 1, 2, \ldots, q$. Equivalently, the simplest form of nonlinear autoregressive (NAR(p)) model is described by

$$y(k) = \Theta\biggl(\sum_{i=1}^{p} a_i y(k-i)\biggr) + e(k) \tag{3.19}$$

and its associated predictor is

$$\hat{y}(k) = \Theta\biggl(\sum_{i=1}^{p} a_i y(k-i)\biggr). \tag{3.20}$$

The associated structures for the predictors described by (3.18) and (3.20) are shown in Figure 3.5. Feedback is present within the NARMA(p, q) predictor, whereas the NAR(p) predictor is an entirely feedforward structure. The structures are simply those of linear filters described in Section 3.6 with the incorporation of a zero-memory nonlinearity.

In control applications, most generally, NARMA(p, q) models also include so-called exogeneous inputs, $u(k-s)$, $s = 1, 2, \ldots, r$, and following the approach of (3.17) and (3.19) the simplest example takes the form

$$y(k) = \Theta\biggl(\sum_{i=1}^{p} a_i y(k-i) + \sum_{j=1}^{q} b_j e(k-j) + \sum_{s=1}^{r} c_s u(k-s)\biggr) + e(k) \tag{3.21}$$

and is termed a nonlinear autoregressive moving average with exogeneous inputs model, NARMAX(p, q, r), with associated predictor

$$\hat{y}(k) = \Theta\biggl(\sum_{i=1}^{p} a_i y(k-i) + \sum_{j=1}^{q} b_j \hat{e}(k-j) + \sum_{s=1}^{r} c_s u(k-s)\biggr), \tag{3.22}$$

which again exploits feedback (Chen and Billings 1989; Siegelmann *et al.* 1997). This is the most straightforward form of nonlinear predictor structure derived from linear filters.

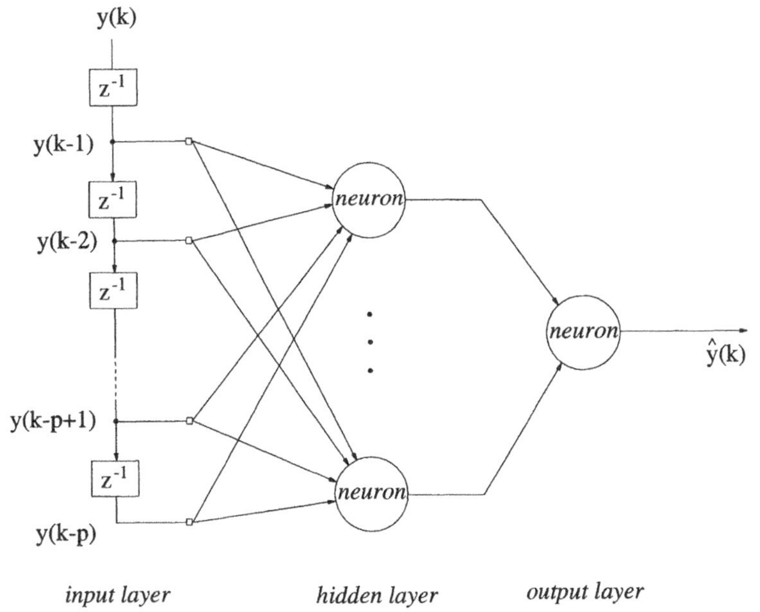

Figure 3.6 Multilayer feedforward neural network

3.8 Feedforward Neural Networks: Memory Aspects

The nonlinearity present in the predictors described by (3.18), (3.20) and (3.22) only appears at the overall output, in the same manner as in the simple neuron depicted in Figure 3.3. These predictors could therefore be referred to as single neuron structures. More generally, however, in neural networks, the nonlinearity is distributed through certain layers, or stages, of processing.

In Figure 3.6 a multilayer feedforward neural network is shown. The measurement samples appear at the input layer, and the output prediction is given from the output layer. To be consistent with the problem of prediction of a single discrete time random signal, only a single output is assumed. In between, there exist so-called hidden layers. Notice the outputs of each layer are only connected to the inputs of the adjacent layer. The nonlinearity inherent in the network is due to the overall action of all the activation functions of the neurons within the structure.

In the problem of prediction, the nature of the inputs to the multilayer feedforward neural network must capture something about the time evolution of the underlying discrete time random signal. The simplest situation is for the inputs to be time-delayed versions of the signal, i.e. $y(k-i)$, $i = 1, 2, \dots, p$, and is commonly termed a tapped delay line or delay space embedding (Mozer 1993). Such a block of inputs provides the network with a short-term memory of the signal. At each time sample, k, the inputs of the network only see the effect of one sample of $y(k)$, and Mozer (1994) terms this a high-resolution memory. The overall predictor can then be represented as

$$\hat{y}(k) = \Phi(y(k-1), y(k-2), \dots, y(k-p)), \tag{3.23}$$

where Φ represents the nonlinear mapping of the neural network.

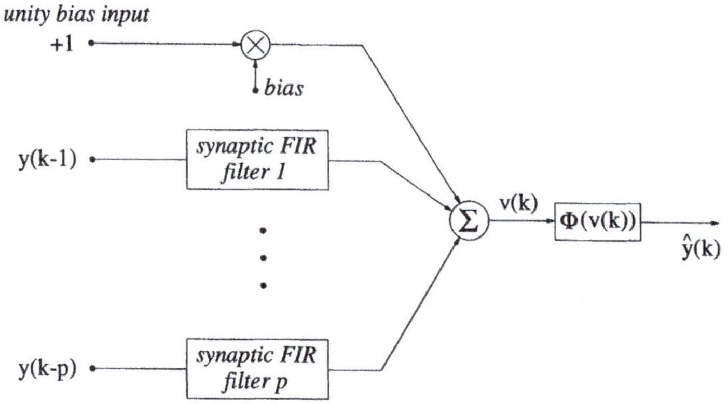

Figure 3.7 Structure of the neuron of a time delay neural network

Other forms of memory for the network include: samples with nonuniform delays, i.e. $y(k-i)$, $i = \tau_1, \tau_2, \dots, \tau_p$; exponential, where each input to the network, denoted $\tilde{y}_i(k)$, $i = 1, 2, \dots, p$, is calculated recursively from $\tilde{y}_i(k) = \mu_i \tilde{y}_i(k-1) + (1-\mu_i) y_i(k)$, where $\mu_i \in [-1, 1]$ is the exponential factor which controls the depth (Mozer 1993) or time spread of the memory and $y_i(k) = y(k-i)$, $i = 1, 2, \dots, p$. A delay line memory is therefore termed high-resolution low-depth, while an exponential memory is low-resolution but high-depth. In continuous time, Principe *et al.* (1993) proposed the Gamma memory, which provided a method to trade resolution for depth. A discrete time version of this memory is described by

$$\tilde{y}_{\mu,j}(k) = \mu \tilde{y}_{\mu,j}(k-1) + (1-\mu)\tilde{y}_{\mu,j-1}(k-1), \tag{3.24}$$

where the index j is included because it is necessary to evaluate (3.24) for $j = 0, 1, \dots, i$, where i is the delay of the particular input to the network and $\tilde{y}_{\mu,-1}(k) = y(k+1)$, for all $k \geqslant 0$, and $\tilde{y}_{\mu,j}(0) = 0$, for all $j \geqslant 0$. The form of the equation is, moreover, a convex mixture. The choice of μ controls the trade-off between depth and resolution; small μ provides low-depth and high-resolution memory, whereas high μ yields high-depth and low-resolution memory.

Restricting the memory in a multilayer feedforward neural network to the input layer may, however, lead to structures with an excessively large number of parameters. Wan (1993) therefore utilises a time-delay network where the memory is integrated within each layer of the network. Figure 3.7 shows the form of a neuron within a time-delay network, in which the multipliers of the basic neuron of Figure 3.3 are replaced by FIR filters to capture the dynamics of the input signals. Networks formed from such neurons are functionally equivalent to networks with only the memory at their input but generally have many fewer parameters, which is beneficial for learning algorithms.

The integration of memory into a multilayer feedforward network yields the structure for nonlinear prediction. It is clear, therefore, that such networks belong to the class of nonlinear filters.

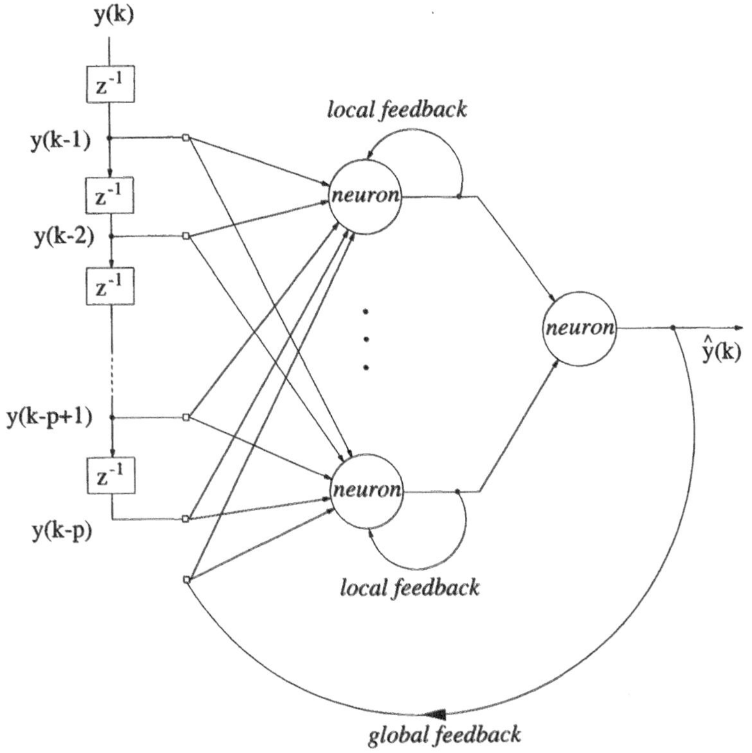

Figure 3.8 Structure of a recurrent neural network with local and global feedback

3.9 Recurrent Neural Networks: Local and Global Feedback

In Figure 3.6, the inputs to the network are drawn from the discrete time signal $y(k)$. Conceptually, it is straightforward to consider connecting the delayed versions of the output, $\hat{y}(k)$, of the network to its input. Such connections, however, introduce feedback into the network and therefore the stability of such networks must be considered, this is a particular focus of later parts of this book. The provision of feedback, with delay, introduces memory to the network and so is appropriate for prediction.

The feedback within recurrent neural networks can be achieved in either a local or global manner. An example of a recurrent neural network is shown in Figure 3.8 with connections for both local and global feedback. The local feedback is achieved by the introduction of feedback within the hidden layer, whereas the global feedback is produced by the connection of the network output to the network input. Interneuron connections can also exist in the hidden layer, but they are not shown in Figure 3.8. Although explicit delays are not shown in the feedback connections, they are assumed to be present within the neurons in order that the network is realisable. The operation of a recurrent neural network predictor that employs global feedback can now be represented by

$$\hat{y}(k) = \Phi(y(k-1), y(k-2), \dots, y(k-p), \hat{e}(k-1), \dots, \hat{e}(k-q)), \tag{3.25}$$

where again $\Phi(\,\cdot\,)$ represents the nonlinear mapping of the neural network and

$$\hat{e}(k-j) = y(k-j) - \hat{y}(k-j), \qquad j = 1, \dots, q.$$

A taxonomy of recurrent neural networks architectures is presented by Tsoi and Back (1997). The choice of structure depends upon the dynamics of the signal, learning algorithm and ultimately the prediction performance. There is, unfortunately, no hard and fast rule as to the best structure to use for a particular problem (Personnaz and Dreyfus 1998).

3.10 State-Space Representation and Canonical Form

The structures in this chapter have been developed on the basis of difference equation representations. Simple nonlinear predictors can be formed by placing a zero-memory nonlinearity within the output stage of a classical linear predictor. In this case, the nonlinearity is restricted to the output stage, as in a single layer neural network realisation. On the other hand, if the nonlinearity is distributed through many layers of weighted interconnections, the concept of neural networks is fully exploited and more powerful nonlinear predictors may ensue. For the purpose of prediction, memory stages may be introduced at the input or within the network. The most powerful approach is to introduce feedback and to unify feedback networks. Nerrand *et al.* (1994) proposed an insightful canonical state-space representation:

> Any feedback network can be cast into a canonical form that consists of a feedforward (static) network:
>
> whose outputs are the outputs of the neurons that have desired values, and the values of the state variables,
>
> whose inputs are the inputs of the network and the values of the state variables, the latter being delayed by one time unit.

Note that in the prediction of a single discrete-time random signal, the network will have only one output neuron with a predicted value. For a dynamic system, such as a recurrent neural network for prediction, the state represents *a set of quantities that summarizes all the information about the past behaviour of the system that is needed to uniquely describe its future behaviour, except for the purely external effects arising from the applied input (excitation)* (Haykin 1999b).

It should be noted that, whereas it is always possible to rewrite a nonlinear input-output model in a state-space representation, an input–output model equivalent to a given state-space model might not exist and, if it does, it is surely of higher order. Under fairly general conditions of observability of a system, however, an equivalent input–output model does exist but it may be of high order. A state-space model is likely to have lower order and require a smaller number of past inputs and, hopefully, a smaller number of parameters. This has fundamental importance when only a limited number of data samples is available. Takens' theorem (Wan 1993) implies that for a wide class of deterministic systems, there exists a diffeomorphism (one-to-one differential mapping) between a finite window of the time series and the underlying

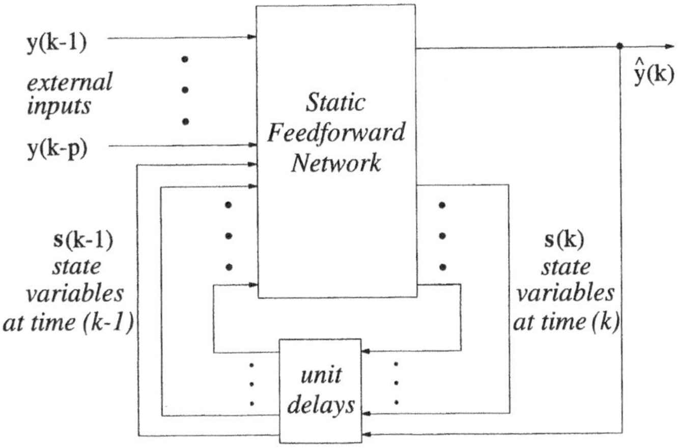

Figure 3.9 Canonical form of a recurrent neural network for prediction

state of the dynamic system which gives rise to the time series. A neural network can therefore approximate this mapping to realise a predictor.

In Figure 3.9, the general canonical form of a recurrent neural network is represented. If the state is assumed to contain N variables, then a state vector is defined as $\boldsymbol{s}(k) = [s_1(k), s_2(k), \dots, s_N(k)]^{\mathrm{T}}$, and a vector of p external inputs is given by $\boldsymbol{y}(k-1) = [y(k-1), y(k-2), \dots, y(k-p)]^{\mathrm{T}}$. The state evolution and output equations of the recurrent network for prediction are given, respectively, by

$$\boldsymbol{s}(k) = \varphi(\boldsymbol{s}(k-1), \boldsymbol{y}(k-1), \hat{y}(k-1)), \tag{3.26}$$

$$\hat{y}(k) = \psi(\boldsymbol{s}(k-1), \boldsymbol{y}(k-1), \hat{y}(k-1)), \tag{3.27}$$

where φ and Ψ represent general classes of nonlinearities. The particular choice of N minimal state variables is not unique, therefore several canonical forms[8] exist. A procedure for the determination of N for an arbitrary recurrent neural network is described by Nerrand *et al.* (1994). The NARMA and NAR predictors described by (3.18) and (3.20), however, follow naturally from the canonical state-space representation because the elements of the state vector are calculated from the inputs and outputs of the network. Moreover, even if the recurrent neural network contains local feedback and memory, it is still possible to convert the network into the above canonical form (Personnaz and Dreyfus 1998).

3.11 Summary

The aim of this chapter has been to show the commonality between the structures of nonlinear filters and neural networks. To this end, the basic building blocks for both structures have been shown to be adders, delayers, multipliers and zero-memory nonlinearities, and the manner in which these elements are interconnected defines

[8] These canonical forms stem from Jordan canonical forms of matrices and companion matrices. Notice that in fact $\hat{y}(k)$ is a state variable but shown separately to emphasise its role as the predicted output.

the particular structure. The theory of linear predictors, for stationary discrete time random signals, which are optimal in the minimum mean square prediction error sense, has been shown to be well established. The structures of linear predictors have also been demonstrated to be established in signal processing and statistical modelling. Nonlinear predictors have then been developed on the basis of defining the dynamics of a discrete time random signal by a nonlinear model. In essence, in their simplest form these predictors have two stages: a weighted linear combination of inputs and/or past outputs, as for linear predictors, and a second stage defined by a zero-memory nonlinearity.

The neuron, the fundamental processing element in neural networks, has been introduced. Multilayer feedforward neural networks have been introduced in which the nonlinearity is distributed throughout the structure. To operate in a prediction mode, some local memory is required either at the input or integral to the network structure. Recurrent neural networks have then been formulated by connecting delayed versions of the global output to the input of a multilayer feedforward structure; or by the introduction of local feedback within the network. A canonical state-space form has been used to represent an arbitrary neural network.

4

Activation Functions Used in Neural Networks

4.1 Perspective

The choice of nonlinear activation function has a key influence on the complexity and performance of artificial neural networks, note the term *neural network* will be used interchangeably with the term *artificial neural network*. The brief introduction to activation functions given in Chapter 3 is therefore extended. Although sigmoidal nonlinear activation functions are the most common choice, there is no strong *a priori* justification why models based on such functions should be preferred to others.

We therefore introduce neural networks as universal approximators of functions and trajectories, based upon the Kolmogorov universal approximation theorem, which is valid for both feedforward and recurrent neural networks. From these universal approximation properties, we then demonstrate the need for a sigmoidal activation function within a neuron. To reduce computational complexity, approximations to sigmoid functions are further discussed. The use of nonlinear activation functions suitable for hardware realisation of neural networks is also considered.

For rigour, we extend the analysis to complex activation functions and recognise that a suitable complex activation function is a Möbius transformation. In that context, a framework for rigorous analysis of some inherent properties of neural networks, such as fixed points, nesting and invertibility based upon the theory of modular groups of Möbius transformations is provided.

All the relevant definitions, theorems and other mathematical terms are given in Appendix B and Appendix C.

4.2 Introduction

A century ago, a set of 23 (originally) unsolved problems in mathematics was proposed by David Hilbert (Hilbert 1901–1902). In his lecture 'Mathematische Probleme' at the second International Congress of Mathematics held in Paris in 1900, he presented 10 of them. These problems were designed to serve as examples for the kinds of problems whose solutions would lead to further development of disciplines in mathematics.

His 13th problem concerned solutions of polynomial equations. Although his original formulation dealt with properties of the solution of the seventh degree algebraic equation,[1] this problem can be restated as: *Prove that there are continuous functions of* n *variables, not representable by a superposition of continuous functions of* $(n-1)$ *variables.* In other words, could a general algebraic equation of a high degree be expressed by sums and compositions of single-variable functions?[2] In 1957, Kolmogorov showed that the conjecture of Hilbert was not correct (Kolmogorov 1957).

Kolmogorov's theorem is a general representation theorem stating that any real-valued continuous function f defined on an n-dimensional cube I^n $(n \geqslant 2)$ can be represented as

$$f(x_1, \dots, x_n) = \sum_{q=1}^{2n+1} \Phi_q\left(\sum_{p=1}^{n} \psi_{pq}(x_p)\right), \tag{4.1}$$

where $\Phi_q(\cdot)$, $q = 1, \dots, 2n+1$, and $\psi_{pq}(\cdot)$, $p = 1, \dots, n$, $q = 1, \dots, 2n+1$, are typically nonlinear continuous functions of one variable.

For a neural network representation, this means that an activation function of a neuron has to be nonlinear to form a universal approximator. This also means that every continuous function of many variables can be represented by a four-layered neural network with two hidden layers and an input and output layer, whose hidden units represent mappings Φ and ψ. However, this does not mean that a network with two hidden layers necessarily provides an accurate representation of function f. In fact, functions ψ_{pq} of Kolmogorov's theorem are quite often highly nonsmooth, whereas for a neural network we want smooth nonlinear activation functions, as is required by gradient-descent learning algorithms (Poggio and Girosi 1990). Vitushkin (1954) showed that there are functions of more than one variable which do not have a representation by superpositions of differentiable functions (Beiu 1998). Important questions about Kolmogorov's representation are therefore existence, constructive proofs and bounds on the size of a network needed for approximation.

Kolmogorov's representation has been improved by several authors. Sprecher (1965) replaced functions ψ_{pq} in the Kolmogorov representation by $\lambda^{pq}\psi_q$, where λ is a constant and ψ_q are monotonic increasing functions which belong to the class of Lipschitz functions. Lorentz (1976) showed that the functions Φ_q can be replaced by only one function Φ. Hecht-Nielsen reformulated this result for MLPs so that they are able to approximate any function. In this case, functions ψ are nonlinear activation functions in hidden layers, whereas functions Φ are nonlinear activation functions in the output layer. The functions Φ and ψ are found, however, to be generally highly nonsmooth. Further, in Katsuura and Sprecher (1994), the function ψ is obtained through a graph that is the limit point of an iterated composition of contraction mappings on their domain.

In applications of neural networks for universal approximation, the existence proof for approximation by neural networks is provided by Kolmogorov's theorem, which

[1] Hilbert conjectured that the roots of the equation $x^7 + ax^3 + bx^2 + cx + 1 = 0$ as functions of coefficients a, b, c are not representable by sums and superpositions of functions of two coefficients, or 'Show the impossibility of solving the general seventh degree equation by functions of two variables.'

[2] For example, function xy is a composition of functions $g(\cdot) = \exp(\cdot)$ and $h(\cdot) = \log(\cdot)$, therefore $xy = e^{\log(x)+\log(y)} = g(h(x) + h(y))$ (Gorban and Wunsch 1998).

in the neural network community was first recognised by Hecht-Nielsen (1987) and Lippmann (1987). The first constructive proof of neural networks as universal approximators was given by Cybenko (1989). Most of the analyses rest on the denseness property of nonlinear functions that approximate the desired function in the space in which the desired function is defined. In Cybenko's results, for instance, if σ is a continuous discriminatory function,[3] then finite sums of the form,

$$g(\boldsymbol{x}) = \sum_{i=1}^{N} w_i \sigma(\boldsymbol{a}_i^{\mathrm{T}} \boldsymbol{x} + b_i), \tag{4.2}$$

where w_i, b_i, $i = 1, \dots, N$, are coefficients, are dense in the space of continuous functions defined on $[0, 1]^n$. Following the classical approach to approximation, this means that given any continuous function f defined on $[0, 1]^N$ and any $\varepsilon > 0$, there is a $g(\boldsymbol{x})$ given by (4.2) for which $|g(\boldsymbol{x}) - f(\boldsymbol{x})| < \varepsilon$ for all $\boldsymbol{x} \in [0, 1]^N$. Cybenko then concludes that any bounded and measurable sigmoidal function is discriminatory (Cybenko 1989), and that a three-layer neural network with a sufficient number of neurons in its hidden layer can represent an arbitrary function (Beiu 1998; Cybenko 1989).

Funahashi (1989) extended this to include sigmoidal functions so that any continuous function is approximately realisable by three-layer networks with bounded and monotonically increasing activation functions within hidden units. Hornik *et al.* (1989) showed that the output function does not have to be continuous, and they also proved that a neural network can approximate simultaneously both a function and its derivative (Hornik *et al.* 1990). Hornik (1990) further showed that the activation function has to be bounded and nonconstant (but not necessarily continuous), Kurkova (1992) revealed the existence of an approximate representation of functions by superposition of nonlinear functions within the constraints of neural networks. Leshno *et al.* (1993) relaxed the condition for the activation function to be 'locally bounded piecewise continuous' (i.e. if and only if the activation function is not a polynomial). This result encompasses most of the activation functions commonly used.

Funahashi and Nakamura (1993), in their article 'Approximation of dynamical systems by continuous time recurrent neural networks', proved that the universal approximation theorem also holds for trajectories and patterns and for recurrent neural networks. Li (1992) also showed that recurrent neural networks are universal approximators. Some recent results, moreover, suggest that 'smaller nets perform better' (Elsken 1999), which recommends recurrent neural networks, since a small-scale RNN has dynamics that can be achieved only by a large scale feedforward neural network.

[3] $\sigma(\cdot)$ is discriminatory if for a Borel measure μ on $[0, 1]^N$,

$$\int_{[0,1]^N} \sigma(\boldsymbol{a}^{\mathrm{T}} \boldsymbol{x} + b) \, \mathrm{d}\mu(\boldsymbol{x}) = 0, \quad \forall \boldsymbol{a} \in \mathbb{R}^N, \quad \forall b \in \mathbb{R},$$

implies that $\mu = 0$. The sigmoids Cybenko considered had limits

$$\sigma(t) = \begin{cases} 0, & t \to -\infty, \\ 1, & t \to \infty. \end{cases}$$

This justifies the use of the logistic function $\sigma(x) = 1/(1 + \mathrm{e}^{-\beta x})$ in neural network applications.

Sprecher (1993) considered the problem of dimensionality of neural networks and demonstrated that the number of hidden layers is independent of the number of input variables N. Barron (1993) described spaces of functions that can be approximated by the relaxed algorithm of Jones using functions computed by single-hidden-layer networks or perceptrons. Attali and Pages (1997) provided an approach based upon the Taylor series expansion. Maiorov and Pinkus have given lower bounds for neural network based approximation (Maiorov and Pinkus 1999). Approximation ability of neural networks has also been rigorously studied in Williamson and Helmke (1995).

Sigmoid neural units usually use a 'bias' or 'threshold' term in computing the activation potential (combination function, net input $\mathrm{net}(k) = \boldsymbol{x}^{\mathrm{T}}(k)\boldsymbol{w}(k)$) of the neural unit. The bias term is a connection weight from a unit with a constant value as shown in Figure 3.3. The bias unit is connected to every neuron in a neural network, the weight of which can be trained just like any other weight in a neural network.

From the geometric point of view, for an MLP with N output units, the operation of the network can be seen as defining an N-dimensional hypersurface in the space spanned by the inputs to the network. The weights define the position of this surface. Without a bias term, all the hypersurfaces would pass through the origin (Mandic and Chambers 2000c), which in turn means that the universal approximation property of neural networks would not hold if the bias was omitted.

A result by Hornik (1993) shows that a sufficient condition for the universal approximation property without biases is that no derivative of the activation function vanishes at the origin, which implies that a fixed nonzero bias can be used instead of a trainable bias.

Why use activation functions?

To introduce nonlinearity into a neural network, we employ nonlinear activation (output) functions. Without nonlinearity, since a composition of linear functions is again a linear function, an MLP would not be functionally different from a linear filter and would not be able to perform nonlinear separation and trajectory learning for nonlinear and nonstationary signals.

Due to the Kolmogorov theorem, almost any nonlinear function is a suitable candidate for an activation function of a neuron. However, for gradient-descent learning algorithms, this function ought to be differentiable. It also helps if the function is bounded.[4] For the output neuron, one should either use an activation function suited to the distribution of desired (target) values, or preprocess the inputs to achieve this goal. If, for instance, the desired values are positive but have no known upper bound, an exponential nonlinear activation function can be used.

It is important to identify classes of functions and processes that can be approximated by artificial neural networks. Similar problems occur in nonlinear circuit theory, where analogue nonlinear devices are used to synthesise desired transfer functions (gyrators, impedance converters), and in digital signal processing where digital filters

[4] The function $f(x) = \mathrm{e}^x$ is a suitable candidate for an activation function and is suitable for unbounded signals. It is also continuously differentiable. However, to control the dynamics, fixed points and invertibility of a neural network, it is desirable to have bounded, 'squashing' activation functions for neurons.

are designed to approximate arbitrarily well any transfer function. Fuzzy sets are also universal approximators of functions and their derivatives (Kreinovich *et al.* 2000; Mitaim and Kosko 1996, 1997).

4.3 Overview

We first explain the requirements of an activation function mathematically. We will then introduce different types of nonlinear activation functions and discuss their properties and realisability. Finally, a complex form of activation functions within the framework of Möbius transformations will be introduced.

4.4 Neural Networks and Universal Approximation

Learning an input–output relationship from examples using a neural network can be considered as the problem of approximating an unknown function $f(x)$ from a set of data points (Girosi and Poggio 1989a). This is why the analysis of neural networks for approximation is important for neural networks for prediction, and also system identification and trajectory tracking. The property of uniform approximation is also found in algebraic and trigonometric polynomials, such as in the case of Weierstrass and Fourier representation, respectively.

A neural activation function $\sigma(\cdot)$ is typically chosen to be a continuous and differentiable nonlinear function that belongs to the class $S = \{\sigma_i \mid i = 1, 2, \ldots, n\}$ of sigmoid[5] functions having the following desirable properties[6]

(i) $\sigma_i \in S$ for $i = 1, \ldots, n$;

(ii) $\sigma_i(x_i)$ is a continuously differentiable function;

(iii) $\sigma_i'(x_i) = \dfrac{\mathrm{d}\sigma_i(x_i)}{\mathrm{d}x_i} > 0$ for all $x_i \in \mathbb{R}$;

(iv) $\sigma_i(\mathbb{R}) = (a_i, b_i)$, $a_i, b_i \in \mathbb{R}$, $a_i \neq b_i$;

(v) $\sigma_i'(x) \to 0$ as $x \to \pm\infty$;

(vi) $\sigma_i'(x)$ takes a global maximal value $\max_{x\in\mathbb{R}} \sigma_i'(x)$ at a unique point $x = 0$;

(vii) a sigmoidal function has only one inflection point, preferably at $x = 0$;

(viii) from (iii), function σ_i is monotonically nondecreasing, i.e. if $x_1 < x_2$ for each $x_1, x_2 \in \mathbb{R} \Rightarrow \sigma_i(x_1) \leqslant \sigma_i(x_2)$;

(ix) σ_i is uniformly Lipschitz, i.e. there exists a constant $L > 0$ such that $\|\sigma_i(x_1) - \sigma_i(x_2)\| \leqslant L\|x_1 - x_2\|$, $\forall x_1, x_2 \in \mathbb{R}$, or in other words

$$\frac{\sigma_i(x_1) - \sigma_i(x_2)}{x_1 - x_2} \leqslant L, \qquad \forall x_1, x_2 \in \mathbb{R}, \qquad x_1 \neq x_2.$$

---

[5] Sigmoid means S-shaped.

[6] The constraints we impose on sigmoidal functions are stricter than the ones commonly employed.

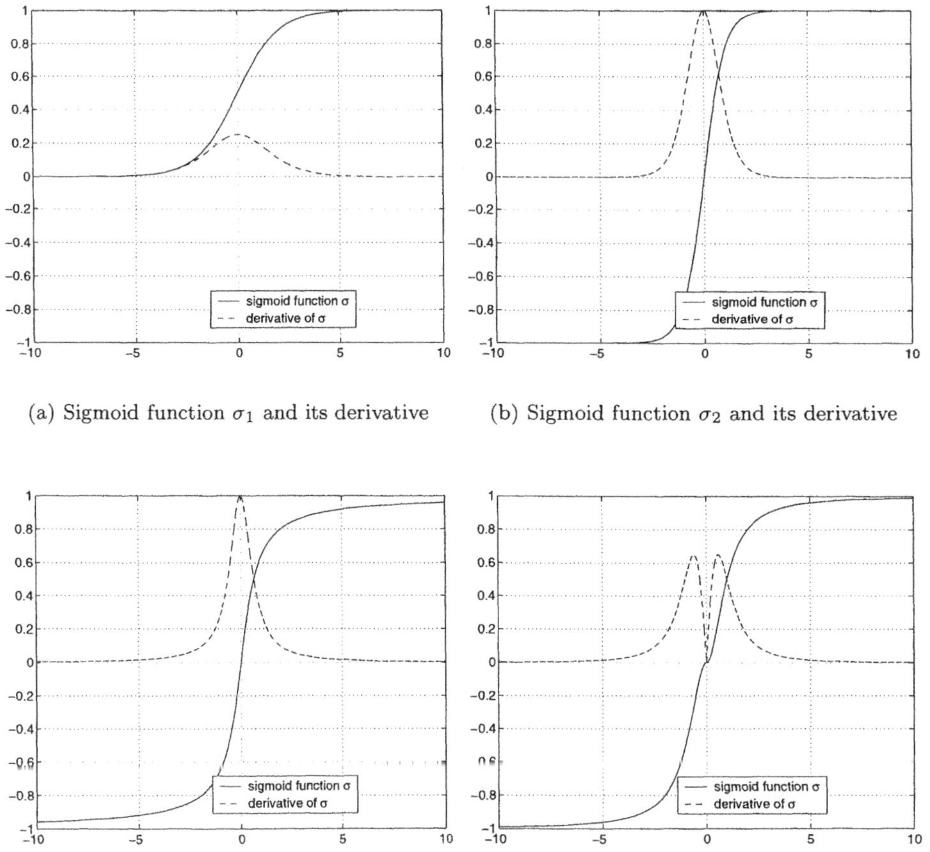

(a) Sigmoid function σ_1 and its derivative

(b) Sigmoid function σ_2 and its derivative

(c) Sigmoid function σ_3 and its derivative

(d) Sigmoid function σ_4 and its derivative

Figure 4.1 Sigmoid functions and their derivatives

We will briefly discuss some of the above requirements. Property (ii) represents continuous differentiability of a sigmoid function, which is important for higher order learning algorithms, which require not only existence of the Jacobian matrix, but also the existence of a Hessian and matrices containing higher-order derivatives. This is also necessary if the behaviour of a neural network is to be described via Taylor series expansion about the current point in the state space of the network. Property (iii) states that a sigmoid should have a positive first derivative, which in turn means that a gradient descent algorithm which is employed for training of a neural network should have gradient vectors pointing towards the bottom of the bowl shaped error performance surface, which is the global minimum of the surface. Property (vi) means that the point around which the first derivative is centred is the origin. This is connected with property (vii) which means that the second derivative of the activation function should change its sign at the origin. Going back to the error performance surface, this means that irrespective of whether the current prediction error is positive or negative, the gradient vector of the network at that point should point downwards. Monotonicity, required by (viii) is useful for uniform convergence of algorithms and in search for fixed points of neural networks. Finally, the Lipschitz condition is connected with the boundedness of an activation function and degenerates into requirements of uniform convergence given by the contraction mapping theorem for $L < 1$.

Surveys of neural transfer functions can be found in Duch and Jankowski (1999) and Cichocki and Unbehauen (1993). Examples of sigmoidal functions are

$$\left.\begin{aligned} \sigma_1(x) &= \frac{1}{1+\mathrm{e}^{-\beta x}}, \qquad \beta \in \mathbb{R}, \\ \sigma_2(x) &= \tanh(\beta x) = \frac{\mathrm{e}^{\beta x} - \mathrm{e}^{-\beta x}}{\mathrm{e}^{\beta x} + \mathrm{e}^{-\beta x}}, \qquad \beta \in \mathbb{R}, \\ \sigma_3(x) &= \frac{2}{\pi} \arctan(\tfrac{1}{2}\pi\beta x), \qquad \beta \in \mathbb{R}, \\ \sigma_4(x) &= \frac{x^2}{1+x^2} \operatorname{sgn}(x), \end{aligned}\right\} \tag{4.3}$$

where $\sigma(x) = \Phi(x)$ as in Chapter 3. For $\beta = 1$, these functions and their derivatives are given in Figure 4.1. The function σ_1, also known as the logistic function,[7] is unipolar, whereas the other three activation functions are bipolar. Two frequently used sigmoid functions in neural networks are σ_1 and σ_2. Their derivatives are also simple to calculate, and are

$$\left.\begin{aligned} \sigma_1'(x) &= \beta\sigma_1(x)(1-\sigma_1(x)), \\ \sigma_2'(x) &= \beta\,\mathrm{sech}^2(x) = \beta(1-\sigma_2^2(x)). \end{aligned}\right\} \tag{4.4}$$

We can easily modify activation functions to have different saturation values. For the logistic function $\sigma_1(x)$, whose saturation values are $(0, 1)$, to obtain saturation values $(-1, 1)$, we perform

$$\sigma_s(x) = \frac{2}{1+\mathrm{e}^{-\beta x}} - 1. \tag{4.5}$$

To modify the input data to fall within the range of an activation function, we can normalise, standardise or rescale the input data, using mean μ, standard deviation std and the minimum and maximum range $R_{\min}$ and $R_{\max}$.[8] Cybenko (1989) has shown that neural networks with a single hidden layer of neurons with sigmoidal functions are

---

[7] The logistic map $\dot{f} = rf(1 - f/K)$ (Strogatz 1994) is used to describe population dynamics, where f is the growth of a population of organisms, r denotes the growth rate and K is the so-called carrying capacity (population cannot grow unbounded). Fixed points of this map in the phase space are 0 and K, hence the population always approaches the carrying capacity. Under these conditions, the graph of $f(t)$ belongs to the class of sigmoid functions.

[8] To normalise the input data to $\mu = 0$ and std $= 1$, we calculate

$$\mu = \frac{\sum_{i=1}^{N} x_i}{N}, \qquad \mathrm{std} = \sqrt{\frac{\sum_{i=1}^{N}(x_i - \mu)^2}{N}},$$

and perform the standardisation of the input data as $\tilde{x}_i = (x_i - \mu)/\mathrm{std}$. To translate data to midrange universal approximators and provided they have enough neurons, can approximate an arbitrary continuous function on a compact set with arbitrary precision. These results do not mean that sigmoidal functions always provide an optimal choice.[9]

Two functions determine the way signals are processed by neurons.

Combination functions. Each processing unit in a neural network performs some mathematical operation on values that are fed into it via synaptic connections (weights) from other units. The resulting value is called the activation potential or 'net input'. This operation is known as a 'combination function', 'activation function' or 'net input'. Any combination function is a net: $\mathbb{R}^N \to \mathbb{R}$ function, and its output is a scalar. Most frequently used combination functions are inner product (linear) combination functions (as in MLPs and RNNs) and Euclidean or Mahalanobis distance combination functions (as in RBF networks).

Activation functions. Neural networks for nonlinear processing of signals map their net input provided by a combination function onto the output of a neuron using a scalar function called a 'nonlinear activation function', 'output function' or sometimes even 'activation function'. The entire functional mapping performed by a neuron (composition of a combination function and a nonlinear activation function) is sometimes called a 'transfer' function of a neuron $\sigma : \mathbb{R}^N \to \mathbb{R}$. Nonlinear activation functions with a bounded range are often called 'squashing' functions, such as the commonly used tanh and logistic functions. If a unit does not transform its net input, it is said to have an 'identity' or 'linear' activation function.[10]

Distance based combination functions (proximity functions) $D(\boldsymbol{x};\boldsymbol{t}) \propto \|\boldsymbol{x}-\boldsymbol{t}\|$, are used to calculate how close $\boldsymbol{x}$ is to a prototype vector $\boldsymbol{t}$. It is also possible to use some combination of the inner product and distance activation functions, for instance in the form $\alpha \boldsymbol{w}^{\mathrm{T}}\boldsymbol{x} + \beta\|\boldsymbol{x}-\boldsymbol{t}\|$ (Duch and Jankowski 1999). Many other functions can be used to calculate the net input, as for instance

$$A(\boldsymbol{x},\boldsymbol{w}) = w_0 + \sum_{i=1}^{N} w_i x_i + w_{N+1}\sum_{i=1}^{N} x_i^2$$

(Ridella *et al.* 1997).

4.5 Other Activation Functions

By the universal approximation theorems, there are many choices of the nonlinear activation function. Therefore, in this section we describe some commonly used application-motivated activation functions of a neuron.

---

0 and standardise to range R, we perform

$$Z = \frac{\max_i\{x_i\} + \min_i\{x_i\}}{R}, \qquad S_x = \max_i\{x_i\} - \min_i\{x_i\}, \qquad x_i^n = \frac{x_i - Z}{S_x/R}.$$

[9] Rational transfer functions (Leung and Haykin 1993) and Gaussian transfer functions also allow NNs to implement universal approximators.

[10] http://www.informatik.uni-freiburg.de/~heinz/faq.html

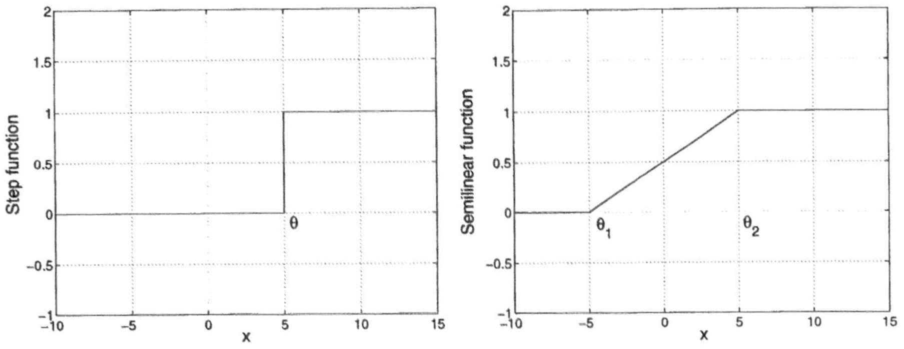

(a) Step activation function

(b) Semilinear activation function

Figure 4.2 Step and semilinear activation function

The hard-limiter Heaviside (step) function was frequently used in the first implementations of neural networks, due to its simplicity. It is given by

$$H(x) = \begin{cases} 0, & x \leqslant \theta, \\ 1, & x > \theta, \end{cases} \tag{4.6}$$

where θ is some threshold. A natural extension of the step function is the multistep function $H_{\mathrm{MS}}(x; \boldsymbol{\theta}) = y_i, \ \theta_i \leqslant x \leqslant \theta_{i+1}$. A variant of this function resembles a staircase $\theta_1 < \theta_2 < \cdots < \theta_N \Leftrightarrow y_1 < y_2 < \cdots < y_N$, and is often called the staircase function. The semilinear function is defined as

$$H_{\mathrm{SL}}(x; \theta_1, \theta_2) = \begin{cases} 0, & x \leqslant \theta_1, \\ (x - \theta_1)/(\theta_2 - \theta_1), & \theta_1 < x \leqslant \theta_2, \\ 1, & x > \theta_2. \end{cases} \tag{4.7}$$

The functions (4.6) and (4.7) are depicted in Figure 4.2. Both the above mentioned functions have discontinuous derivatives, preventing the use of gradient-based training procedures. Although they are, strictly speaking, S-shaped, we do not use them for neural networks for real-time processing, and this is why we restricted ourselves to differentiable functions in our nine requirements that a suitable activation function should satisfy. With the development of neural network theory, these discontinuous functions were later generalised to logistic functions, leading to the *graded response neurons*, which are suitable for gradient-based training. Indeed, the logistic function

$$\sigma(x) = \frac{1}{1 + \mathrm{e}^{-\beta x}} \tag{4.8}$$

degenerates into the step function (4.6), as $\beta \to \infty$.

Many other activation functions have been designed for special purposes. For instance, a modified activation function which enables single layer perceptrons to solve

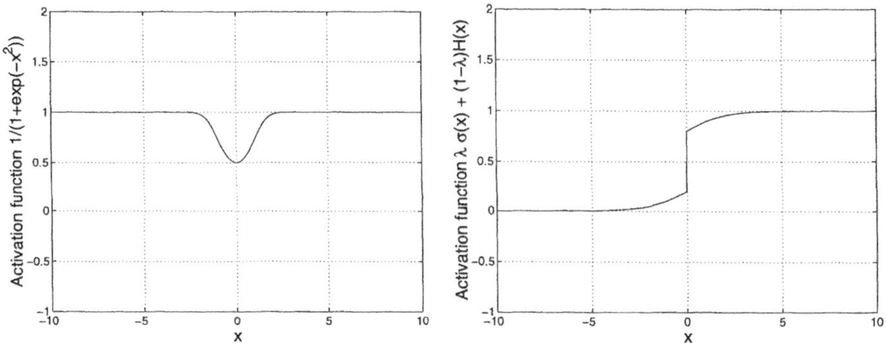

(a) The function (4.9) (b) The function (4.10) for $\lambda = 0.4$

Figure 4.3 Other activation functions

some linearly inseparable problems has been proposed in Zhang and Sarhadi (1993) and takes the form,

$$f(x) = \frac{1}{1 + \mathrm{e}^{-(x^2 + \mathrm{bias})}}. \tag{4.9}$$

The function (4.9) is differentiable and therefore a network based upon this function can be trained using gradient descent methods. The square operation in the exponential term of the function enables individual neurons to perform limited nonlinear classification. This activation function has been employed for image segmentation (Zhang and Sarhadi 1993). There have been efforts to combine two or more forms of commonly used functions to obtain an improved activation function. For instance, a function defined by

$$f(x) = \lambda\sigma(x) + (1 - \lambda)H(x), \tag{4.10}$$

where $\sigma(x)$ is a sigmoid function, $H(x)$ is a hard-limiting function and $0 \leqslant \lambda \leqslant 1$, has been used in Jones (1990). The function (4.10) is a weighted sum of functions σ and H. The functions (4.9) and (4.10) are depicted in Figure 4.3.

Another possibility is to use a linear combination of sigmoid functions instead of a single sigmoid function as an activation function of a neuron. A sigmoid packet f is therefore defined as a linear combination of a set of sigmoid functions with different amplitudes h, slopes β and biases b (Peng *et al.* 1998). This function is defined as

$$f(x) = \sum_{n=1}^{N} h_n \sigma_n = \sum_{n=1}^{N} \frac{h_n}{1 + \mathrm{e}^{-\beta_n x + b_n}}. \tag{4.11}$$

During the learning phase, all parameters (h, β, b) can be adjusted for adaptive shape-refining. Intuitively, a Gaussian-shaped activation function can be, for instance, approximated by a difference of two sigmoids, as shown in Figure 4.4. Other options include spline neural networks[11] (Guarnieri *et al.* 1999; Vecci *et al.* 1997) and wavelet

[11] Splines are piecewise polynomials (often cubic) that are smooth and can retain the 'squashing property'.

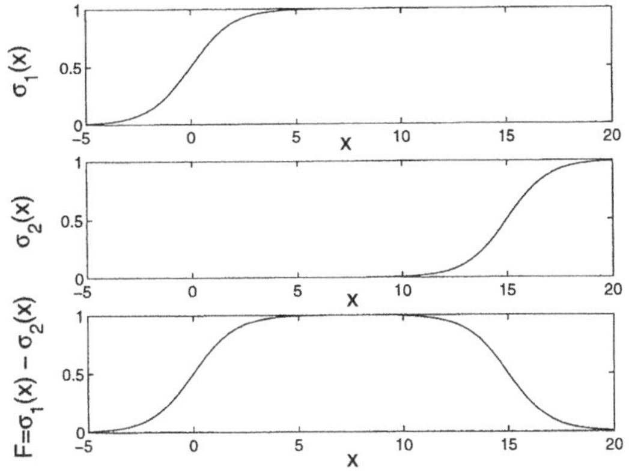

Figure 4.4 Approximation capabilities of sigmoid functions

based neural networks (Zhang *et al.* 1995), where the structure of the network is similar to the RBF, except that the RBFs are replaced by orthonormal scaling functions that are not necessarily radial-symmetric.

For neural systems operating on chaotic input signals, the most commonly used activation function is a sinusoidal function. Another activation function that is often used in order to detect chaos in the input signal is the so-called saturated-modulus function given by (Dogaru *et al.* 1996; Nakagawa 1996)

$$\varphi(x) = \begin{cases} |x|, & |x| \leqslant 1, \\ 1, & |x| > 1. \end{cases} \tag{4.12}$$

This activation function ensures chaotic behaviour even for a very small number of neurons within the network. This function corresponds to the rectifying operation used in electronic instrumentation and is therefore called a saturated modulus or saturated rectifier function.

4.6 Implementation Driven Choice of Activation Functions

When neurons of a neural network are realised in hardware, due to the limitation of processing power and available precision, activation functions can be significantly different from their ideal forms (Al-Ruwaihi 1997; Yang *et al.* 1998). Implementations of nonlinear activation functions of neurons proposed by various authors are based on a look-up table, McLaurin polynomial approximation, piecewise linear approximation or stepwise linear approximation (Basaglia *et al.* 1995; Murtagh and Tsoi 1992). These approximations require more iterations of the learning algorithm to converge as compared with standard sigmoids.

For neurons based upon look-up tables, samples of a chosen sigmoid are put into a ROM or RAM to store the desired activation function. Alternatively, we use simplified activation functions that approximate the chosen activation function and are not demanding regarding processor time and memory. Thus, for instance, for the logistic function, its derivative can be expressed as $\sigma'(x) = \sigma(x)(1 - \sigma(x))$, which is simple

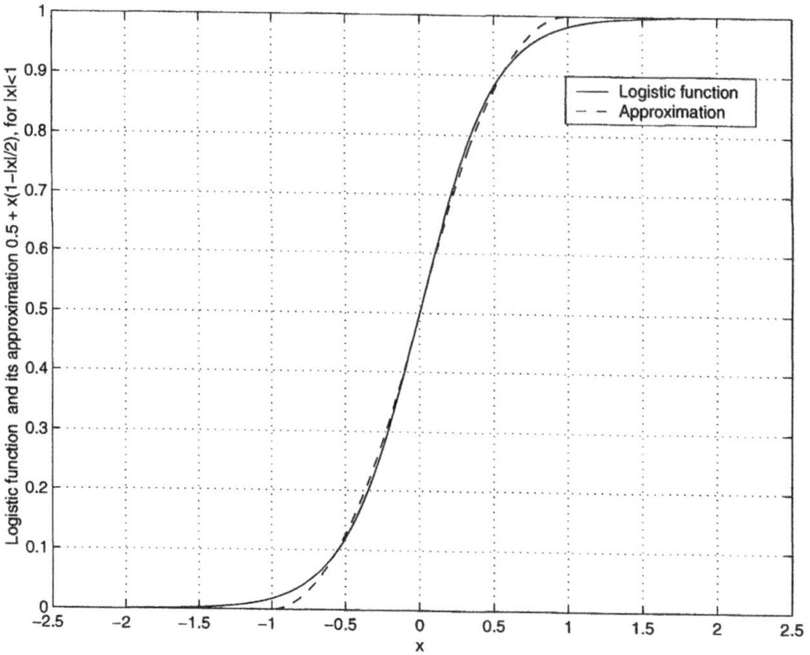

Figure 4.5 Logistic function and its approximation

to calculate. The logistic function can be approximated using

$$f(x) = \begin{cases} 0, & x \leqslant -1, \\ 0.5 + x(1 - |x|/2), & -1 < x < 1, \\ 1, & x \geqslant 1. \end{cases} \tag{4.13}$$

The maximal absolute deviation between this function and the logistic function is less than 0.02.[12] The function (4.13) is compared with the logistic function and shown in Figure 4.5. There are several other approximations. To save on computational complexity, we can approximate sigmoid functions with a series of straight lines, i.e. by a piecewise-linear functions. Another sigmoid was proposed by David Elliott (Elliott 1993)

$$\sigma(x) = \frac{x}{1 + |x|} \tag{4.14}$$

with derivative

$$\sigma' = \frac{1}{(1 + |x|)^2} = (1 - |\sigma|)^2,$$

which is also easy to calculate. The function (4.14) and its derivative are shown in Figure 4.6.

In a digital VLSI implementation of an MLP, the computation of the activation function of each neuron is performed using a look-up table (LUT), i.e. a RAM or

[12] http://www.dontveter.com/bpr/bpr.html

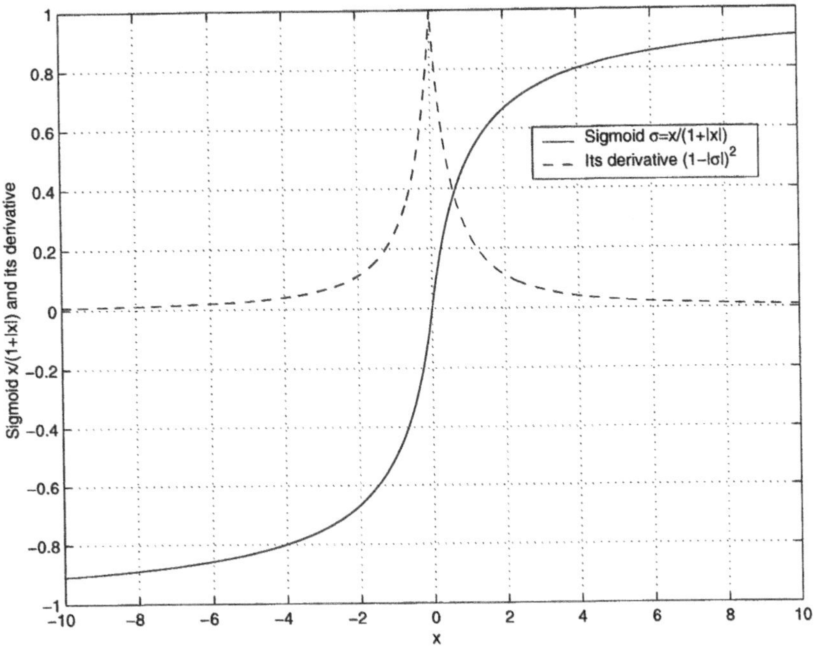

Figure 4.6 Sigmoid function and its derivative

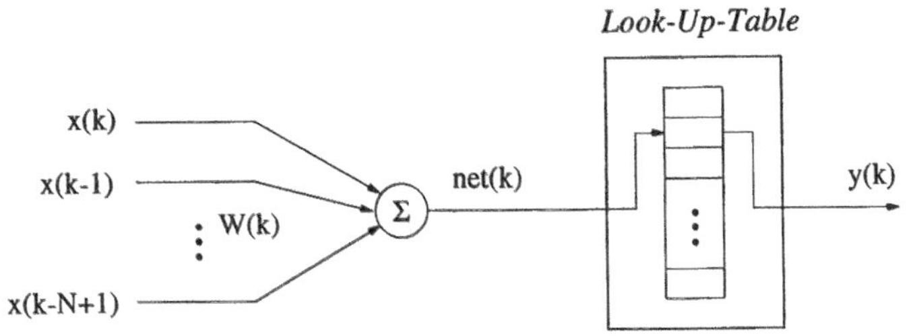

Figure 4.7 LUT neuron

ROM memory which is addressed in some way (Piazza *et al.* 1993). An adaptive LUT based neuron is depicted in Figure 4.7.

Although sigmoidal functions are a typical choice for MLPs, several other functions have been considered. Recently, the use of polynomial activation functions has been proposed (Chon and Cohen 1997; Piazza *et al.* 1992; Song and Manry 1993). Networks with polynomial neurons have been shown to be isomorphic to Volterra filters (Chon and Cohen 1997; Song and Manry 1993). However, calculating a polynomial activation

$$f(x) = a_0 + a_1(x) + \cdots + a_M x^M \tag{4.15}$$

for every neuron and every time instant is extremely computationally demanding and is unlikely to be acceptable for real-time applications. Since their calculation is much slower than simple arithmetic operations, other sigmoidal functions might be useful for hardware implementations of neural networks for online applications. An overview of such functions is given in Duch and Jankowski (1999).

4.7 MLP versus RBF Networks

MLP- and RBF-based neural networks are the two most commonly used types of feed-forward networks. A fundamental difference between the two is the way in which hidden units combine values at their inputs. MLP networks use inner products, whereas RBFs use Euclidean or Mahalanobis distance as a combination function. An RBF is given by

$$f(\boldsymbol{x}) = \sum_{i=1}^{N} c_i G(\boldsymbol{x}; \boldsymbol{t}_i), \tag{4.16}$$

where $G(\,\cdot\,)$ is a basis function, c_i, $i = 1, \dots, N$, are coefficients, $\boldsymbol{t}_i$, $i = 1, \dots, N$, are centres of radial bases, and $\boldsymbol{x}$ is the input vector.

Both multilayer perceptrons and RBFs have good approximation properties and are related for normalised inputs. In fact, an MLP network can always simulate a Gaussian RBF network, whereas the converse is true only for certain values of the bias parameter (Poggio and Girosi 1990; Yee *et al.* 1999).

4.8 Complex Activation Functions

Recent results suggest that despite the existence of the universal approximation property, approximation by real-valued neural networks might not be continuous (Kainen *et al.* 1999) for some standard types of neural networks, such as Heaviside perceptrons or Gaussian radial basis functions.[13] For many functions there is not a best approximation in $\mathbb{R}$. However, there is always a unique best approximation in $\mathbb{C}$.

Many apparently real-valued mathematical problems can be better understood if they are embedded in the complex plane. Every variable is then represented by a complex number $z = x + \mathrm{j}\,y$, where x and y are real numbers and $\mathrm{j} = \sqrt{-1}$. Example problems cast into the complex plane include the analysis of transfer functions and polynomial equations. This has motivated researchers to generalise neural networks to the complex plane (Clarke 1990). Concerning the hardware realisation, the complex weights of the neural network represent impedance as opposed to resistance in real-valued networks.

If we again consider the approximation,

$$f(x) = \sum_{i=1}^{N} c_i \sigma(x - a_i), \tag{4.17}$$

where σ is a sigmoid function, different choices of σ will give different realisations of f. An extensive analysis of this problem is given in Helmke and Williamson (1995) and

[13] Intuitively, since a measure of the quality of an approximation is a distance function, for instance, the $\mathcal{L}_2$ distance given by $(\int_a^b (f(x) - g(x))^2 \, dx)^{1/2}$, there might occur a case where we have to calculate an integral which is not possible to be calculated within the field of real numbers $\mathbb{R}$, but which is easy to calculate in the field of complex numbers $\mathbb{C}$ – recall the function e^{-x^2}.

Williamson and Helmke (1995). Going back to elementary function approximation, if $\sigma(x) = x^{-1}$, then (4.17) represents a partial fraction expansion of a rational function f. The coefficients a_i and c_i are, respectively, poles and residuals of (4.17). Notice, however, that both a_i and c_i are allowed to be complex.[14]

A complex sigmoid is naturally obtained by analytic continuation of a real-valued sigmoid onto the complex plane. In order to extend a gradient-based learning algorithm for complex signals, the employed activation function should be analytic. Using analytic continuation to extend an activation function to the complex plane, however, has a consequence that, by the Liouville Theorem (Theorem C.1.4), the only bounded differentiable functions defined on the entire complex plane are constant functions. For commonly used activation functions, however, the singularities occur in a limited set.[15] For the logistic function

$$\sigma(z) = \frac{1}{1 + \mathrm{e}^{-z}} = u + \mathrm{j}v,$$

if z approaches any value in the set $\{0 \pm \mathrm{j}(2n+1)\pi,\ n \in \mathbb{Z}\}$, then $|\sigma(z)| \to \infty$. Similar conditions for the tanh are $\{0 \pm \mathrm{j}((2n+1)/2)\pi,\ n \in \mathbb{Z}\}$, whereas for e^{-z^2}, we have singularities for $z = 0 + \mathrm{j}y$ (Georgiou and Koutsougeras 1992).

Hence, a function obtained by an analytic continuation to the complex plane is generally speaking not an appropriate activation function. Generalising the discussion for real activation functions, properties that a function $\sigma(z) = u(x, y) + \mathrm{j}v(x, y)$ should satisfy so that it represents a suitable activation function in the complex plane are (Georgiou and Koutsougeras 1992)

- u and v are nonlinear in x and y;
- $\sigma(z)$ is bounded $\Rightarrow$ u and v are bounded;
- $\sigma(z)$ has bounded partial derivatives u_x, u_y, v_x, v_y, which satisfy the Cauchy–Riemann conditions (Mathews and Howell 1997);
- $\sigma(z)$ is not entire (not a constant).

Regarding fixed point iteration and global asymptotic stability of neural networks, which will be discussed in more detail in Chapter 7, complex-valued neural networks can generate the dynamics

$$z \leftarrow \Phi(z). \tag{4.18}$$

Functions of the form $cz(1-z)$, for instance, give rise to the Mandelbrot and Julia sets (Clarke 1990; Devaney 1999; Strogatz 1994). A single complex neuron with a feedback connection is thus capable of performing complicated discriminations and generation of nonlinear behaviour.

---

[14] Going back to transfer function approximation in signal processing, functions of the type (4.17) are able to approximate a Butterworth function of any degree if (4.17) is allowed to have complex coefficients (such as in the case of an RLC realisation). On the other hand, functions with real coefficients (an RC network) cannot approximate a Butterworth function whose order is $\geqslant 2$.

[15] The exponential function $\exp : \mathbb{C} \to \mathbb{C} \setminus \{0\}$ maps the set $\{z = a + (2k+1)\mathrm{j}\pi\}$, $a \in \mathbb{R}$, $k \in \mathbb{Z}$ onto the negative real axis, which determines singularities of complex sigmoids.

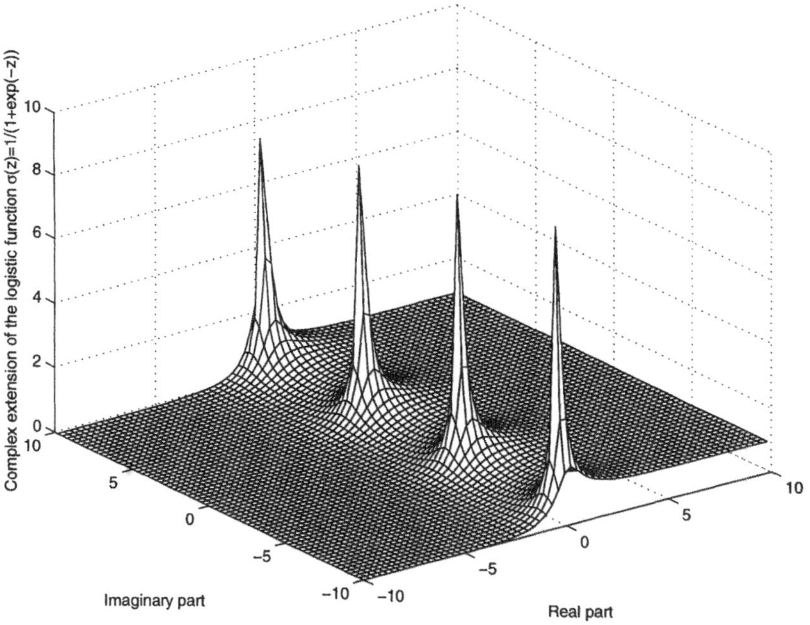

Figure 4.8 Complex extension of the logistic function $\sigma(z) = 1/(1 + \mathrm{e}^{-z})$

To provide conditions on the capability of complex neural networks to approximate nonlinear dynamics, a density theorem for complex MLPs with nonanalytic activation function and a hidden layer is proved in Arena *et al.* (1998b). The often cited denseness conditions are, as pointed out by Cotter (1990), special cases of the Stone–Weierstrass theorem.

In the context of learning algorithms, Leung and Haykin (1991) developed their complex backpropagation algorithm considering the following activation function,

$$f(z) = \frac{1}{1 + \mathrm{e}^{-z}} : \mathbb{C}^N \to \mathbb{C}, \tag{4.19}$$

whose magnitude is shown in Figure 4.8. This complex extension of the logistic function has singularities due to the complex exponential in the denominator of (4.19). It is safe to use (4.19) if the inputs are scaled to the range of the complex logistic function where it is analytic. In Benvenuto and Piazza (1992), the following activation function is proposed,

$$\sigma(z) = \sigma(x) + \mathrm{j}\sigma(y), \tag{4.20}$$

where $\sigma(z)$ is a two-dimensional extension of a standard sigmoid. The magnitude of this function is shown in Figure 4.9. The function (4.20) is not analytic and bounded on $\mathbb{C}$. It is, however, discriminatory, and linear combinations of functions of this type are dense in $\mathbb{C}$ (Arena *et al.* 1998a).

Another proposed complex sigmoid is (Benvenuto and Piazza 1992)

$$\sigma(z) = \frac{2c_1}{1 + \mathrm{e}^{-c_2 z}} - c_1, \tag{4.21}$$

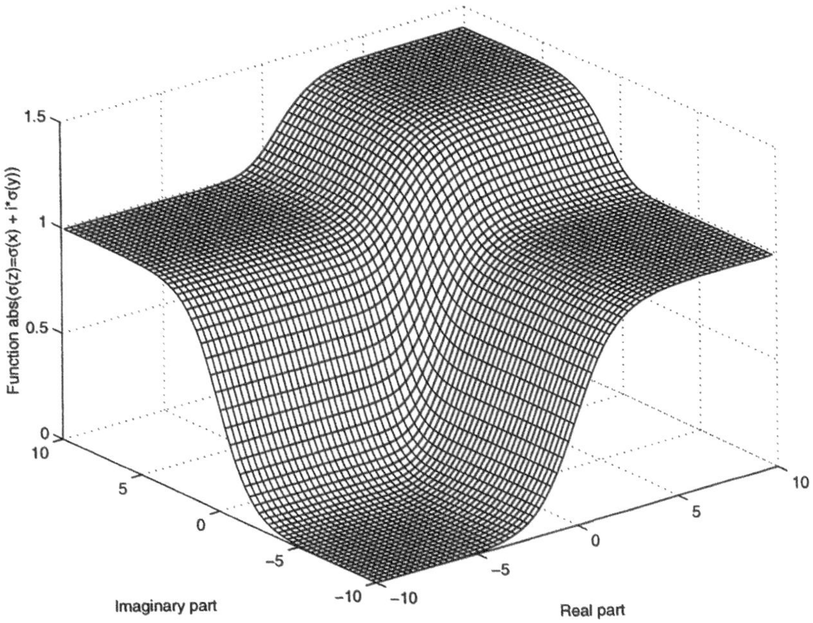

Figure 4.9 Complex sigmoid $\sigma(z) = \sigma(z_r) + \mathrm{j}\sigma(z_i)$

where c_1, c_2 are suitable parameters. The derivative of this function is

$$\sigma'(z) = \frac{c_2}{2c_1}(c_1^2 - \sigma^2(z)). \tag{4.22}$$

Other work on complex backpropagation was proposed in Kim and Guest (1990).

A suitable complex activation function would have the property that an excitation near zero would remain close to zero, and large excitations would be mapped into bounded outputs. One such function is given by (Clarke 1990)

$$\sigma(z) = \frac{(\cos\theta + \mathrm{j}\sin\theta)(z - s)}{1 - \alpha^* z}, \tag{4.23}$$

where θ is a rotation angle and α is a complex constant of magnitude less than one. The operator $(\cdot)^*$ denotes complex conjugation; the sign of the imaginary part of the asterisked variable is changed. This function is a conformal mapping of the unit disc in the complex plane onto itself and is unique. Further, σ maps large complex numbers into $-1/\alpha$ and thus satisfies the above criteria. The one flaw in σ is a singularity at $z = 1/\alpha$, but in view of Liouville's theorem this is unavoidable. The magnitude plot of this function is shown in Figure 4.10.

A simple function that satisfies all the above properties is (Georgiou and Koutsougeras 1992)

$$f(z) = \frac{z}{c + (1/r)|z|}, \tag{4.24}$$

where c and r are real positive constants. This function maps any point in the complex plane onto the open disc $\{z : |z| < r\}$, as shown in Figure 4.11.

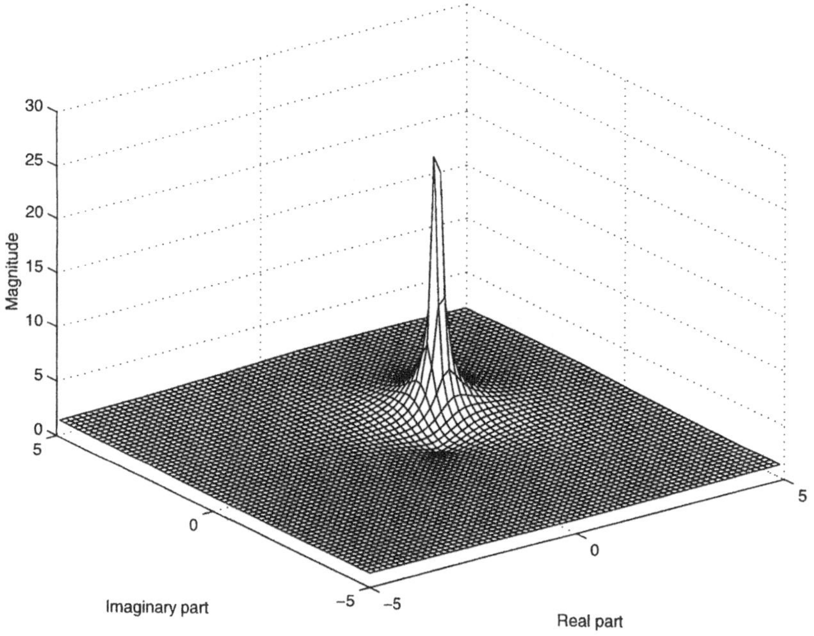

Figure 4.10 A complex activation function

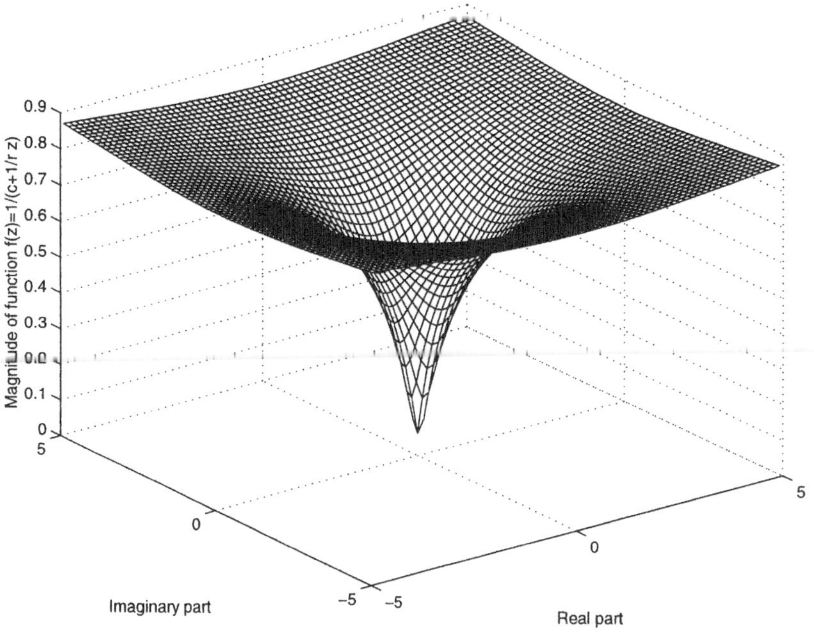

Figure 4.11 Magnitude of the function (4.24)

4.9 Complex Valued Neural Networks as Modular Groups of Compositions of Möbius Transformations

We next offer a different perspective upon some inherent properties of neural networks, such as fixed points, nesting and invertibility, by exposing the representations of neural networks in the framework of Möbius transformations. This framework includes the consideration of complex weights and inputs to the network together with complex sigmoidal activation functions.

4.9.1 Möbius Transformation

Observation 4.9.1.

(i) *The map $g : \mathbb{C} \to \mathbb{C}$, with $g(z) = \mathrm{e}^z$ is holomorphic on $\mathbb{C}$.*

(ii) *The complex sigmoidal activation function $f(z)$ of a neuron in a neural network is holomorphic and conformal.*

Definition 4.9.2 (Möbius mapping (Apostol 1997)). Let a, b, c and d denote four complex constants with the restriction that $ad \neq bc$. The function

$$w = f(z) = \frac{az + b}{cz + d} \tag{4.25}$$

is called a Möbius transformation, bilinear transformation or linear fractional transformation.

The condition $ad \neq bc$ is necessary, since for complex variables z_1 and z_2,

$$f(z_1) - f(z_2) = \frac{(ad - bc)(z_1 - z_2)}{(cz_1 + d)(cz_2 + d)},$$

which is constant for $ad - bc = 0$. The Möbius transformation is analytic everywhere except for a pole at $z = -d/c$, and also one-to-one and onto a half-plane, and vice versa, which means that its inverse is also a Möbius transformation.

Remark 4.9.3. The Möbius transformation does not determine the coefficients a, b, c, d uniquely, i.e. if $\varphi \in \mathbb{C} \setminus \{0\}$, then coefficients φa, φb, φc, φd correspond to the same transformation.

In addition, every Möbius transformation (except $f(z) = z$) has one or two fixed points[16] z^*, such that $f(z^*) = z^*$ (Apostol 1997).

4.9.2 Activation Functions and Möbius Transformations

Equating the coefficients associated with the powers of z in the transfer function of a sigmoid activation function, to those in a general form of the Möbius transformation (4.25), i.e.

$$\frac{1 - \mathrm{e}^{-\beta\,\mathrm{net}}}{1 + \mathrm{e}^{-\beta\,\mathrm{net}}} = \frac{az + b}{cz + d} \tag{4.26}$$

[16] See Definition 4.9.7.

we see that the tanh function satisfies the conditions of a Möbius transformation for $z = \mathrm{e}^{-\beta\,\mathrm{net}}$ and (Mandic 2000c)

$$a = -1, \qquad b = 1, \qquad c = 1, \qquad d = 1. \tag{4.27}$$

Also, the condition $ad - bc \neq 0$ is satisfied. Therefore, the hyperbolic tangent activation function is a Möbius transformation and holomorphic. So too is the logistic function, for which $a = 0$, $b = 1$, $c = 1$, $d = 1$ and $ad - bc \neq 0$.

Proposition 4.9.4 (see Dumitras and Lazarescu 1997). *The sigmoidal transformation $f(z)$ performed by a neuron in a neural network on a complex signal $z = \alpha + \mathrm{j}\beta$ is a Möbius transformation.*

To analyse an N-dimensional nested nonlinear system, as exhibited by a feedforward and recurrent neural network with hidden neurons, we use the notion of a modular group (Apostol 1997), which is explained in detail in Appendix C.1.1.

Example 4.9.5. Show that the transfer function between neurons in consecutive layers within a general neural network belongs to a modular group Γ of compositions of Möbius transformations.

Solution. Without loss in generality let us consider only two neurons from consecutive layers. Notice that their nonlinear functions are nested. Let us denote the functions performed by these two neurons by

$$H_1(z) = \frac{a_1 z + b_1}{c_1 z + d_1} \quad \text{and} \quad H_2(z) = \frac{a_2 z + b_2}{c_2 z + d_2}.$$

Then their composition (transfer function from a layer to a layer) is $H_1(H_2(z)) = H_1 \circ H_2$ and can be expressed as

$$\begin{aligned} H_1 \circ H_2 &= \left(a_1 \frac{a_2 z + b_2}{c_2 z + d_2} + b_1\right) \bigg/ \left(c_1 \frac{a_2 z + b_2}{c_2 z + d_2} + d_1\right) \\ &= \frac{(a_1 a_2 + b_1 c_2) z + a_1 b_2 + b_1 d_2}{(a_2 c_1 + c_2 d_1) z + b_2 c_1 + d_1 d_2}, \end{aligned} \tag{4.28}$$

which belongs to group Γ. Also if the Möbius mappings performed by H_1 and H_2 are, respectively, described by matrices

$$M_1 = \begin{bmatrix} a_1 & b_1 \\ c_1 & d_1 \end{bmatrix} \quad \text{and} \quad M_2 = \begin{bmatrix} a_2 & b_2 \\ c_2 & d_2 \end{bmatrix},$$

then the composition $H_1 \circ H_2$ is described by

$$\begin{aligned} H(z) = H_1(z) \circ H_2(z) \quad &\Leftrightarrow \quad (M_1 \times M_2) z = \left(\begin{bmatrix} a_1 & b_1 \\ c_1 & d_1 \end{bmatrix} \times \begin{bmatrix} a_2 & b_2 \\ c_2 & d_2 \end{bmatrix}\right) z \\ &= \begin{bmatrix} a_1 a_2 + b_1 c_2 & a_1 b_2 + b_1 d_2 \\ a_2 c_1 + c_2 d_1 & b_2 c_1 + d_1 d_2 \end{bmatrix} z = M z, \end{aligned} \tag{4.29}$$

which again belongs to the group Γ. □

Observation 4.9.6 (see Mandic 2000b,c). *The global input–output relationship in a neural network can be considered as a Möbius transformation.*

4.9.3 Existence and Uniqueness of Fixed Points in a Complex Neural Network via Theory of Modular Groups

Definition 4.9.7. A point x^* which is mapped onto itself under a map G so that $x^* = G(x^*)$ is called a fixed point.

Since by the Brower fixed point theorem (Zeidler 1986), a continuous function on a convex, closed and bounded set has a fixed point, so too Möbius transformations have fixed points. Therefore, it suffices to investigate the characteristics of the nonlinear activation function to obtain fixed points for a general neural network based nonlinear system. In that case, both the external input vector to the system $\boldsymbol{x}(k)$ and the parameter vector $\boldsymbol{w}(k)$ are assumed to form a time-invariant part of the variable net.

In Zakeri (1996), it is proven that a proper holomorphic map in the complex plane is uniquely determined up to post-composition with a Möbius transformation by its critical points. To connect fixed (critical) points and nested Möbius transformations, which occur in RNN-based architectures, we use the following observation.

Observation 4.9.8 (see Mandic 2000b). *For a given architecture, fixed points of the complex activation function employed in a neural network determine the fixed points of the neural network considered.*

Existence. The existence conditions for a fixed point of a sigmoid activation function are provided by recognising that a sigmoid activation function is a Möbius transformation (Proposition 4.9.4). Since every Möbius map has one or two fixed points (Mathews and Howell 1997), so too does a sigmoid activation function of a neuron. By Observation 4.9.6, a global input–output relationship in a neural network can be considered in the framework of compositions of Möbius transformations, and hence has fixed points.

Uniqueness. By Theorem C.1.5, transfer functions of two neurons in consecutive layers can be described by matrices $M_1 = T^{n_1}ST^{n_2}S\cdots ST^{n_k}$ and $M_2 = T^{m_1}ST^{n_2}S\cdots ST^{m_l}$. Using the property described in Zakeri (1996), the composition of their activation functions (nested nonlinearities) represented by the product $M_1 \times M_2$ has critical points defined by M_1 and M_2.

As a consequence of complex activation functions being Möbius transformations, we also conclude that (Mandic 2000c)

- a general neural network has an inverse (the existence of an inverse);
- the inverse of a neural network is not necessarily unique (the uniqueness of an inverse).

Since the input–output relationship in a complex neural network can be considered in the framework of Möbius transformations (which has an inverse, which is again a Möbius transformation), a general complex neural network has an inverse (*existence*). On the other hand, Möbius transformations are uniformly determined up to a multiplication by a constant. This means that there is more than one solution to the inverse problem (*uniqueness*).

4.10 Summary

An overview of nonlinear activation functions used in neural networks has been provided. We have started from the problem of function approximation and trajectory learning and evaluated neural networks suitable for these problems. Properties of neural networks realised in hardware have also been addressed. For rigour, the analysis has been extended to neural networks with complex activation functions, for which we have built a unified framework via modular groups of Möbius transformations.

Existence and uniqueness conditions of fixed points and invertibility of such mappings have been derived. These results apply both for the general input–output relationship in a neural network as well as for a single neuron. This analysis provides a strong mathematical background for further analysis of neural networks for adaptive filtering and prediction.

5

Recurrent Neural Networks Architectures

5.1 Perspective

In this chapter, the use of neural networks, in particular recurrent neural networks, in system identification, signal processing and forecasting is considered. The ability of neural networks to model nonlinear dynamical systems is demonstrated, and the correspondence between neural networks and block-stochastic models is established. Finally, further discussion of recurrent neural network architectures is provided.

5.2 Introduction

There are numerous situations in which the use of linear filters and models is limited. For instance, when trying to identify a saturation type nonlinearity, linear models will inevitably fail. This is also the case when separating signals with overlapping spectral components.

Most real-world signals are generated, to a certain extent, by a nonlinear mechanism and therefore in many applications the choice of a nonlinear model may be necessary to achieve an acceptable performance from an adaptive predictor. Communications channels, for instance, often need nonlinear equalisers to achieve acceptable performance. The choice of model has crucial importance[1] and practical applications have shown that nonlinear models can offer a better prediction performance than their linear counterparts. They also reveal rich dynamical behaviour, such as limit cycles, bifurcations and fixed points, that cannot be captured by linear models (Gershenfeld and Weigend 1993).

By *system* we consider the actual underlying physics[2] that generate the data, whereas by *model* we consider a mathematical description of the system. Many variations of mathematical models can be postulated on the basis of datasets collected from observations of a system, and their suitability assessed by various performance

[1] System identification, for instance, consists of choice of the model, model parameter estimation and model validation.

[2] Technically, the notions of *system* and *process* are equivalent (Pearson 1995; Sjöberg *et al.* 1995).

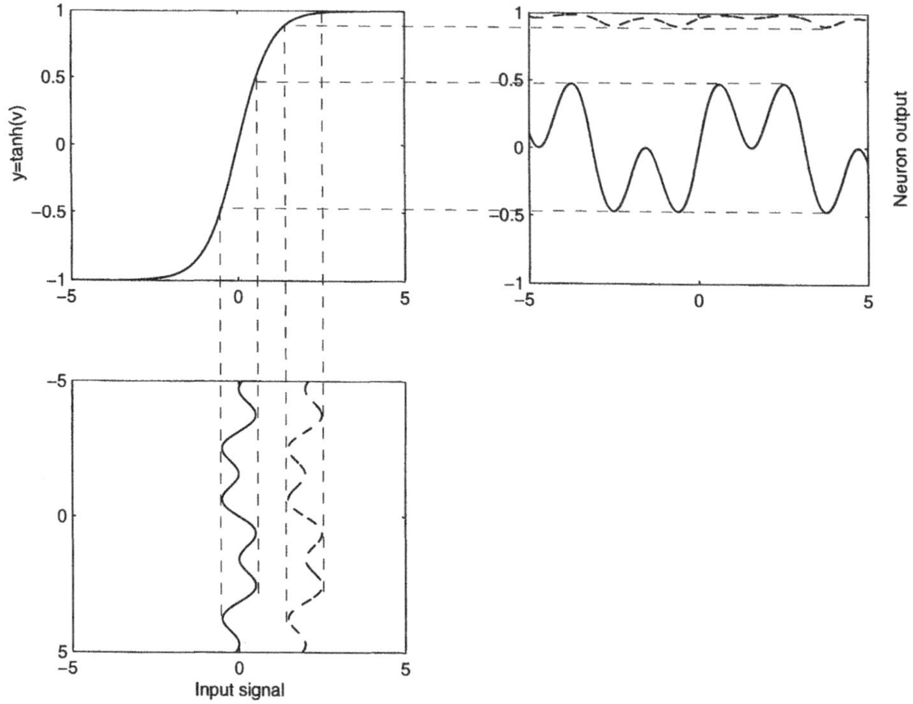

Figure 5.1 Effects of $y = \tanh(v)$ nonlinearity in a neuron model upon two example inputs

metrics. Since it is not possible to characterise nonlinear systems by their impulse response, one has to resort to less general models, such as homomorphic filters, morphological filters and polynomial filters. Some of the most frequently used polynomial filters are based upon Volterra series (Mathews 1991), a nonlinear analogue of the linear impulse response, threshold autoregressive models (TAR) (Priestley 1991) and Hammerstein and Wiener models. The latter two represent structures that consist of a linear dynamical model and a static zero-memory nonlinearity. An overview of these models can be found in Haber and Unbehauen (1990). Notice that for nonlinear systems, the ordering of the modules within a modular structure[3] plays an important role.

To illustrate some important features associated with nonlinear neurons, let us consider a squashing nonlinear activation function of a neuron, shown in Figure 5.1. For two identical mixed sinusoidal inputs with different offsets, passed through this nonlinearity, the output behaviour varies from amplifying and slightly distorting the input signal (solid line in Figure 5.1) to attenuating and considerably nonlinearly distorting the input signal (broken line in Figure 5.1). From the viewpoint of system theory, neural networks represent nonlinear maps, mapping one metric space to another.

[3] To depict this, for two modules performing nonlinear functions $H_1 = \sin(x)$ and $H_2 = e^x$, we have $H_1(H_2(x)) \neq H_2(H_1(x))$ since $\sin(e^x) \neq e^{\sin(x)}$. This is the reason to use the term *nesting* rather than cascading in modular neural networks.

Nonlinear system modelling has traditionally focused on Volterra–Wiener analysis. These models are nonparametric and computationally extremely demanding. The Volterra series expansion is given by

$$y(k) = h_0 + \sum_{i=0}^{N} h_1(i)x(k-i) + \sum_{i=0}^{N}\sum_{j=0}^{N} h_2(i,j)x(k-i)x(k-j) + \cdots \tag{5.1}$$

for the representation of a causal system. A nonlinear system represented by a Volterra series is completely characterised by its Volterra kernels h_i, $i = 0,1,2,\ldots$. The Volterra modelling of a nonlinear system requires a great deal of computation, and mostly second- or third-order Volterra systems are used in practice.

Since the Volterra series expansion is a Taylor series expansion with memory, they both fail when describing a system with discontinuities, such as

$$y(k) = A\,\mathrm{sgn}(x(k)), \tag{5.2}$$

where $\mathrm{sgn}(\,\cdot\,)$ is the signum function.

To overcome this difficulty, nonlinear parametric models of nonlinear systems, termed NARMAX, that are described by nonlinear difference equations, have been introduced (Billings 1980; Chon and Cohen 1997; Chon *et al.* 1999; Connor 1994). Unlike the Volterra–Wiener representation, the NARMAX representation of nonlinear systems offers compact representation.

The NARMAX model describes a system by using a nonlinear functional dependence between lagged inputs, outputs and/or prediction errors. A polynomial expansion of the transfer function of a NARMAX neural network does not comprise of delayed versions of input and output of order higher than those presented to the network. Therefore, the input of an insufficient order will result in undermodelling, which complies with Takens' embedding theorem (Takens 1981).

Applications of neural networks in forecasting, signal processing and control require treatment of dynamics associated with the input signal. Feedforward networks for processing of dynamical systems tend to capture the dynamics by including past inputs in the input vector. However, for dynamical modelling of complex systems, there is a need to involve feedback, i.e. to use recurrent neural networks. There are various configurations of recurrent neural networks, which are used by Jordan (1986) for control of robots, by Elman (1990) for problems in linguistics and by Williams and Zipser (1989a) for nonlinear adaptive filtering and pattern recognition. In Jordan's network, past values of network outputs are fed back into hidden units, in Elman's network, past values of the outputs of hidden units are fed back into themselves, whereas in the Williams–Zipser architecture, the network is fully connected, having one hidden layer.

There are numerous modular and hybrid architectures, combining linear adaptive filters and neural networks. These include the pipelined recurrent neural network and networks combining recurrent networks and FIR adaptive filters. The main idea here is that the linear filter captures the linear 'portion' of the input process, whereas a neural network captures the nonlinear dynamics associated with the process.

5.3 Overview

The basic modes of modelling, such as *parametric, nonparametric, white box, black box* and *grey box* modelling are introduced. Afterwards, the dynamical richness of neural models is addressed and feedforward and recurrent modelling for noisy time series are compared. Block-stochastic models are introduced and neural networks are shown to be able to represent these models. The chapter concludes with an overview of recurrent neural network architectures and recurrent neural networks for NARMAX modelling.

5.4 Basic Modes of Modelling

The notions of parametric, nonparametric, black box, grey box and white box modelling are explained. These can be used to categorise neural network algorithms, such as the direct gradient computation, *a posteriori* and normalised algorithms. The basic idea behind these approaches to modelling is *not to estimate what is already known.* One should, therefore, utilise prior knowledge and knowledge about the physics of the system, when selecting the neural network model prior to parameter estimation.

5.4.1 Parametric versus Nonparametric Modelling

A review of nonlinear input–output modelling techniques is given in Pearson (1995). Three classes of input–output models are *parametric, nonparametric* and *semiparametric* models. We next briefly address them.

- *Parametric* modelling assumes a fixed structure for the model. The model identification problem then simplifies to estimating a finite set of parameters of this fixed model. This estimation is based upon the prediction of real input data, so as to best match the input data dynamics. An example of this technique is the broad class of ARIMA/NARMA models. For a given structure of the model (NARMA for instance) we recursively estimate the parameters of the chosen model.

- *Nonparametric* modelling seeks a particular model structure from the input data. The actual model is not known beforehand. An example taken from nonparametric regression is that we look for a model in the form of $y(k) = f(x(k))$ without knowing the function $f(\,\cdot\,)$ (Pearson 1995).

- *Semiparametric* modelling is the combination of the above. Part of the model structure is completely specified and known beforehand, whereas the other part of the model is either not known or loosely specified.

Neural networks, especially recurrent neural networks, can be employed within estimators of all of the above classes of models. Closely related to the above concepts are white, grey and black box modelling techniques.

5.4.2 White, Grey and Black Box Modelling

To understand and analyse real-world physical phenomena, various mathematical models have been developed. Depending on some *a priori* knowledge about the process, data and model, we differentiate between three fairly general modes of modelling. The idea is to distinguish between three levels of prior knowledge, which have been 'colour-coded'. An overview of the white, grey and black box modelling techniques can be found in Aguirre (2000) and Sjöberg *et al.* (1995).

Given data gathered from planet movements, then Kepler's gravitational laws might well provide the initial framework in building a mathematical model of the process. This mode of modelling is referred to as *white box* modelling (Aguirre 2000), underlying its fairly deterministic nature. Static data are used to calculate the parameters, and to do that the underlying physical process has to be understood. It is therefore possible to build a *white box* model entirely from physical insight and prior knowledge. However, the underlying physics are generally not completely known, or are too complicated and often one has to resort to other types of modelling.

The exact form of the input–output relationship that describes a real-world system is most commonly unknown, and therefore modelling is based upon a chosen set of known functions. In addition, if the model is to approximate the system with an arbitrary accuracy, the set of chosen nonlinear continuous functions must be dense. This is the case with polynomials. In this light, neural networks can be viewed as another mode of functional representations. *Black box* modelling therefore assumes no previous knowledge about the system that produces the data. However, the chosen network structure belongs to architectures that are known to be flexible and have performed satisfactorily on similar problems. The aim hereby is to find a function F that approximates the process y based on the previous observations of process y_{PAST} and input u, as

$$y = F(y_{\mathrm{PAST}}, u). \tag{5.3}$$

This 'black box' establishes a functional dependence between the input and output, which can be either linear or nonlinear. The downside is that it is generally not possible to learn about the true physical process that generates the data, especially if a linear model is used. Once the training process is complete, a neural network represents a *black box*, nonparametric process model. Knowledge about the process is embedded in the values of the network parameters (i.e. synaptic weights).

A natural compromise between the two previous models is so-called *grey box* modelling. It is obtained from *black box* modelling if some information about the system is known *a priori*. This can be a probability density function, general statistics of the process data, impulse response or attractor geometry. In Sjöberg *et al.* (1995), two subclasses of *grey box* models are considered: *physical* modelling, where a model structure is built upon understanding of the underlying physics, as for instance the state-space model structure; and *semiphysical* modelling, where, based upon physical insight, certain nonlinear combinations of data structures are suggested, and then estimated by *black box* methodology.

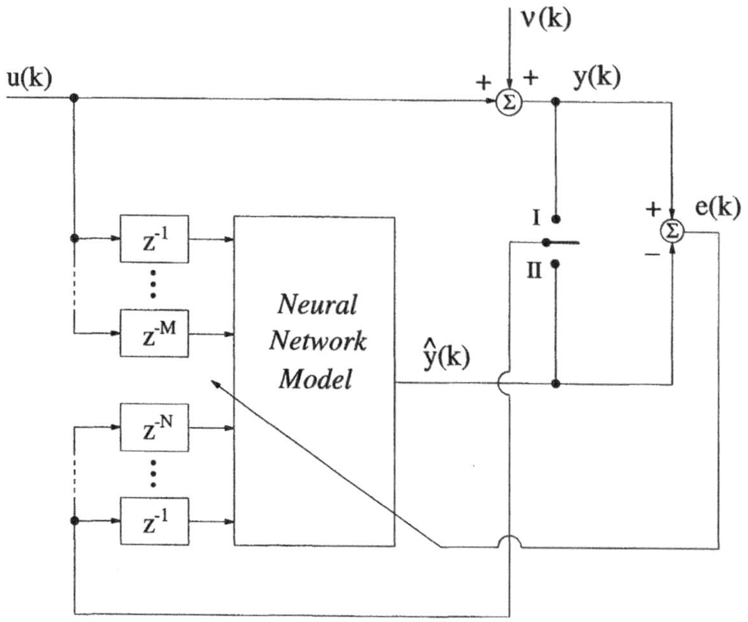

Figure 5.2 Nonlinear prediction configuration using a neural network model

5.5 NARMAX Models and Embedding Dimension

For neural networks, the number of input nodes specifies the dimension of the network input. In practice, the true state of the system is not observable and the mathematical model of the system that generates the dynamics is not known. The question arises: is the sequence of measurements $\{y(k)\}$ sufficient to reconstruct the nonlinear system dynamics? Under some regularity conditions, Takens' (1981) and Mane's (1981) embedding theorems establish this connection. To ensure that the dynamics of a nonlinear process estimated by a neural network are fully recovered, it is convenient to use Takens' embedding theorem (Takens 1981), which states that to obtain a faithful reconstruction of the system dynamics, the *embedding dimension* d must satisfy

$$d \geqslant 2D + 1, \tag{5.4}$$

where D is the dimension of the system attractor. Takens' embedding theorem (Takens 1981; Wan 1993) establishes a diffeomorphism between a finite window of the time series

$$[y(k-1), y(k-2), \dots, y(k-N)] \tag{5.5}$$

and the underlying state of the dynamic system which generates the time series. This implies that a nonlinear regression

$$y(k) = g[y(k-1), y(k-2), \dots, y(k-N)] \tag{5.6}$$

can model the nonlinear time series. An important feature of the delay-embedding theorem due to Takens (1981) is that it is physically implemented by delay lines.

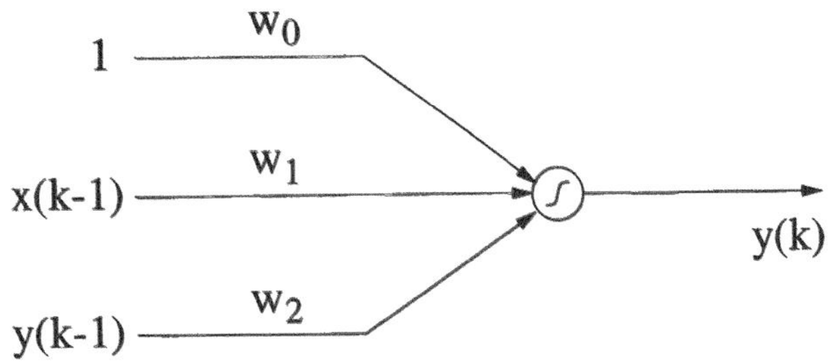

Figure 5.3 A NARMAX recurrent perceptron with $p = 1$ and $q = 1$

There is a deep connection between time-lagged vectors and underlying dynamics. Delay vectors are not just a representation of a state of the system, their length is the key to recovering the full dynamical structure of a nonlinear system. A general starting point would be to use a network for which the input vector comprises delayed inputs and outputs, as shown in Figure 5.2. For the network in Figure 5.2, both the input and the output are passed through delay lines, hence indicating the NARMAX character of this network. The switch in this figure indicates two possible modes of learning which will be explained in Chapter 6.

5.6 How Dynamically Rich are Nonlinear Neural Models?

To make an initial step toward comparing neural and other nonlinear models, we perform a Taylor series expansion of the sigmoidal nonlinear activation function of a single neuron model as (Billings *et al.* 1992)

$$\Phi(v(k)) = \frac{1}{1+e^{-\beta v(k)}} = \frac{1}{2} + \frac{\beta}{4}v(k) - \frac{\beta^3}{48}v^3(k) + \frac{\beta^5}{480}v^5(k) - \frac{17\beta^7}{80\,640}v^7(k) + \cdots. \quad (5.7)$$

Depending on the steepness β and the activation potential $v(k)$, the polynomial representation (5.7) of the transfer function of a neuron exhibits a complex nonlinear behaviour.

Let us now consider a NARMAX recurrent perceptron with $p = 1$ and $q = 1$, as shown in Figure 5.3, which is a simple example of recurrent neural networks. Its mathematical description is given by

$$y(k) = \Phi(w_1 x(k-1) + w_2 y(k-1) + w_0). \quad (5.8)$$

Expanding (5.8) using (5.7) yields

$$y(k) = \tfrac{1}{2} + \tfrac{1}{4}[w_1 x(k-1) + w_2 y(k-1) + w_0] - \tfrac{1}{48}[w_1 x(k-1) + w_2 y(k-1) + w_0]^3 + \cdots, \quad (5.9)$$

where $\beta = 1$. Expression (5.9) illustrates the dynamical richness of squashing activation functions. The associated dynamics, when represented in terms of polynomials are quite complex. Networks with more neurons and hidden layers will produce more complicated dynamics than those in (5.9). Following the same approach, for a general recurrent neural network, we obtain (Billings *et al.* 1992)

$$\begin{aligned} y(k) = c_0 + c_1x(k-1) + c_2y(k-1) + c_3x^2(k-1) \\ + c_4y^2(k-1) + c_5x(k-1)y(k-1) + c_6x^3(k-1) \\ + c_7y^3(k-1) + c_8x^2(k-1)y(k-1) + \cdots . \end{aligned} \quad (5.10)$$

Equation (5.10) does not comprise delayed versions of input and output samples of order higher than those presented to the network. If the input vector were of an insufficient order, undermodelling would result, which complies with Takens' embedding theorem. Therefore, when modelling an unknown dynamical system or tracking unknown dynamics, it is important to concentrate on the embedding dimension of the network. Representation (5.10) also models an offset (mean value) c_0 of the input signal.

5.6.1 Feedforward versus Recurrent Networks for Nonlinear Modelling

The choice of which neural network to employ to represent a nonlinear physical process depends on the dynamics and complexity of the network that is best for representing the problem in hand. For instance, due to feedback, recurrent networks may suffer from instability and sensitivity to noise. Feedforward networks, on the other hand, might not be powerful enough to capture the dynamics of the underlying nonlinear dynamical system. To illustrate this problem, we resort to a simple IIR (ARMA) linear system described by the following first-order difference equation

$$z(k) = 0.5z(k-1) + 0.1x(k-1). \quad (5.11)$$

The system (5.11) is stable, since the pole of its transfer function is at 0.5, i.e. within the unit circle in the z-plane. However, in a noisy environment, the output $z(k)$ is corrupted by noise $e(k)$, so that the noisy output $y(k)$ of system (5.11) becomes

$$y(k) = z(k) + e(k), \quad (5.12)$$

which will affect the quality of estimation based on this model. This happens because the noise terms accumulate during recursions[4] (5.11) as

$$y(k) = 0.5y(k-1) + 0.1x(k-1) + e(k) - 0.5e(k-1). \quad (5.13)$$

An equivalent FIR (MA) representation of the same filter (5.11), using the method of long division, gives

$$z(k) = 0.1x(k-1) + 0.05x(k-2) + 0.025x(k-3) + 0.0125x(k-4) + \cdots \quad (5.14)$$

and the representation of a noisy system now becomes

$$y(k) = 0.1x(k-1) + 0.05x(k-2) + 0.025x(k-3) + 0.0125x(k-4) + \cdots + e(k). \quad (5.15)$$

[4] Notice that if the noise $e(k)$ is zero mean and white it appears coloured in (5.13), i.e. correlated with previous outputs, which leads to biased estimates.

Clearly, the noise in (5.15) is not correlated with the previous outputs and the estimates are unbiased.[5] The price to pay, however, is the infinite length of the exact representation of (5.11).

A similar principle applies to neural networks. In Chapter 6 we address the modes of learning in neural networks and discuss the bias/variance dilemma for recurrent neural networks.

5.7 Wiener and Hammerstein Models and Dynamical Neural Networks

Under relatively mild conditions,[6] the output signal of a nonlinear model can be considered as a combination of outputs from some suitable submodels. The structure identification, model validation and parameter estimation based upon these submodels are more convenient than those of the whole model. Block oriented stochastic models consist of static nonlinear and dynamical linear modules. Such models often occur in practice, examples of which are

- the Hammerstein model, where a zero-memory nonlinearity is followed by a linear dynamical system characterised by its transfer function $H(z) = N(z)/D(z)$;
- the Wiener model, where a linear dynamical system is followed by a zero-memory nonlinearity.

5.7.1 Overview of Block-Stochastic Models

The definitions of certain stochastic models are given by the

1. Wiener system
$$y(k) = g(H(z^{-1})u(k)), \tag{5.16}$$
where $u(k)$ is the input to the system, $y(k)$ is the output,
$$H(z^{-1}) = \frac{C(z^{-1})}{D(z^{-1})}$$
is the z-domain transfer function of the linear component of the system and $g(\cdot)$ is a nonlinear function;

2. Hammerstein system
$$y(k) = H(z^{-1})g(u(k)); \tag{5.17}$$

3. Uryson system, defined by
$$y(k) = \sum_{i=1}^{M} H_i(z^{-1})g_i(u(k)). \tag{5.18}$$

---

[5] Under the usual assumption that the external additive noise $e(k)$ is not correlated with the input signal $x(k)$.

[6] A finite degree polynomial steady-state characteristic.

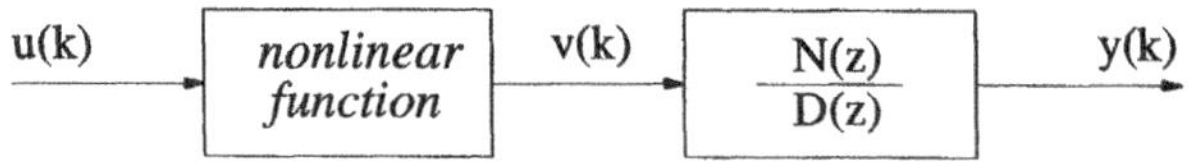

(a) The Hammerstein stochastic model

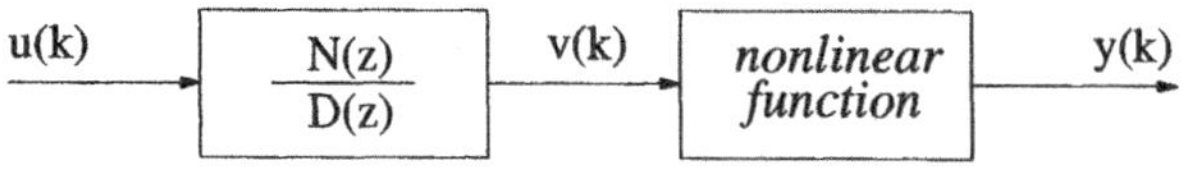

(b) The Wiener stochastic model

Figure 5.4 Nonlinear stochastic models used in control and signal processing

Theoretically, there are finite size neural systems with dynamic synapses which can represent all of the above. Moreover, some modular neural architectures, such as the PRNN (Haykin and Li 1995), are able to represent block-cascaded Wiener–Hammerstein systems described by (Mandic and Chambers 1999c)

$$y(k) = \Phi_N(H_N(z^{-1})\Phi_{N-1}(H_{N-1}(z^{-1})\cdots\Phi_1(H_1(z^{-1})u(k)))) \tag{5.19}$$

and

$$y(k) = H_N(z^{-1})\Phi_N(H_{N-1}(z^{-1})\Phi_{N-1}\cdots\Phi_1(H_1(z^{-1}u(k)))) \tag{5.20}$$

under certain constraints relating the size of networks and order of block-stochastic models. Due to its parallel nature, however, a general Uryson model is not guaranteed to be representable this way.

5.7.2 *Connection Between Block-Stochastic Models and Neural Networks*

Block diagrams of Wiener and Hammerstein systems are shown in Figure 5.4. The nonlinear function from Figure 5.4(a) can be generally assumed to be a polynomial,[7] i.e.

$$v(k) = \sum_{i=0}^{M} \lambda_i u^i(k). \tag{5.21}$$

The Hammerstein model is a conventional parametric model, usually used to represent processes with nonlinearities involved with the process inputs, as shown in Figure 5.4(a). The equation describing the output of a SISO Hammerstein system corrupted with additive output noise $\eta(k)$ is

$$y(k) = \Phi[u(k-1)] + \sum_{i=2}^{\infty} h_i \Phi[u(k-i)] + \nu(k), \tag{5.22}$$

where Φ is a nonlinear function which is continuous. Other requirements are that the linear dynamical subsystem is stable. This network is shown in Figure 5.5.

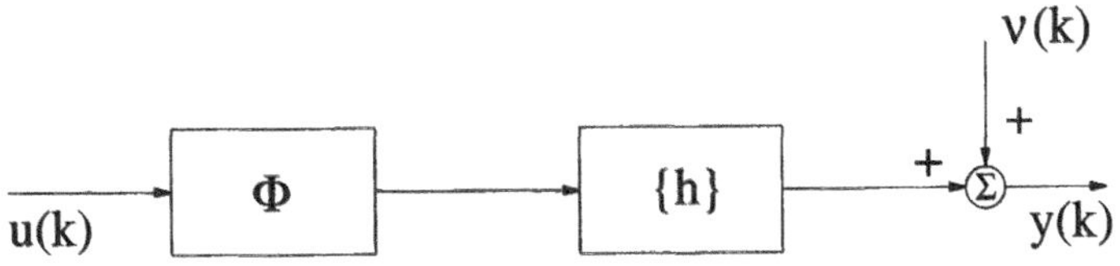

Figure 5.5 Discrete-time SISO Hammerstein model with observation noise

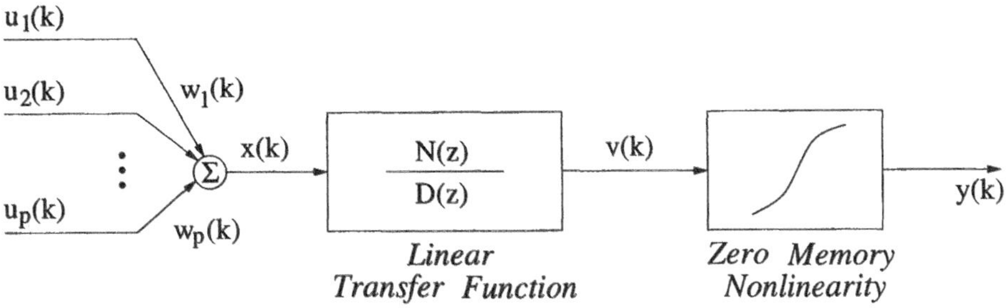

Figure 5.6 Dynamic perceptron

Neural networks with locally distributed dynamics (LDNN) can be considered as locally recurrent networks with global feedforward features. An example of these networks is the *dynamical multilayer perceptron* (DMLP) which consists of dynamical neurons and is shown in Figure 5.6. The model of this dynamic perceptron is described by

$$\left.\begin{aligned} y(k) &= \Phi(v(k)), \\ v(k) &= \sum_{i=0}^{\deg(N(z))} n_i(k)x(k-i) + 1 + \sum_{j=1}^{\deg(D(z))} d_j(k)v(k-j), \\ x(k) &= \sum_{l=1}^{p} w_l(k)u_l(k), \end{aligned}\right\} \tag{5.23}$$

where n_i and d_i denote, respectively, the coefficients of the polynomials in $N(z)$ and $D(z)$ and '1' is included for a possible bias input. From Figure 5.6, the transfer function between $y(k)$ and $x(k)$ represents a Wiener system. Hence, combinations of dynamical perceptrons (such as a recurrent neural network) are able to represent block-stochastic Wiener–Hammerstein models. Gradient-based learning rules can be developed for a recurrent neural network representing block-stochastic models. Both the Wiener and Hammerstein models can exhibit a more general structure, as shown in Figure 5.7, for the Hammerstein model. Wiener and Hammerstein models can be combined to produce more complicated block-stochastic models. A representative of these models is the Wiener–Hammerstein model, shown in Figure 5.8. This figure shows a Wiener stochastic model, followed by a linear dynamical system represented by its transfer function $H_2(z) = N_2(z)/D_2(z)$, hence building a Wiener–Hammerstein

[7] By the Weierstrass theorem, polynomials can approximate arbitrarily well any nonlinear function, including sigmoid functions.

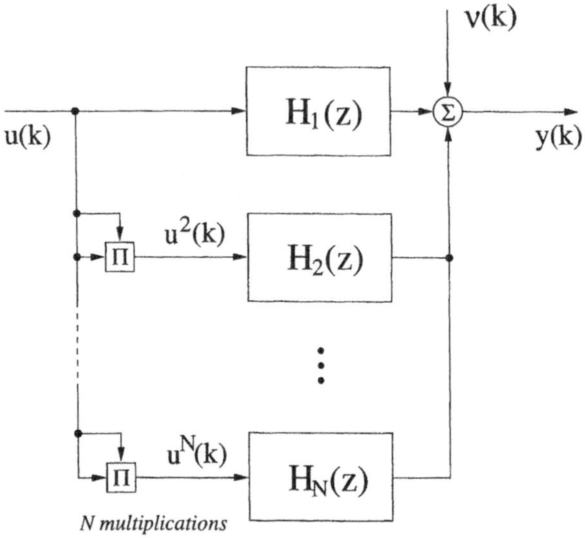

Figure 5.7 Generalised Hammerstein model

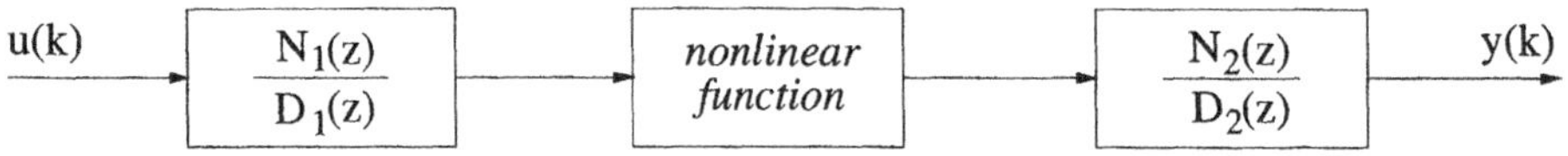

Figure 5.8 The Wiener–Hammerstein model

block-stochastic system. In practice, we can build complicated block cascaded systems this way.

Wiener and Hammerstein systems are frequently used to compensate each other (Kang *et al.* 1998). This includes finding an inverse of the first module in the combination. If these models are represented by neural networks, Chapter 4 provides a general framework for uniqueness, existence and convergence of inverse neural models. The following example from Billings and Voon (1986) shows that the Wiener model can be represented by a NARMA model, which, in turn can be modelled by a recurrent neural network.

Example 5.7.1. The Wiener model

$$\left.\begin{aligned} w(k) &= 0.8w(k-1) + 0.4u(k-1), \\ y(k) &= w(k) + w^3(k) + e(k), \end{aligned}\right\} \tag{5.24}$$

was identified as

$$\begin{aligned} y(k) = {} & 0.7578y(k-1) + 0.3891u(k-1) - 0.037\,23y^2(k-1) \\ & + 0.3794y(k-1)u(k-1) + 0.0684u^2(k-1) + 0.1216y(k-1)u^2(k-1) \\ & + 0.0633u^3(k-1) - 0.739e(k-1) - 0.368u(k-1)e(k-1) + e(k), \end{aligned} \tag{5.25}$$

which is a NARMA model, and hence can be realised by a recurrent neural network.

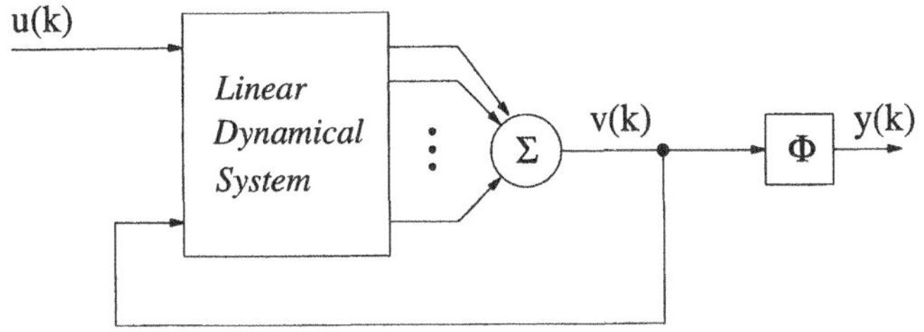

(a) Activation feedback scheme

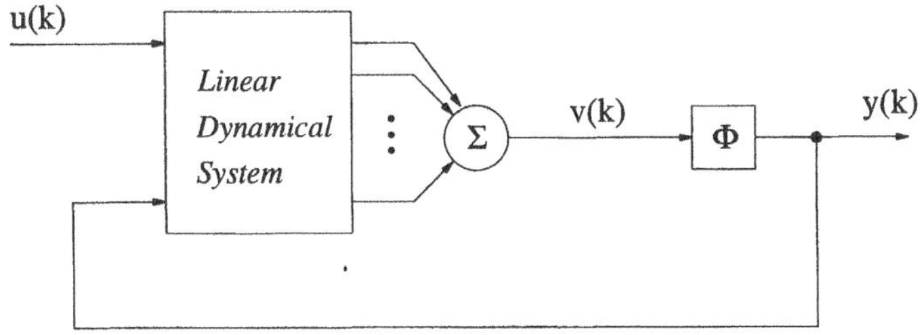

(b) Output feedback scheme

Figure 5.9 Recurrent neural network architectures

5.8 Recurrent Neural Network Architectures

Two straightforward ways to include recurrent connections in neural networks are *activation feedback* and *output feedback*, as shown, respectively, in Figure 5.9(a) and Figure 5.9(b). These schemes are closely related to the state space representation of neural networks. A comprehensive and insightful account of canonical forms and state space representation of general neural networks is given in Nerrand *et al.* (1993) and Dreyfus and Idan (1998). In Figure 5.9, the blocks labelled 'linear dynamical systems' comprise of delays and multipliers, hence providing linear combination of their input signals. The output of a neuron shown in Figure 5.9(a) can be expressed as

$$\left.\begin{aligned} v(k) &= \sum_{i=0}^{M} w_{u,i}(k)u(k-i) + \sum_{j=1}^{N} w_{v,j}(k)v(k-j), \\ y(k) &= \Phi(v(k)), \end{aligned}\right\} \tag{5.26}$$

where $w_{u,i}$ and $w_{v,j}$ correspond to the weights associated with u and v, respectively.

The transfer function of a neuron shown in Figure 5.9(b) can be expressed as

$$\left.\begin{aligned} v(k) &= \sum_{i=0}^{M} w_{u,i}(k)u(k-i) + \sum_{j=1}^{N} w_{y,j}(k)y(k-j), \\ y(k) &= \Phi(v(k)), \end{aligned}\right\} \tag{5.27}$$

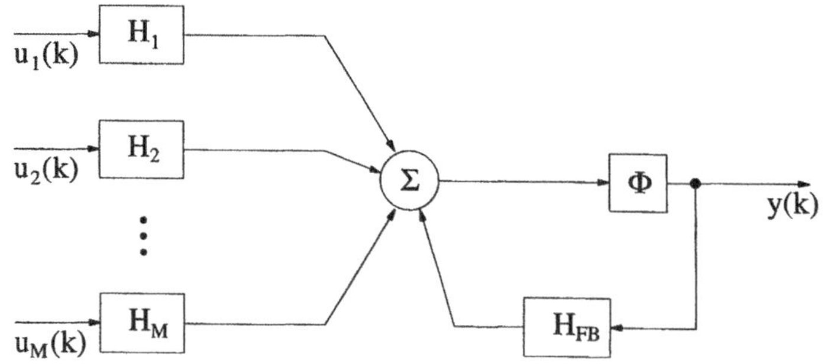

Figure 5.10 General LRGF architecture

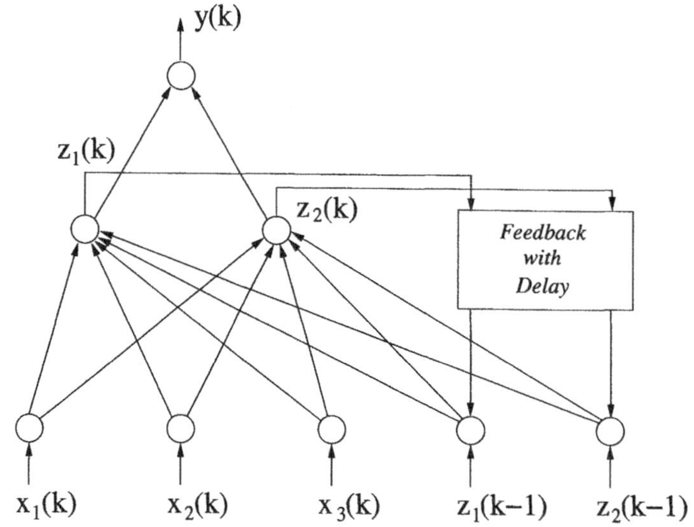

Figure 5.11 An example of Elman recurrent neural network

where $w_{y,j}$ correspond to the weights associated with the delayed outputs. A comprehensive account of types of synapses and short-term memories in dynamical neural networks is provided by Mozer (1993).

The networks mentioned so far exhibit a locally recurrent architecture, but when connected into a larger network, they have a feedforward structure. Hence they are referred to as locally recurrent–globally feedforward (LRGF) architectures. A general LRGF architecture is shown in Figure 5.10. This architecture allows for the dynamic synapses both within the input (represented by $H_1, \dots, H_M$) and the output feedback (represented by $H_{\rm FB}$), hence comprising some of the aforementioned schemes.

The Elman network is a recurrent network with a hidden layer, a simple example of which is shown in Figure 5.11. This network consists of an MLP with an additional input which consists of delayed state space variables of the network. Even though it contains feedback connections, it is treated as a kind of MLP. The network shown in

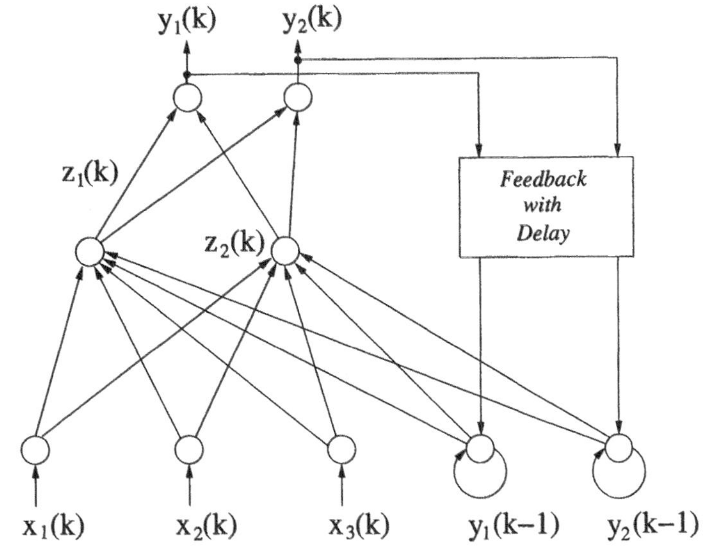

Figure 5.12 An example of Jordan recurrent neural network

Figure 5.12 is an example of the Jordan network. It consists of a multilayer perceptron with one hidden layer and a feedback loop from the output layer to an additional input called the context layer. In the context layer, there are self-recurrent loops. Both Jordan and Elman networks are structurally locally recurrent globally feedforward (LRGF), and are rather limited in including past information.

A network with a rich representation of past outputs, which will be extensively considered in this book, is a fully connected recurrent neural network, known as the Williams–Zipser network (Williams and Zipser 1989a), shown in Figure 5.13. We give a detailed introduction to this architecture. This network consists of three layers: the input layer, the processing layer and the output layer. For each neuron i, $i = 1, 2, \ldots, N$, the elements u_j, $j = 1, 2, \ldots, p + N + 1$, of the input vector to a neuron $\boldsymbol{u}$ (5.31), are weighted, then summed to produce an internal activation function of a neuron v (5.30), which is finally fed through a nonlinear activation function Φ (5.28), to form the output of the ith neuron y_i (5.29). The function Φ is a monotonically increasing sigmoid function with slope β, as for instance the logistic function,

$$\Phi(v) = \frac{1}{1 + \mathrm{e}^{-\beta v}}. \tag{5.28}$$

At the time instant k, for the ith neuron, its weights form a $(p + N + 1) \times 1$ dimensional weight vector $\boldsymbol{w}_i^{\mathrm{T}}(k) = [w_{i,1}(k), \ldots, w_{i,p+N+1}(k)]$, where p is the number of external inputs, N is the number of feedback connections and $(\cdot)^{\mathrm{T}}$ denotes the vector transpose operation. One additional element of the weight vector $\boldsymbol{w}$ is the bias input weight. The feedback consists of the delayed output signals of the RNN. The following equations fully describe the RNN from Fig-

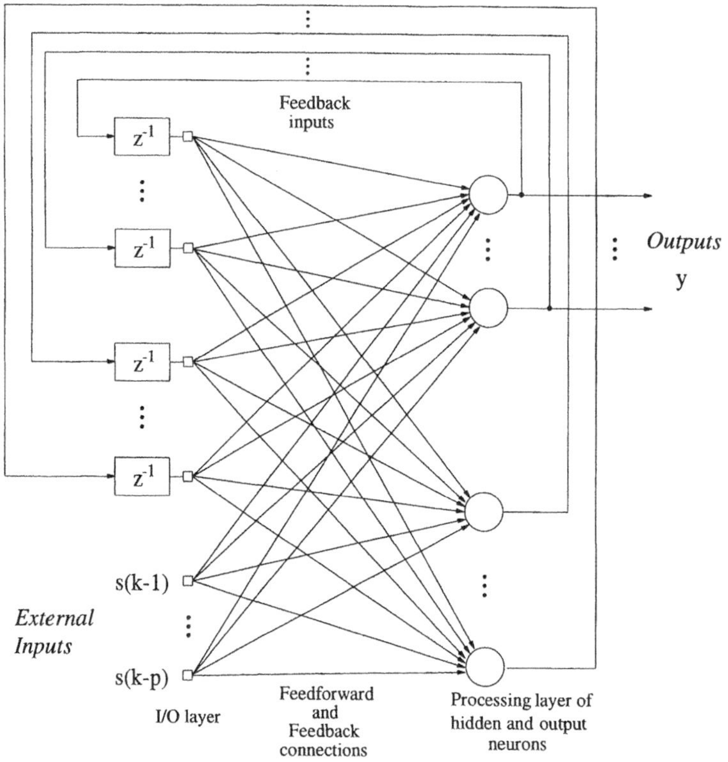

Figure 5.13 A fully connected recurrent neural network

ure 5.13,

$$y_i(k) = \Phi(v_i(k)), \qquad i = 1, 2, \dots, N, \tag{5.29}$$

$$v_i(k) = \sum_{l=1}^{p+N+1} w_{i,l}(k) u_l(k), \tag{5.30}$$

$$u_i^{\mathrm{T}}(k) = [s(k-1), \dots, s(k-p), 1, y_1(k-1), y_2(k-1), \dots, y_N(k-1)], \tag{5.31}$$

where the $(p + N + 1) \times 1$ dimensional vector $\boldsymbol{u}$ comprises both the external and feedback inputs to a neuron, as well as the unity valued constant bias input.

5.9 Hybrid Neural Network Architectures

These networks consist of a cascade of a neural network and a linear adaptive filter. If a neural network is considered as a complex adaptable nonlinearity, then hybrid neural networks resemble Wiener and Hammerstein stochastic models. An example of these networks is given in Khalaf and Nakayama (1999), for prediction of noisy time series. A neural subpredictor is cascaded with a linear FIR predictor, hence making a hybrid predictor. The block diagram of this type of neural network architecture is

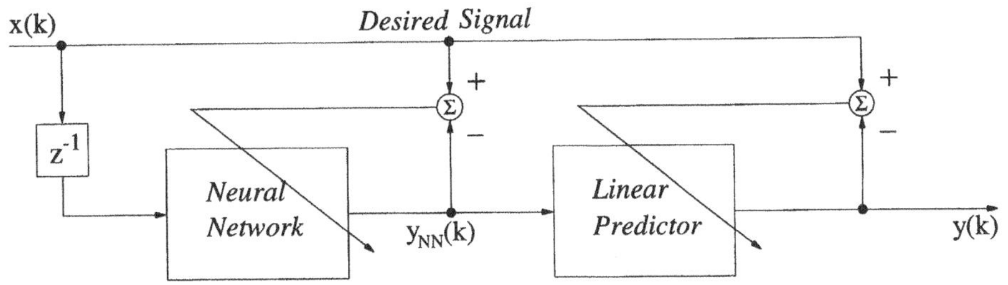

Figure 5.14 A hybrid neural predictor

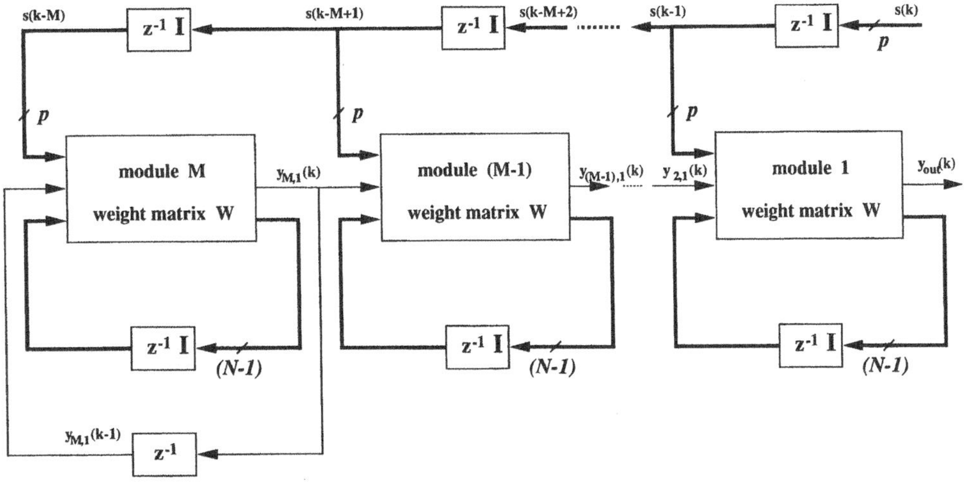

Figure 5.15 Pipelined recurrent neural network

given in Figure 5.14. The neural network from Figure 5.14 can be either a feedforward neural network or a recurrent neural network.

Another example of hybrid structures is the so-called pipelined recurrent neural network (PRNN), introduced by Haykin and Li (1995) and shown in Figure 5.15. It consists of a modular nested structure of small-scale fully connected recurrent neural networks and a cascaded FIR adaptive filter. In the PRNN configuration, the M modules, which are FCRNNs, are connected as shown in Figure 5.15. The cascaded linear filter is omitted. The description of this network follows the approach from Mandic *et al.* (1998) and Baltersee and Chambers (1998). The uppermost module of the PRNN, denoted by M, is simply an FCRNN, whereas in modules $(M-1,\dots,1)$, the only difference is that the feedback signal of the output neuron within module m, denoted by $y_{m,1}$, $m = 1,\dots,M-1$, is replaced with the appropriate output signal $y_{m+1,1}$, $m = 1,\dots,M-1$, from its left neighbour module $m+1$. The $(p \times 1)$-dimensional external signal vector $\boldsymbol{s}^{\mathrm{T}}(k) = [s(k),\dots,s(k-p+1)]$ is delayed by m time steps $(z^{-m}\boldsymbol{I})$ before feeding the module m, where z^{-m}, $m = 1,\dots,M$, denotes the m-step time delay operator and $\boldsymbol{I}$ is the $(p \times p)$-dimensional identity matrix. The weight vectors $\boldsymbol{w}_n$ of each neuron n, are embodied in an $(p+N+1) \times N$ dimensional weight matrix $\boldsymbol{W}(k) = [\boldsymbol{w}_1(k),\dots,\boldsymbol{w}_N(k)]$, with N being the number of neurons in each module. All the modules operate using the same weight matrix $\boldsymbol{W}$. The overall output signal of the PRNN is $y_{\text{out}}(k) = y_{1,1}(k)$, i.e. the output of the first neuron of the first module. A full mathematical description of the PRNN is given in the following equations:

$$y_{i,n}(k) = \Phi(v_{i,n}(k)), \tag{5.32}$$

$$v_{i,n}(k) = \sum_{l=1}^{p+N+1} w_{n,l}(k)u_{i,l}(k), \tag{5.33}$$

$$\boldsymbol{u}_i^{\mathrm{T}}(k) = [s(k-i), \ldots, s(k-i-p+1), 1, \\ y_{i+1,1}(k), y_{i,2}(k-1), \ldots, y_{i,N}(k-1)] \quad \text{for } 1 \leqslant i \leqslant M-1, \tag{5.34}$$

$$\boldsymbol{u}_M^{\mathrm{T}}(k) = [s(k-M), \ldots, s(k-M-p+1), 1, \\ y_{M,1}(k-1), y_{M,2}(k-1), \ldots, y_{M,N}(k-1)] \quad \text{for } i = M. \tag{5.35}$$

At the time step k for each module i, $i = 1, \ldots, M$, the one-step forward prediction error $e_i(k)$ associated with a module is then defined as a difference between the desired response of that module $s(k-i+1)$, which is actually the next incoming sample of the external input signal, and the actual output of the ith module $y_{i,1}(k)$ of the PRNN, i.e.

$$e_i(k) = s(k-i+1) - y_{i,1}(k), \qquad i = 1, \ldots, M. \tag{5.36}$$

Thus, the overall cost function of the PRNN becomes a weighted sum of all squared error signals,

$$E(k) = \sum_{i=1}^{M} \lambda^{i-1} e_i^2(k), \tag{5.37}$$

where $e_i(k)$ is defined in Equation (5.36) and λ, $\lambda \in (0, 1]$, is a forgetting factor.

Other architectures combining linear and nonlinear blocks include the so-called 'sandwich' structure which was used for estimation of Hammerstein systems (Ibnkahla *et al.* 1998). The architecture used was a linear–nonlinear–linear combination.

5.10 Nonlinear ARMA Models and Recurrent Networks

A general NARMA(p, q) recurrent network model can be expressed as (Chang and Hu 1997)

$$\hat{x}(k) = \Phi\biggl(\sum_{i=1}^{p} w_{1,i}(k)x(k-i) + w_{1,p+1}(k) + \sum_{j=p+2}^{p+q+1} w_{1,j}(k)\hat{e}(k+j-2-p-q) \\ + \sum_{l=p+q+2}^{p+q+N} w_{1,l}(k)y_{l-p-q}(k-1)\biggr). \tag{5.38}$$

A realisation of this model is shown in Figure 5.16. The NARMA(p, q) scheme shown in Figure 5.16 is a common Williams–Zipser type recurrent neural network, which

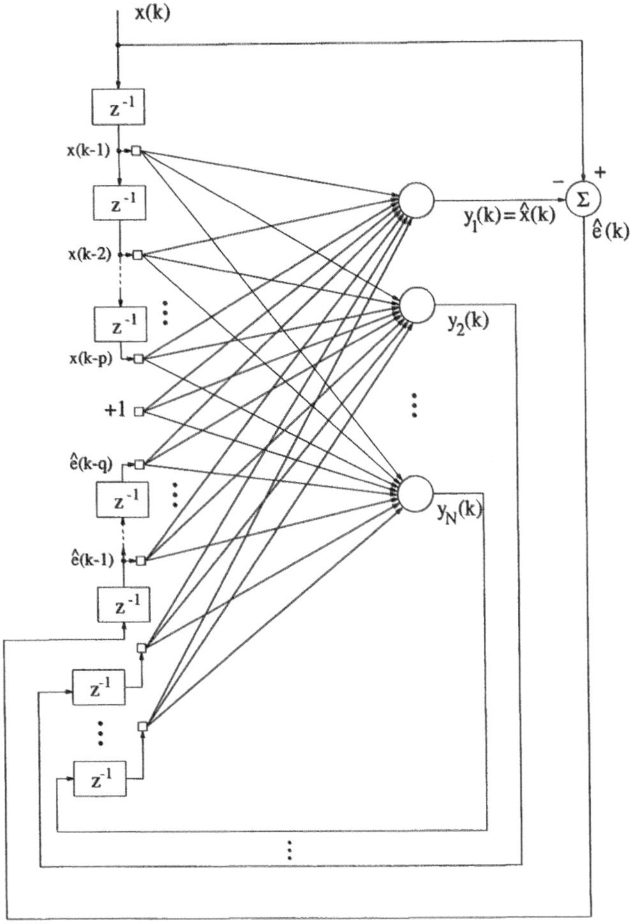

Figure 5.16 Alternative recurrent NARMA(p, q) network

consists of only two layers, the output layer of output and hidden neurons $y_1, \ldots, y_N$, and the input layer of feedforward and feedback signals

$$x(k-1), \ldots, x(k-p), +1, \hat{e}(k-1), \ldots, \hat{e}(k-q), y_2(k-1), \ldots, y_N(k-1).$$

The nonlinearity in this case is determined by both the nonlinearity associated with the output neuron of the recurrent neural network and nonlinearities in hidden neurons.

The inputs to this network, given in (5.38), however, comprise the prediction error terms (residuals) $\hat{e}(k-1), \ldots, \hat{e}(k-q)$, which make the learning in such networks difficult. Namely, the well-known real-time recurrent learning (RTRL) algorithm (Haykin 1994; Williams and Zipser 1989a) was derived to minimise the instantaneous squared prediction error $\hat{e}(k)$, and hence cannot be applied directly to the RNN realisations of the NARMA(p, q) network, as shown above, since the inputs to the network comprise the delayed prediction error terms $\{\hat{e}\}$. It is therefore desirable to find another equivalent representation of the NARMA(p, q) network, which would be more suited for the RTRL-based learning.

If, for the sake of clarity, we denote the predicted values $\hat{x}$ by y, i.e. to match the notation common in RNNs with the NARMA(p, q) theory, and have $y_1(k) = \hat{x}(k)$, and keep the symbol x for the exact values of the input signal being predicted, the NARMA network from (5.38), can be approximated further as (Connor 1994)

$$\begin{aligned} y_1(k) &= h(x(k-1), x(k-2), \dots, x(k-p), \hat{e}(k-1), \hat{e}(k-2), \dots, \hat{e}(k-q)) \\ &= h(x(k-1), x(k-2), \dots, x(k-p), (x(k-1) - y_1(k-1)), \dots \\ &\qquad \dots, (x(k-q) - y_1(k-q))) \\ &= H(x(k-1), x(k-2), \dots, x(k-p), y_1(k-1), \dots, y_1(k-q)). \end{aligned} \tag{5.39}$$

In that case, the scheme shown in Figure 5.16 should be redrawn, remaining topologically the same, with y_1 replacing the corresponding $\hat{e}$ terms among the inputs to the network.

On the other hand, the alternative expression for the conditional mean predictor, depicted in Figure 5.16 can be written as

$$\begin{aligned} \hat{x}(k) = \Phi\Bigg(\sum_{i=1}^{p} w_{1,i}(k)x(k-i) + w_{1,p+1}(k) + \sum_{j=p+2}^{p+q+1} w_{1,j}(k)\hat{x}(k+j-2-p-q) \\ + \sum_{l=p+q+2}^{p+q+N} w_{1,l}(k)y_{l-p-q}(k-1)\Bigg) \end{aligned} \tag{5.40}$$

or, bearing in mind (5.39), the notation used earlier (Haykin and Li 1995; Mandic *et al.* 1998) for the examples on the prediction of speech, i.e. $x(k) = s(k)$, and that $y_1(k) = \hat{s}(k)$,

$$\begin{aligned} \hat{s}(k) = \Phi\Bigg(\sum_{i=1}^{p} w_{1,i}(k)s(k-i) + w_{1,p+1}(k) + \sum_{j=p+2}^{p+q+1} w_{1,j}(k)y_1(k+j-2-p-q) \\ + \sum_{l=p+q+2}^{p+q+N} w_{1,l}(k)y_{l-p-q}(k-1)\Bigg), \end{aligned} \tag{5.41}$$

which is the common RNN lookalike notation. This scheme offers a simpler solution to the NARMA(p, q) problem, as compared to the previous one, since the only nonlinear function used is the activation function of a neuron Φ, while the set of signals being processed is the same as in the previous scheme. Furthermore, the scheme given in (5.41) and depicted in Figure 5.17 resembles the basic ARMA structure.

Li (1992) has shown that the recurrent network of (5.41) with a sufficiently large number of neurons and appropriate weights can be found by performing the RTRL algorithm such that the sum of squared prediction errors $E < \delta$ for an arbitrary $\delta > 0$. In other words, $\|s - \hat{s}\|_D < \delta$, where $\|\cdot\|_D$ denotes the $\mathcal{L}_2$ norm with respect to the training set D. Moreover, this scheme, shown also in Figure 5.17, fits into the well-known learning strategies, such as the RTRL algorithm, which recommends this

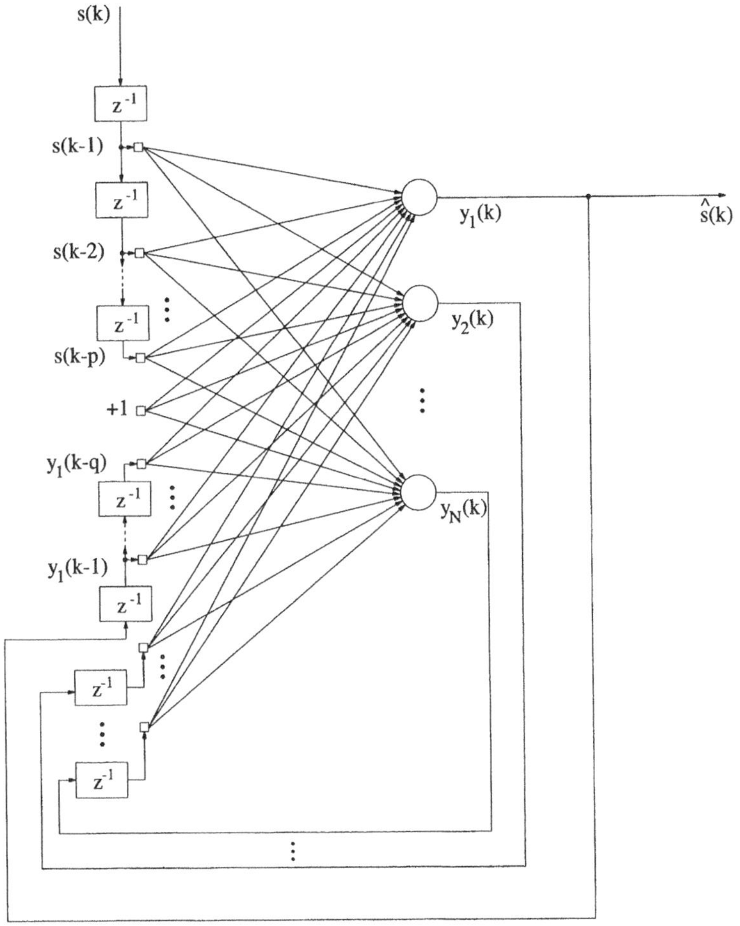

Figure 5.17 Recurrent NARMA(p, q) implementation of prediction model

scheme for NARMA/NARMAX nonlinear prediction applications (Baldi and Atiya 1994; Draye *et al.* 1996; Kosmatopoulos *et al.* 1995; McDonnell and Waagen 1994; Nerrand *et al.* 1994; Wu and Niranjan 1994).

5.11 Summary

A review of recurrent neural network architectures in the fields of nonlinear dynamical modelling, system identification, control, signal processing and forecasting has been provided. A relationship between neural network models and NARMA/NARMAX models, as well as Wiener and Hammerstein structures has been established. Particular attention has been devoted to the fully connected recurrent neural network and its use in NARMA/NARMAX modelling has been highlighted.

6

Neural Networks as Nonlinear Adaptive Filters

6.1 Perspective

Neural networks, in particular recurrent neural networks, are cast into the framework of nonlinear adaptive filters. In this context, the relation between recurrent neural networks and polynomial filters is first established. Learning strategies and algorithms are then developed for neural adaptive system identifiers and predictors. Finally, issues concerning the choice of a neural architecture with respect to the bias and variance of the prediction performance are discussed.

6.2 Introduction

Representation of nonlinear systems in terms of NARMA/NARMAX models has been discussed at length in the work of Billings and others (Billings 1980; Chen and Billings 1989; Connor 1994; Nerrand *et al.* 1994). Some cognitive aspects of neural nonlinear filters are provided in Maass and Sontag (2000). Pearson (1995), in his article on nonlinear input–output modelling, shows that block oriented nonlinear models are a subset of the class of Volterra models. So, for instance, the Hammerstein model, which consists of a static nonlinearity $f(\cdot)$ applied at the output of a linear dynamical system described by its z-domain transfer function $H(z)$, can be represented[1] by the Volterra series.

In the previous chapter, we have shown that neural networks, be they feedforward or recurrent, cannot generate time delays of an order higher than the dimension of the input to the network. Another important feature is the capability to generate subharmonics in the spectrum of the output of a nonlinear neural filter (Pearson 1995). The key property for generating subharmonics in nonlinear systems is recursion, hence, recurrent neural networks are necessary for their generation. Notice that, as

[1] Under the condition that the function f is analytic, and that the Volterra series can be thought of as a generalised Taylor series expansion, then the coefficients of the model (6.2) that do not vanish are $h_{i,j,\dots,z} \neq 0 \Leftrightarrow i = j = \cdots = z$.

pointed out in Pearson (1995), block-stochastic models are, generally speaking, not suitable for this application.

In Hakim *et al.* (1991), by using the Weierstrass polynomial expansion theorem, the relation between neural networks and Volterra series is established, which is then extended to a more general case and to continuous functions that cannot be expanded via a Taylor series expansion.[2] Both feedforward and recurrent networks are characterised by means of a Volterra series and vice versa.

Neural networks are often referred to as 'adaptive neural networks'. As already shown, adaptive filters and neural networks are formally equivalent, and neural networks, employed as nonlinear adaptive filters, are generalisations of linear adaptive filters. However, in neural network applications, they have been used mostly in such a way that the network is first trained on a particular training set and subsequently used. This approach is not an online adaptive approach, which is in contrast with linear adaptive filters, which undergo continual adaptation.

Two groups of learning techniques are used for training recurrent neural networks: a direct gradient computation technique (used in nonlinear adaptive filtering) and a recurrent backpropagation technique (commonly used in neural networks for offline applications). The real-time recurrent learning (RTRL) algorithm (Williams and Zipser 1989a) is a technique which uses direct gradient computation, and is used if the network coefficients change slowly with time. This technique is essentially an LMS learning algorithm for a nonlinear IIR filter. It should be noticed that, with the same computation time, it might be possible to unfold the recurrent neural network into the corresponding feedforward counterparts and hence to train it by backpropagation. The backpropagation through time (BPTT) algorithm is such a technique (Werbos 1990).

Some of the benefits involved with neural networks as nonlinear adaptive filters are that no assumptions concerning Markov property, Gaussian distribution or additive measurement noise are necessary (Lo 1994). A neural filter would be a suitable choice even if mathematical models of the input process and measurement noise are not known (black box modelling).

6.3 Overview

We start with the relationship between Volterra and bilinear filters and neural networks. Recurrent neural networks are then considered as nonlinear adaptive filters and neural architectures for this case are analysed. Learning algorithms for online training of recurrent neural networks are developed inductively, starting from corresponding algorithms for linear adaptive IIR filters. Some issues concerning the problem of vanishing gradient and bias/variance dilemma are finally addressed.

6.4 Neural Networks and Polynomial Filters

It has been shown in Chapter 5 that a small-scale neural network can represent high-order nonlinear systems, whereas a large number of terms are required for an equiv-

---

[2] For instance nonsmooth functions, such as $|x|$.

alent Volterra series representation. For instance, as already shown, after performing a Taylor series expansion for the output of a neural network depicted in Figure 5.3, with input signals $u(k-1)$ and $u(k-2)$, we obtain

$$\begin{aligned} y(k) = c_0 + c_1 u(k-1) + c_2 u(k-2) + c_3 u^2(k-1) + c_4 u^2(k-2) \\ + c_5 u(k-1)u(k-2) + c_6 u^3(k-1) + c_7 u^3(k-2) + \cdots, \end{aligned} \tag{6.1}$$

which has the form of a general Volterra series, given by

$$y(k) = h_0 + \sum_{i=0}^{N} h_1(i)x(k-i) + \sum_{i=0}^{N}\sum_{j=0}^{N} h_2(i,j)x(k-i)x(k-j) + \cdots, \tag{6.2}$$

Representation by a neural network is therefore more compact. As pointed out in Schetzen (1981), Volterra series are not suitable for modelling saturation type nonlinear functions and systems with nonlinearities of a high order, since they require a very large number of terms for an acceptable representation. The order of Volterra series and complexity of kernels $h(\cdot)$ increase exponentially with the order of the delay in system (6.2). This problem restricts practical applications of Volterra series to small-scale systems.

Nonlinear system identification, on the other hand, has been traditionally based upon the Kolmogorov approximation theorem (neural network existence theorem), which states that a neural network with a hidden layer can approximate an arbitrary nonlinear system. Kolmogorov's theorem, however, is not that relevant in the context of networks for learning (Girosi and Poggio 1989b). The problem is that inner functions in Kolmogorov's formula (4.1), although continuous, have to be highly nonsmooth. Following the analysis from Chapter 5, it is straightforward that multilayered and recurrent neural networks have the ability to approximate an arbitrary nonlinear system, whereas Volterra series fail even for simple saturation elements.

Another convenient form of nonlinear system is the bilinear (truncated Volterra) system described by

$$y(k) = \sum_{j=1}^{N-1} c_j y(k-j) + \sum_{i=0}^{N-1}\sum_{j=1}^{N-1} b_{i,j} y(k-j)x(k-i) + \sum_{i=0}^{N-1} a_i x(k-i). \tag{6.3}$$

Despite its simplicity, this is a powerful nonlinear model and a large class of nonlinear systems (including Volterra systems) can be approximated arbitrarily well using this model. Its functional dependence (6.3) shows that it belongs to a class of general recursive nonlinear models. A recurrent neural network that realises a simple bilinear model is depicted in Figure 6.1. As seen from Figure 6.1, multiplicative input nodes (denoted by '$\times$') have to be introduced to represent the bilinear model. Bias terms are omitted and the chosen neuron is linear.

Example 6.4.1. Show that the recurrent network shown in Figure 6.1 realises a bilinear model. Also show that this network can be described in terms of NARMAX models.

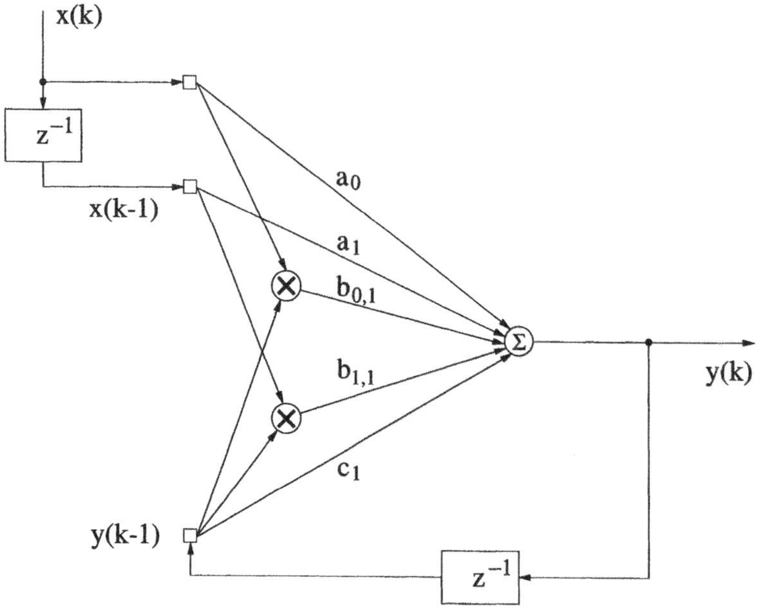

Figure 6.1 Recurrent neural network representation of the bilinear model

Solution. The functional description of the recurrent network depicted in Figure 6.1 is given by

$$y(k) = c_1 y(k-1) + b_{0,1} x(k) y(k-1) + b_{1,1} x(k-1) y(k-1) + a_0 x(k) + a_1 x(k-1), \quad (6.4)$$

which belongs to the class of bilinear models (6.3). The functional description of the network from Figure 6.1 can also be expressed as

$$y(k) = F(y(k-1), x(k), x(k-1)), \quad (6.5)$$

which is a NARMA representation of model (6.4). □

Example 6.4.1 confirms the duality between Volterra, bilinear, NARMA/NARMAX and recurrent neural models. To further establish the connection between Volterra series and a neural network, let us express the activation potential of nodes of the network as

$$\mathrm{net}_i(k) = \sum_{j=0}^{M} w_{i,j} x(k-j), \quad (6.6)$$

where $\mathrm{net}_i(k)$ is the activation potential of the ith hidden neuron, $w_{i,j}$ are weights and $x(k-j)$ are inputs to the network. If the nonlinear activation functions of neurons are expressed via an Lth-order polynomial expansion[3] as

$$\Phi(\mathrm{net}_i(k)) = \sum_{l=0}^{L} \xi_{il} \, \mathrm{net}_i^l(k), \quad (6.7)$$

[3] Using the Weierstrass theorem, this expansion can be arbitrarily accurate. However, in practice we resort to a moderate order of this polynomial expansion.

then the neural model described in (6.6) and (6.7) can be related to the Volterra model (6.2). The actual relationship is rather complicated, and Volterra kernels are expressed as sums of products of the weights from input to hidden units, weights associated with the output neuron, and coefficients ξ_{il} from (6.7). Chon *et al.* (1998) have used this kind of relationship to compare the Volterra and neural approach when applied to processing of biomedical signals.

Hence, to avoid the difficulty of excessive computation associated with Volterra series, an input–output relationship of a nonlinear predictor that computes the output in terms of past inputs and outputs may be introduced as[4]

$$\hat{y}(k) = F(y(k-1), \dots, y(k-N), u(k-1), \dots, u(k-M)), \tag{6.8}$$

where $F(\cdot)$ is some nonlinear function. The function F may change for different input variables or for different regions of interest. A NARMAX model may therefore be a correct representation only in a region around some operating point. Leontaritis and Billings (1985) rigorously proved that a discrete time nonlinear time invariant system can always be represented by model (6.8) in the vicinity of an equilibrium point provided that

- the response function of the system is finitely realisable, and
- it is possible to linearise the system around the chosen equilibrium point.

As already shown, some of the other frequently used models, such as the bilinear polynomial filter, given by (6.3), are obviously cases of a simple NARMAX model.

6.5 Neural Networks and Nonlinear Adaptive Filters

To perform nonlinear adaptive filtering, tracking and system identification of nonlinear time-varying systems, there is a need to introduce dynamics in neural networks. These dynamics can be introduced via recurrent neural networks, which are the focus of this book.

The design of linear filters is conveniently specified by a frequency response which we would like to match. In the nonlinear case, however, since a transfer function of a nonlinear filter is not available in the frequency domain, one has to resort to different techniques. For instance, the design of nonlinear filters may be thought of as a nonlinear constrained optimisation problem in Fock space (deFigueiredo 1997).

In a recurrent neural network architecture, the feedback brings the delayed outputs from hidden and output neurons back into the network input vector $\boldsymbol{u}(k)$, as shown in Figure 5.13. Due to gradient learning algorithms, which are sequential, these delayed outputs of neurons represent filtered data from the previous discrete time instant. Due to this 'memory', at each time instant, the network is presented with the raw,

---

[4] As already shown, this model is referred to as the NARMAX model (nonlinear ARMAX), since it resembles the linear model

$$\hat{y}(k) = a_0 + \sum_{j=1}^{N} a_j y(k-j) + \sum_{i=1}^{M} b_i u(k-i).$$

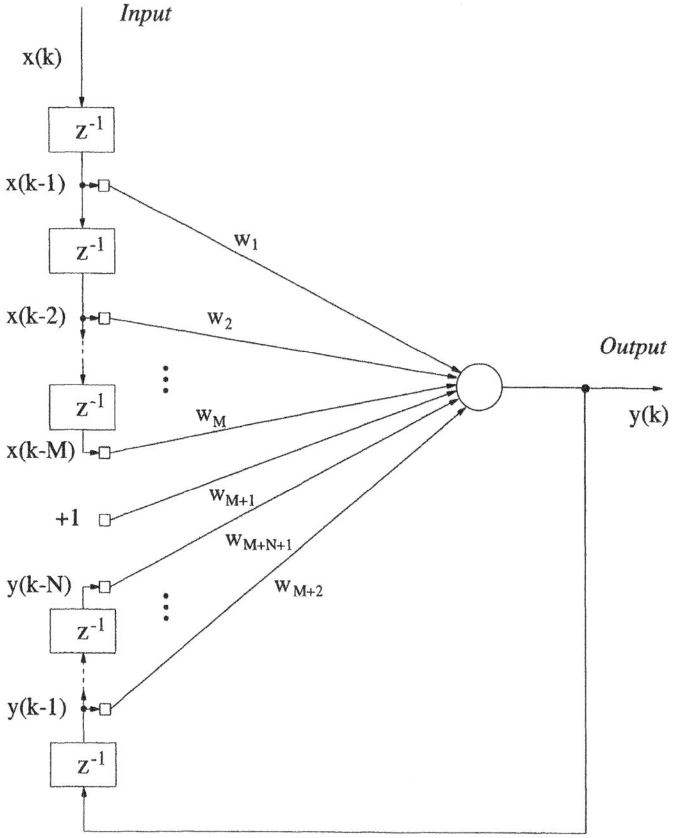

Figure 6.2 NARMA recurrent perceptron

possibly noisy, external input data $s(k), s(k-1), \dots, s(k-M)$ from Figure 5.13 and Equation (5.31), and filtered data $y_1(k-1), \dots, y_N(k-1)$ from the network output. Intuitively, this filtered input history helps to improve the processing performance of recurrent neural networks, as compared with feedforward networks. Notice that the history of past outputs is never presented to the learning algorithm for feedforward networks. Therefore, a recurrent neural network should be able to process signals corrupted by additive noise even in the case when the noise distribution is varying over time.

On the other hand, a nonlinear dynamical system can be described by

$$\boldsymbol{u}(k+1) = \Phi(\boldsymbol{u}(k)) \tag{6.9}$$

with an observation process

$$y(k) = \varphi(\boldsymbol{u}(k)) + \epsilon(k), \tag{6.10}$$

where $\epsilon(k)$ is observation noise (Haykin and Principe 1998). Takens' embedding theorem (Takens 1981) states that the geometric structure of system (6.9) can be recovered

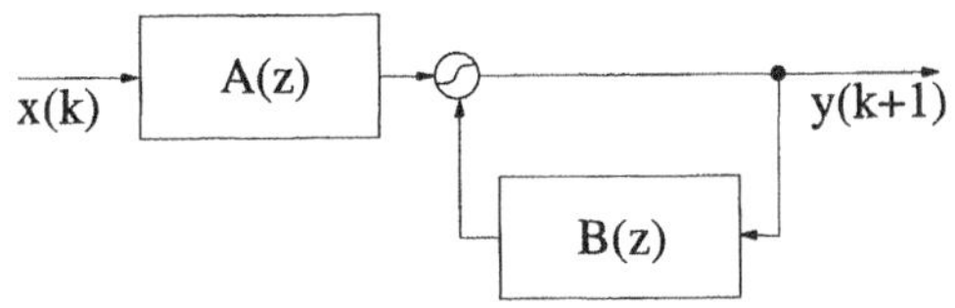

(a) A recurrent nonlinear neural filter

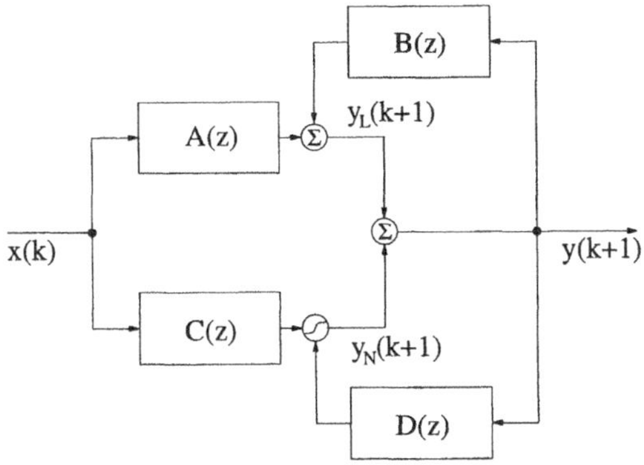

(b) A recurrent linear/nonlinear neural filter structure

Figure 6.3 Nonlinear IIR filter structures

from the sequence $\{y(k)\}$ in a D-dimensional space spanned by[5]

$$\boldsymbol{y}(k) = [y(k), y(k-1), \dots, y(k-(D-1))] \tag{6.11}$$

provided that $D \geqslant 2d+1$, where d is the dimension of the state space of system (6.9). Therefore, one advantage of NARMA models over FIR models is the parsimony of NARMA models, since an upper bound on the order of a NARMA model is twice the order of the state (phase) space of the system being analysed.

The simplest recurrent neural network architecture is a recurrent perceptron, shown in Figure 6.2. This is a simple, yet effective architecture. The equations which describe the recurrent perceptron shown in Figure 6.2 are

$$\left.\begin{aligned} y(k) &= \Phi(v(k)), \\ v(k) &= \boldsymbol{u}^{\mathrm{T}}(k)\boldsymbol{w}(k), \end{aligned}\right\} \tag{6.12}$$

where $\boldsymbol{u}(k) = [x(k-1), \dots, x(k-M), 1, y(k-1), \dots, y(k-N)]^{\mathrm{T}}$ is the input vector, $\boldsymbol{w}(k) = [w_1(k), \dots, w_{M+N+1}(k)]^{\mathrm{T}}$ is the weight vector and $(\cdot)^{\mathrm{T}}$ denotes the vector transpose operator.

[5] Model (6.11) is in fact a NAR/NARMAX model.

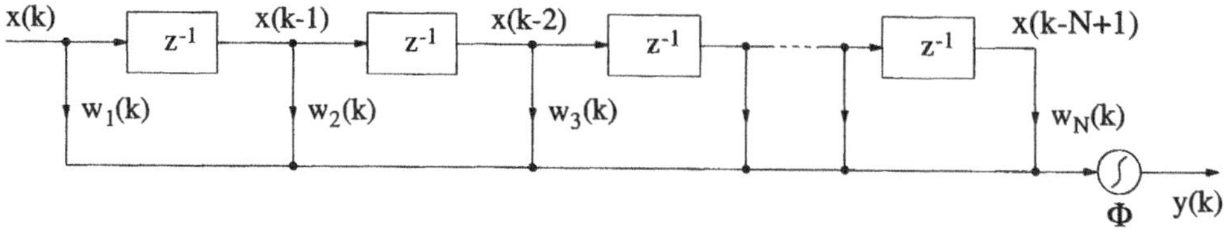

Figure 6.4 A simple nonlinear adaptive filter

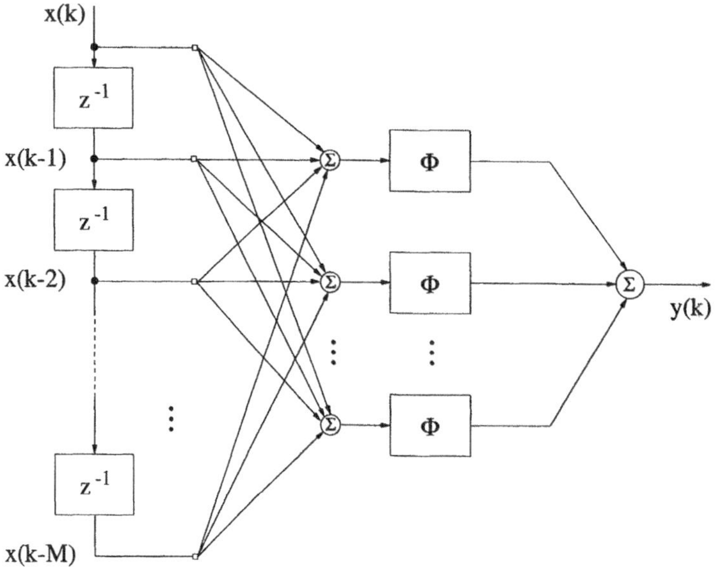

Figure 6.5 Fully connected feedforward neural filter

A recurrent perceptron is a recursive adaptive filter with an arbitrary output function as shown in Figure 6.3. Figure 6.3(a) shows the recurrent perceptron structure as a nonlinear infinite impulse response (IIR) filter. Figure 6.3(b) depicts the parallel linear/nonlinear structure, which is one of the possible architectures. These structures stem directly from IIR filters and are described in McDonnell and Waagen (1994), Connor (1994) and Nerrand *et al.* (1994). Here, $A(z)$, $B(z)$, $C(z)$ and $D(z)$ denote the z-domain linear transfer functions. The general structure of a fully connected, multilayer neural feedforward filter is shown in Figure 6.5 and represents a generalisation of a simple nonlinear feedforward perceptron with dynamic synapses, shown in Figure 6.4. This structure consists of an input layer, layer of hidden neurons and an output layer. Although the output neuron shown in Figure 6.5 is linear, it could be nonlinear. In that case, attention should be paid that the dynamic ranges of the input signal and output neuron match.

Another generalisation of a fully connected recurrent neural filter is shown in Figure 6.6. This network consists of nonlinear neural filters as depicted in Figure 6.5, applied to both the input and output signal, the outputs of which are summed together. This is a fairly general structure which resembles the architecture of a lin-

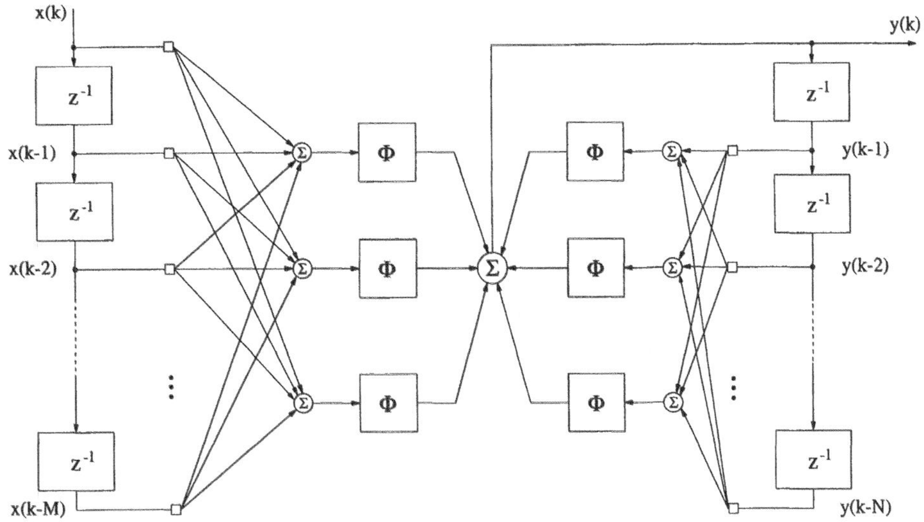

Figure 6.6 Fully connected recurrent neural filter

ear IIR filter and is the extension of the NARMAX recurrent perceptron shown in Figure 6.2.

Narendra and Parthasarathy (1990) provide deep insight into structures of neural networks for identification of nonlinear dynamical systems. Due to the duality between system identification and prediction, the same architectures are suitable for prediction applications. From Figures 6.3–6.6, we can identify four general architectures of neural networks for prediction and system identification. These architectures come as combinations of linear/nonlinear parts from the architecture shown in Figure 6.6, and for the nonlinear prediction configuration are specified as follows.

(i) The output $y(k)$ is a linear function of previous outputs and a nonlinear function of previous inputs, given by

$$y(k) = \sum_{j=1}^{N} a_j(k)y(k-j) + F(u(k-1), u(k-2), \dots, u(k-M)), \tag{6.13}$$

where $F(\cdot)$ is some nonlinear function. This architecture is shown in Figure 6.7(a).

(ii) The output $y(k)$ is a nonlinear function of past outputs and a linear function of past inputs, given by

$$y(k) = F(y(k-1), y(k-2), \dots, y(k-N)) + \sum_{i=1}^{M} b_i(k)u(k-i). \tag{6.14}$$

This architecture is depicted in Figure 6.7(b).

(iii) The output $y(k)$ is a nonlinear function of both past inputs and outputs. The functional relationship between the past inputs and outputs can be expressed

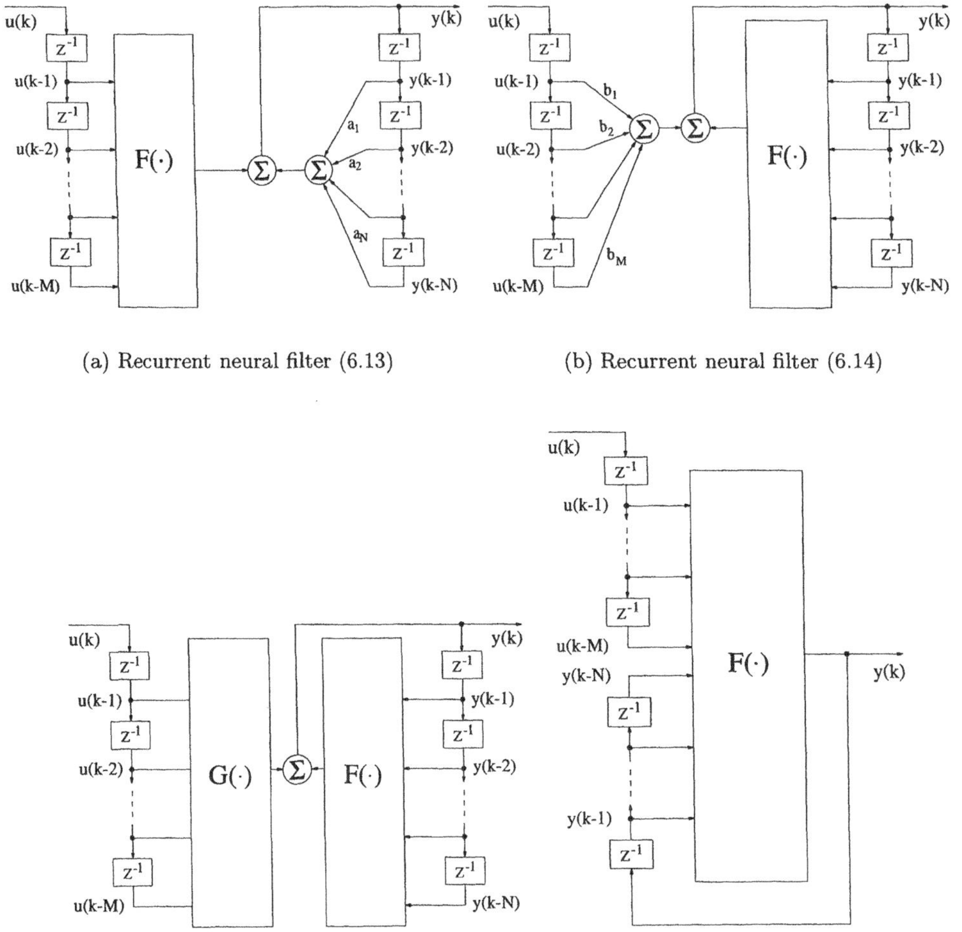

(a) Recurrent neural filter (6.13)

(b) Recurrent neural filter (6.14)

(c) Recurrent neural filter (6.15)

(d) Recurrent neural filter (6.16)

Figure 6.7 Architectures of recurrent neural networks as nonlinear adaptive filters

in a separable manner as

$$y(k) = F(y(k-1), \dots, y(k-N)) + G(u(k-1), \dots, u(k-M)). \tag{6.15}$$

This architecture is depicted in Figure 6.7(c).

(iv) The output $y(k)$ is a nonlinear function of past inputs and outputs, as

$$y(k) = F(y(k-1), \dots, y(k-N), u(k-1), \dots, u(k-M)). \tag{6.16}$$

This architecture is depicted in Figure 6.7(d) and is most general.

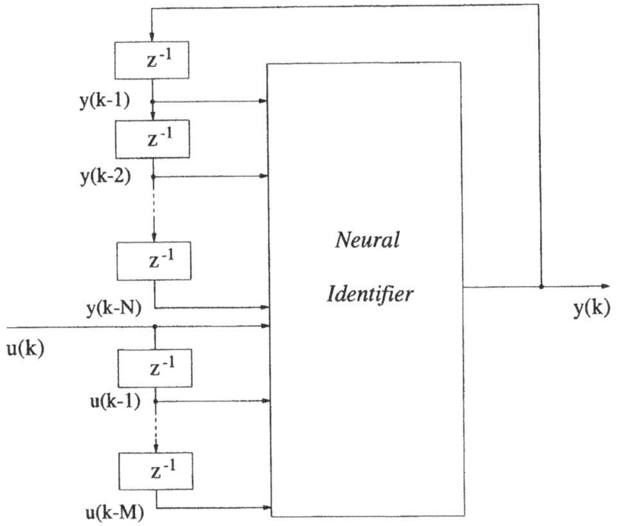

Figure 6.8 NARMA type neural identifier

6.6 Training Algorithms for Recurrent Neural Networks

A natural error criterion, upon which the training of recurrent neural networks is based, is in the form of the accumulated squared prediction error over the whole dataset, given by

$$E(k) = \frac{1}{M+1} \sum_{m=0}^{M} \lambda(m) e^2(k-m), \tag{6.17}$$

where M is the length of the dataset and $\lambda(m)$ are weights associated with a particular instantaneous error $e(k-m)$. For stationary signals, usually $\lambda(m) = 1$, $m = 1, 2, \dots, M$, whereas in the nonstationary case, since the statistics change over time, it is unreasonable to take into account the whole previous history of the errors. For this case, a forgetting mechanism is usually employed, whereby $0 < \lambda(m) < 1$. Since many real-world signals are nonstationary, online learning algorithms commonly use the squared instantaneous error as an error criterion, i.e.

$$E(k) = \tfrac{1}{2} e^2(k). \tag{6.18}$$

Here, the coefficient $\frac{1}{2}$ is included for convenience in the derivation of the algorithms.

6.7 Learning Strategies for a Neural Predictor/Identifier

A NARMA/NARMAX type neural identifier is depicted in Figure 6.8. When considering a neural predictor, the only difference is the position of the neural module within the system structure, as shown in Chapter 2. There are two main training strategies to estimate the weights of the neural network shown in Figure 6.8. In the first approach, the links between the real system and the neural identifier are as depicted in Figure 6.9. During training, the configuration shown in Figure 6.9 can be

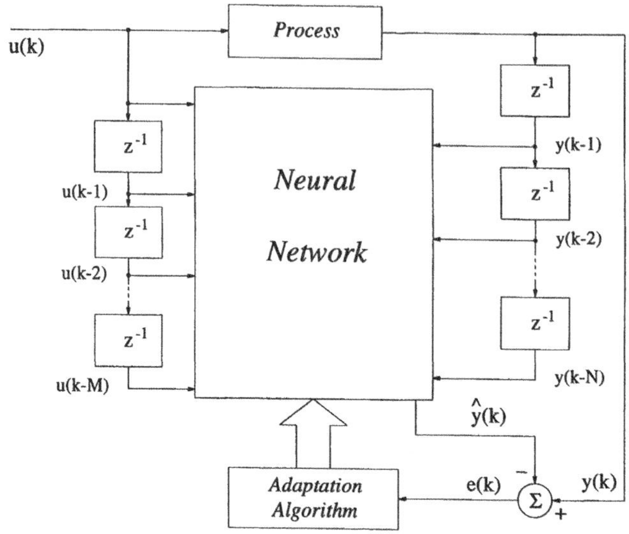

Figure 6.9 The nonlinear series–parallel (teacher forcing) learning configuration

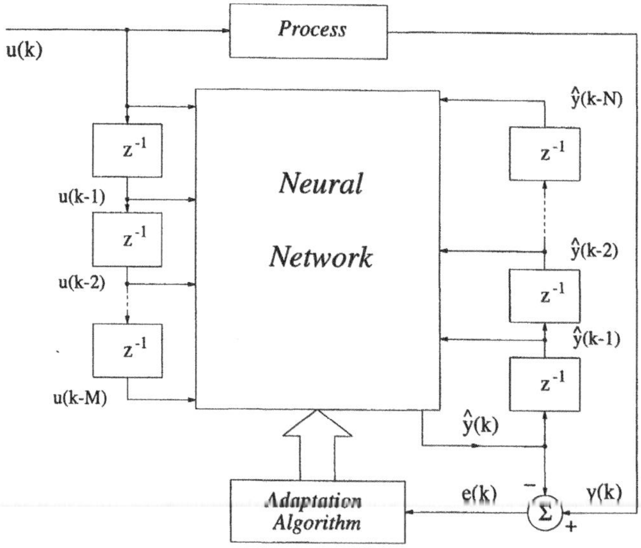

Figure 6.10 The nonlinear parallel (supervised) learning configuration

described by

$$\hat{y}(k) = f(u(k), \ldots, u(k-M), y(k-1), \ldots, y(k-N)), \tag{6.19}$$

which is referred to as the *nonlinear series–parallel model* (Alippi and Piuri 1996; Qin *et al.* 1992). In this configuration, the desired signal $y(k)$ is presented to the network, which produces biased estimates (Narendra 1996).

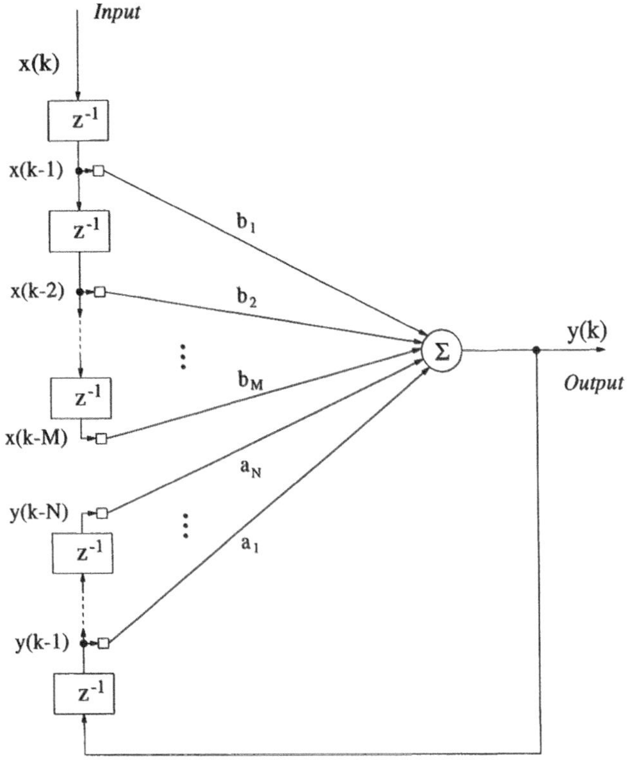

Figure 6.11 Adaptive IIR filter

To overcome such a problem, a training configuration depicted in Figure 6.10 may be considered. This configuration is described by

$$\hat{y}(k) = f(u(k), \dots, u(k-M), \hat{y}(k-1), \dots, \hat{y}(k-N)). \tag{6.20}$$

Here, the previous estimated outputs $\hat{y}(k)$ are fed back into the network. It should be noticed that these two configurations require different training algorithms. The configuration described by Figure 6.10 and Equation (6.20) is known as the *nonlinear parallel model* (Alippi and Piuri 1996; Qin *et al.* 1992) and requires the use of a recursive training algorithm, such as the RTRL algorithm.

The nonlinear prediction configuration using a recurrent neural network is shown in Figure 5.2, where the signal to be predicted $u(k)$ is delayed through a tap delay line and fed into a neural predictor.

6.7.1 Learning Strategies for a Neural Adaptive Recursive Filter

To introduce learning strategies for recurrent neural networks, we start from corresponding algorithms for IIR adaptive filters. An IIR adaptive filter can be thought of as a recurrent perceptron from Figure 6.2, for which the neuron is linear, i.e. it performs only summation instead of both summation and nonlinear mapping. An IIR adaptive filter in the prediction configuration is shown in Figure 6.11. A comprehensive account of adaptive IIR filters is given in Regalia (1994).

Two classes of adaptive learning algorithms used for IIR systems are the *equation error* and *output error* algorithms (Shynk 1989). In the equation error configuration, the desired signal $d(k)$ is fed back into the adaptive filter, whereas in the output error configuration, the signals that are fed back are the estimated outputs $\hat{y}(k)$.

6.7.2 Equation Error Formulation

The output $y_{\mathrm{EE}}(k)$ of the equation error IIR filter strategy is given by

$$y_{\mathrm{EE}}(k) = \sum_{i=1}^{N} a_i(k)d(k-i) + \sum_{j=1}^{M} b_j(k)x(k-j), \tag{6.21}$$

where $\{a_i(k)\}$ and $\{b_j(k)\}$ are adjustable coefficients which correspond, respectively, to the feedback and input signals. Since the functional relationship (6.21) does not comprise of delayed outputs $y_{\mathrm{EE}}(k)$, this filter *does not* have feedback and the output $y_{\mathrm{EE}}(k)$ depends *linearly* on its coefficients. This means that the learning algorithm for this structure is in fact a kind of LMS algorithm for an FIR structure with inputs $\{d(k)\}$ and $\{x(k)\}$. A more compact expression for filter (6.21) is given by

$$y_{\mathrm{EE}}(k) = A(k, z^{-1})d(k) + B(k, z^{-1})x(k), \tag{6.22}$$

where

$$A(k, z^{-1}) = \sum_{i=1}^{N} a_i(k)z^{-i} \quad \text{and} \quad B(k, z^{-1}) = \sum_{j=1}^{M} b_j(k)z^{-j}.$$

The equation error $e_{\mathrm{EE}}(k) = d(k) - y_{\mathrm{EE}}(k)$ can be expressed as

$$e_{\mathrm{EE}}(k) = d(k) - [1 - A(k, z^{-1})]d(k) - B(k, z^{-1})x(k), \tag{6.23}$$

whereby the name of the method is evident. Since $e_{\mathrm{EE}}(k)$ is a linear function of coefficients $\{a_i(k)\}$ and $\{b_j(k)\}$, the error performance surface in this case is quadratic, with a single global minimum. In the presence of noise, however, this minimum is in a different place from in the output error formulation.

6.7.3 Output Error Formulation

The output $y_{\mathrm{OE}}(k)$ of the output error learning strategy is given by

$$y_{\mathrm{OE}}(k) = \sum_{i=1}^{N} a_i(k)y_{\mathrm{OE}}(k-i) + \sum_{j=1}^{M} b_j(k)x(k-j). \tag{6.24}$$

A more compact form of Equation (6.24) can be expressed as

$$y_{\mathrm{OE}}(k) = \frac{B(k, z^{-1})}{1 - A(k, z^{-1})}x(k). \tag{6.25}$$

The output error $e_{\mathrm{OE}}(k) = d(k) - y_{\mathrm{OE}}(k)$ is the difference between the teaching signal $d(k)$ and output $y_{\mathrm{OE}}(k)$, hence the name of the method. The output $y_{\mathrm{OE}}(k)$ is a function of the coefficients and past outputs, and so too is the error $e_{\mathrm{OE}}(k)$. As a consequence, the error performance surface for this strategy has potentially multiple local minima, especially if the order of the model is smaller than the order of the process.

Notice that the equation error can be expressed as a filtered version of the output error as

$$e_{\mathrm{EE}}(k) = [1 - A(k, z^{-1})]e_{\mathrm{OE}}(k). \tag{6.26}$$

6.8 Filter Coefficient Adaptation for IIR Filters

We first present the coefficient adaptation algorithm for an *output error* IIR filter. In order to derive the relations for filter coefficient adaptation, let us define the gradient $\nabla_{\Theta}(E(k))$ for the instantaneous cost function $E(k) = \frac{1}{2}e^2(k)$ as

$$\nabla_{\Theta} E(k) = \frac{\partial E(k)}{\partial \Theta(k)} = e_{\mathrm{OE}}(k)\nabla_{\Theta} e_{\mathrm{OE}}(k) = -e_{\mathrm{OE}}(k)\nabla_{\Theta} y_{\mathrm{OE}}(k), \tag{6.27}$$

where $\Theta(k) = [b_1(k), \dots, b_M(k), a_1(k), \dots, a_N(k)]^{\mathrm{T}}$. The gradient vector consists of partial derivatives of the output with respect to filter coefficients

$$\nabla_{\Theta} y_{\mathrm{OE}}(k) = \left[\frac{\partial y_{\mathrm{OE}}(k)}{\partial b_1(k)}, \dots, \frac{\partial y_{\mathrm{OE}}(k)}{\partial b_M(k)}, \frac{\partial y_{\mathrm{OE}}(k)}{\partial a_1(k)}, \dots, \frac{\partial y_{\mathrm{OE}}(k)}{\partial a_N(k)}\right]^{\mathrm{T}}. \tag{6.28}$$

To derive the coefficient update equations, notice that the inputs $\{x(k)\}$ are independent from the feedback coefficients $a_i(k)$. Now, take the derivatives of both sides of (6.24) first with respect to $a_i(k)$ and then with respect to $b_j(k)$ to obtain

$$\left.\begin{aligned} \frac{\partial y_{\mathrm{OE}}(k)}{\partial a_i(k)} &= y_{\mathrm{OE}}(k-i) + \sum_{m=1}^{N} a_m(k)\frac{\partial y_{\mathrm{OE}}(k-m)}{\partial a_i(k)}, \\ \frac{\partial y_{\mathrm{OE}}(k)}{\partial b_j(k)} &= x(k-j) + \sum_{m=1}^{N} a_m(k)\frac{\partial y_{\mathrm{OE}}(k-m)}{\partial b_j(k)}. \end{aligned}\right\} \tag{6.29}$$

There is a difficulty in practical applications of this algorithm, since the partial derivatives in Equation (6.29) are with respect to current values of $a_m(k)$ and $b_m(k)$, which makes Equation (6.29) nonrecursive. Observe that if elements of Θ were independent of $\{y(k-i)\}$, then the gradient calculation would be identical to the FIR case. However, we have delayed samples of $y_{\mathrm{OE}}(k)$ involved in calculation of $\Theta(k)$ and an approximation to algorithm (6.29), known as the pseudolinear regression algorithm (PRA), is used. It is reasonable to assume that with a sufficiently small learning rate η, the coefficients will adapt slowly, i.e.

$$\Theta(k) \approx \Theta(k-1) \approx \cdots \approx \Theta(k-N). \tag{6.30}$$

The previous approximation is particularly good for N small. From (6.29) and (6.30), we finally have the equations for LMS IIR gradient adaptation,

$$\left.\begin{aligned}\frac{\partial y_{\mathrm{OE}}(k)}{\partial a_i(k)} &\approx y_{\mathrm{OE}}(k-i) + \sum_{m=1}^{N} a_m(k)\frac{\partial y_{\mathrm{OE}}(k-m)}{\partial a_i(k-m)},\\ \frac{\partial y_{\mathrm{OE}}(k)}{\partial b_j(k)} &\approx x(k-j) + \sum_{m=1}^{N} a_m(k)\frac{\partial y_{\mathrm{OE}}(k-m)}{\partial b_j(k-m)}.\end{aligned}\right\} \tag{6.31}$$

The partial derivatives

$$\frac{\partial y_{\mathrm{OE}}(k-m)}{\partial a_i(k-m)} \quad \text{and} \quad \frac{\partial y_{\mathrm{OE}}(k-m)}{\partial b_j(k-m)}$$

admit computation in a recursive fashion. For more details see Treichler (1987), Regalia (1994) and Shynk (1989).

To express this algorithm in a more compact form, let us introduce the weight vector $\boldsymbol{w}(k)$ as

$$\boldsymbol{w}(k) = [b_1(k), b_2(k), \dots, b_M(k), a_1(k), a_2(k), \dots, a_N(k)]^{\mathrm{T}} \tag{6.32}$$

and the IIR filter input vector $\boldsymbol{u}(k)$ as

$$\boldsymbol{u}(k) = [x(k-1), \dots, x(k-M), y(k-1), \dots, y(k-N)]^{\mathrm{T}}. \tag{6.33}$$

With this notation, we have, for instance, $w_2(k) = b_2(k)$, $w_{M+2}(k) = a_2(k)$, $u_M(k) = x(k-M)$ or $u_{M+1}(k) = y(k-1)$. Now, Equation (6.31) can be rewritten in a compact form as

$$\frac{\partial y_{\mathrm{OE}}(k)}{\partial w_i(k)} \approx u_i(k) + \sum_{m=1}^{N} w_{m+M}(k)\frac{\partial y_{\mathrm{OE}}(k-m)}{\partial w_i(k-m)}. \tag{6.34}$$

If we denote

$$\pi_i(k) = \frac{\partial y_{\mathrm{OE}}(k)}{\partial w_i(k)}, \qquad i = 1, \dots, M+N,$$

then (6.34) becomes

$$\pi_i(k) \approx u_i(k) + \sum_{m=1}^{N} w_{m+M}(k)\pi_i(k-m). \tag{6.35}$$

Finally, the weight update equation for a linear IIR adaptive filter can be expressed as

$$\boldsymbol{w}(k+1) = \boldsymbol{w}(k) + \eta(k)e(k)\boldsymbol{\pi}(k), \tag{6.36}$$

where $\boldsymbol{\pi}(k) = [\pi_1(k), \dots, \pi_{M+N}(k)]^{\mathrm{T}}$.

The adaptive IIR filter in a system identification configuration for the output error formulation is referred to as a model reference adaptive system (MRAS) in the control literature.

6.8.1 Equation Error Coefficient Adaptation

The IIR filter input vector $\boldsymbol{u}(k)$ for *equation error* adaptation can be expressed as

$$\boldsymbol{u}(k) = [x(k-1), \dots, x(k-M), d(k-1), \dots, d(k-N)]^{\mathrm{T}}. \tag{6.37}$$

Both the external input vector $\boldsymbol{x}(k) = [x(k-1), \dots, x(k-M)]^{\mathrm{T}}$ and the vector of teaching signals $\boldsymbol{d}(k) = [d(k-1), \dots, d(k-N)]^{\mathrm{T}}$ are not generated through the filter and are independent of the filter weights. Therefore, the weight adaptation for an *equation error* IIR adaptive filter can be expressed as

$$\boldsymbol{w}(k+1) = \boldsymbol{w}(k) + \eta(k)e(k)\boldsymbol{u}(k), \tag{6.38}$$

which is identical to the formula for adaptation of FIR adaptive filters. In fact, an *equation error* IIR adaptive filter can be thought of as a dual input FIR adaptive filter.

6.9 Weight Adaptation for Recurrent Neural Networks

The output of a recurrent perceptron, shown in Figure 6.2, with weight vectors $\boldsymbol{w}_a(k) = [w_{M+2}, \dots, w_{M+N+1}]^{\mathrm{T}}$ and $\boldsymbol{w}_b(k) = [w_1, w_2, \dots, w_M, w_{M+1}]^{\mathrm{T}}$, which comprise weights associated with delayed outputs and inputs, respectively, is given by

$$\left.\begin{aligned} y(k) &= \Phi(\mathrm{net}(k)), \\ \mathrm{net}(k) &= \sum_{j=1}^{M} w_j(k)x(k-j) + w_{M+1}(k) + \sum_{m=1}^{N} w_{m+M+1}(k)y(k-m), \end{aligned}\right\} \tag{6.39}$$

where $w_i(k) \in \boldsymbol{w}(k) = [\boldsymbol{w}_b^{\mathrm{T}}(k), \boldsymbol{w}_a^{\mathrm{T}}(k)]^{\mathrm{T}}$, $i = 1, \dots, M+N+1$. The instantaneous output error in this case is given by

$$e(k) = d(k) - y(k), \tag{6.40}$$

where $d(k)$ denotes the desired (teaching) signal, whereas the cost function is $E = \frac{1}{2}e^2(k)$. In order to obtain the weight vector $\boldsymbol{w}(k+1)$, we have to calculate the gradient $\nabla_{\boldsymbol{w}}E(k)$ and the weight update vector $\Delta\boldsymbol{w}(k)$ for which the elements $\Delta w_i(k)$, $i = 1, \dots, M+N+1$, are

$$\Delta w_i(k) = -\eta \frac{\partial E(k)}{\partial w_i(k)} = -\eta e(k) \frac{\partial e(k)}{\partial w_i(k)} = +\eta e(k) \frac{\partial y(k)}{\partial w_i(k)}. \tag{6.41}$$

From (6.39), we have

$$\frac{\partial y(k)}{\partial w_i(k)} = \Phi'(\mathrm{net}(k)) \frac{\partial\, \mathrm{net}(k)}{\partial w_i(k)}. \tag{6.42}$$

Following the analysis provided for IIR adaptive filters (6.34), we see that the partial derivatives of outputs with respect to weights form a recursion. Thus, we have

$$\frac{\partial y(k)}{\partial w_i(k)} \approx \Phi'(\mathrm{net}(k)) \left[u_i(k) + \sum_{m=1}^{N} w_{m+M+1}(k) \frac{\partial y(k-m)}{\partial w_i(k-m)} \right], \tag{6.43}$$

where vector $\boldsymbol{u}(k) = [x(k), \ldots, x(k-M), 1, y(k), \ldots, y(k-N)]^{\mathrm{T}}$ comprises the set of all input signals to a recurrent perceptron, including the delayed inputs, delayed outputs and bias, and $i = 1, \ldots, M+N+1$. If we introduce notation

$$\pi_i(k) = \frac{\partial y(k)}{\partial w_i(k)},$$

then (6.43) can be rewritten as

$$\pi_i(k) = \Phi'(\text{net}(k))\left[u_i(k) + \sum_{m=1}^{N} w_{m+M+1}(k)\pi_i(k-m)\right]. \tag{6.44}$$

In control theory, coefficients $\pi_i(k)$ are called *sensitivities*. It is convenient to assume zero initial conditions for sensitivities (6.44) (Haykin 1994), i.e.

$$\pi_i(0) = 0, \qquad i = 1, \ldots, M+N+1.$$

The analysis presented so far is the basis of the real-time recurrent learning (RTRL) algorithm. The derivation of this online direct-gradient algorithm for a general recurrent neural network is more involved and is given in Appendix D.

Finally, the weight update equation for a nonlinear adaptive filter in the form of a recurrent perceptron can be expressed as

$$\boldsymbol{w}(k+1) = \boldsymbol{w}(k) + \eta(k)e(k)\boldsymbol{\pi}(k), \tag{6.45}$$

where $\boldsymbol{\pi}(k) = [\pi_1(k), \ldots, \pi_{M+N+1}(k)]^{\mathrm{T}}$. In order to calculate vector $\boldsymbol{\pi}(k)$, we have to store the following matrix:

$$\boldsymbol{\Pi}(k) = \begin{bmatrix} \pi_1(k-1) & \pi_2(k-1) & \cdots & \pi_{M+N+1}(k-1) \\ \pi_1(k-2) & \pi_2(k-2) & \cdots & \pi_{M+N+1}(k-2) \\ \vdots & \vdots & \ddots & \vdots \\ \pi_1(k-N) & \pi_2(k-N) & \cdots & \pi_{M+N+1}(k-N) \end{bmatrix}. \tag{6.46}$$

The learning procedure described above is the so-called *supervised learning* (or *output error* learning) algorithm for a recurrent perceptron.

6.9.1 *Teacher Forcing Learning for a Recurrent Perceptron*

The input vector $\boldsymbol{u}(k)$ for *teacher forced* adaptation of a recurrent perceptron can be expressed as

$$\boldsymbol{u}(k) = [x(k-1), \ldots, x(k-M), 1, d(k-1), \ldots, d(k-N)]^{\mathrm{T}}. \tag{6.47}$$

The analysis of this algorithm is analogous to that presented in Section 6.8.1. Hence, the weight adaptation for a *teacher forced* recurrent perceptron can be expressed as

$$\boldsymbol{w}(k+1) = \boldsymbol{w}(k) + \eta(k)e(k)\Phi'(\text{net}(k))\boldsymbol{u}(k), \tag{6.48}$$

which is identical to the formula for adaptation of dual-input nonlinear FIR adaptive filters.

6.9.2 *Training Process for a NARMA Neural Predictor*

Algorithms for training of recurrent neural networks have been extensively studied since the late 1980s. The real-time recurrent learning (RTRL) algorithm (Robinson and Fallside 1987; Williams and Zipser 1989a) enabled training of simple RNNs, whereas Pineda provided recurrent backpropagation (RBP) (Pineda 1987, 1989). RTRL-based training of the RNN employed as a nonlinear adaptive filter is based upon minimising the instantaneous squared error at the output of the first neuron of the RNN (Williams and Zipser 1989a), which can be expressed as $\min(e^2(k)) = \min([s(k) - y_1(k)]^2)$, where $e(k)$ denotes the error at the output of the RNN and $s(k)$ is the teaching signal. It is an *output error* algorithm. The correction $\Delta \boldsymbol{W}(k)$ to the weight matrix $\boldsymbol{W}(k)$ of the RNN for prediction is calculated as

$$\Delta \boldsymbol{W}(k) = -\eta \frac{\partial E(k)}{\partial \boldsymbol{W}(k)} = \eta e(k) \frac{\partial y_1(k)}{\partial \boldsymbol{W}(k)}, \tag{6.49}$$

which turns out to be based upon a recursive calculation of the gradients of the outputs of the neurons (Mandic *et al.* 1998; Williams and Zipser 1989a). A detailed gradient descent training process (RTRL) for RNNs is given in Appendix D.

Similarly to the analysis for IIR filters and recurrent perceptrons, in order to make the algorithm run in real time, an approximation has to be made, namely that for a small learning rate η, the following approximation,

$$\frac{\partial y_i(k-m)}{\partial \boldsymbol{W}(k)} \approx \frac{\partial y_i(k-m)}{\partial \boldsymbol{W}(k-m)}, \qquad i, m = 1, \dots, N, \tag{6.50}$$

holds for slowly time-varying input statistics.

Another frequently used algorithm for training recurrent neural networks is a variant of the extended Kalman filter algorithm called the linearised recursive least-squares (LRLS) algorithm (Baltersee and Chambers 1998; Mandic *et al.* 1998). Its derivation is rather mathematically involved and is given in Appendix D. This algorithm is related to the previously mentioned gradient-based algorithms, and it modifies both the weights and the states of the network on an equal basis.

6.10 The Problem of Vanishing Gradients in Training of Recurrent Neural Networks

Recently, several empirical studies have shown that when using gradient-descent learning algorithms, it might be difficult to learn simple temporal behaviour with long time dependencies (Bengio *et al.* 1994; Mozer 1993), i.e. those problems for which the output of a system at time instant k depends on network inputs presented at times $\tau \ll k$. Bengio *et al.* (1994) analysed learning algorithms for systems with long time dependencies and showed that for gradient-based training algorithms, the information about the gradient contribution K steps in the past vanishes for large K. This effect is referred to as the problem of *vanishing gradient*, which partially explains why gradient descent algorithms are not very suitable to estimate systems and signals with long time dependencies. For instance, common recurrent neural networks encounter problems when learning information with long time dependencies, which is a problem in prediction of nonlinear and nonstationary signals.

The forgetting behaviour experienced in neural networks is formalised in Definition 6.10.1 (Frasconi *et al.* 1992).

Definition 6.10.1 (forgetting behaviour). A recurrent network exhibits forgetting behaviour if

$$\lim_{K\to\infty} \frac{\partial z_i(k)}{\partial z_j(k-K)} = 0 \qquad \forall k \in \mathcal{K},\ i \in \mathcal{O},\ j \in \mathcal{I}, \tag{6.51}$$

where z are state variables, $\mathcal{I}$ denotes the set of input neurons, $\mathcal{O}$ denotes the set of output neurons and $\mathcal{K}$ denotes the time index set.

A state space representation of recurrent NARX neural networks can be expressed as

$$z_i(k+1) = \begin{cases} \Phi(\boldsymbol{u}(k), \boldsymbol{z}(k)), & i = 1, \\ z_i(k), & i = 2, \dots, N, \end{cases} \tag{6.52}$$

where the output $y(k) = z_1(k)$ and z_i, $i = 1, 2, \dots, N$, are state variables of a recurrent neural network. To represent mathematically the problem of vanishing gradients, recall that the weight update for gradient-based methods for a neural network with one output neuron can be expressed as

$$\Delta\boldsymbol{w}(k) = \eta e(k)\left(\frac{\partial y(k)}{\partial \boldsymbol{w}(k)}\right) = \eta e(k)\sum_i\left(\frac{\partial y(k)}{\partial z_i(k)}\frac{\partial z_i(k)}{\partial \boldsymbol{w}(k)}\right). \tag{6.53}$$

Expanding Equation (6.53) and using the chain rule, we have

$$\Delta\boldsymbol{w}(k) = \eta e(k)\sum_i\left(\frac{\partial y_i(k)}{\partial z_i(k)}\sum_{l=1}^{k}\frac{\partial z_i(k)}{\partial z_i(k-l)}\frac{\partial z_i(k-l)}{\partial \boldsymbol{w}(k-l)}\right). \tag{6.54}$$

Partial derivatives of the state space variables $\partial z_i(k)/\partial z_i(l)$ from (6.54) build a Jacobian matrix $\boldsymbol{J}(k, k-l)$,[6] which is given by

$$\boldsymbol{J}(k) = \begin{bmatrix} \dfrac{\partial y(k)}{\partial z_1(k)} & \dfrac{\partial y(k)}{\partial z_2(k)} & \cdots & \dfrac{\partial y(k)}{\partial z_N(k)} \\ 1 & 0 & \cdots & 0 \\ 0 & 1 & \cdots & 0 \\ \vdots & \vdots & \ddots & \vdots \\ 0 & 0 & \cdots & 0 \end{bmatrix} \tag{6.55}$$

If all the eigenvalues of Jacobian (6.55) are inside the unit circle, then the corresponding transition matrix of $\boldsymbol{J}(k)$ is an exponentially decreasing function of k. As a consequence, the states of the network will remain within a set defined by a hyperbolic attractor, and the adaptive system will not be able to escape from this fixed point.

[6] Notice that the Jacobian (6.55) represents a companion matrix. The stability of these matrices is analysed in Mandic and Chambers (2000d) and will be addressed in Chapter 7.

Hence, a small perturbation in the weight vector $\boldsymbol{w}$ affects mostly the near past. This means that even if there was a weight update $\Delta \boldsymbol{w}(k)$ that would move the current point in the state space of the network from a present attractor, the gradient $\nabla_{\boldsymbol{w}} E(k)$ would not carry this information, due to the effects of *vanishing gradient*. Due to this effect, the network for which the dynamics are described above, is not able to estimate/represent long term dependencies in the input signal/system (Haykin 1999b).

Several approaches have been suggested to circumvent the problem of vanishing gradient in training RNNs. Most of them rest upon embedding memory in neural networks, whereas several propose improved learning algorithms, such as the extended Kalman filter algorithm (Mandic *et al.* 1998), Newton type algorithms, annealing algorithms (Mandic and Chambers 1999c; Rose 1998) and *a posteriori* algorithms (Mandic and Chambers 1998c). The deterministic annealing approach, for instance, offers (Rose 1998) (i) the ability to avoid local minima, (ii) applicability to many different structures/architectures, (iii) the ability to minimise the cost function even when its gradients tend to vanish for time indices from the distant past.

Embedded memory is particularly significant in recurrent NARX and NARMAX neural networks (Lin *et al.* 1997). This embedded memory can help to speed up propagation of gradient information, and hence help to reduce the effect of vanishing gradient (Lin *et al.* 1996). There are various methods to introduce memory and temporal information into neural networks. These include (Kim 1998) (i) creating a spatial representation of temporal pattern, (ii) putting time delays into the neurons or their connections, (iii) employing recurrent connections, (iv) using neurons with activations that sum inputs over time, (v) using some combination of these. The PRNN, for instance, uses the combination of (i) and (iii).

6.11 Learning Strategies in Different Engineering Communities

Learning strategies classification for a broad class of adaptive systems is given in Nerrand *et al.* (1994), where learning algorithms are classified as directed, semidirected and unidirected. *Directed* algorithms are suitable mostly for the modelling of noiseless dynamical systems, or for systems with noise added to the state variables of the black-box model of the system, whereas *unidirected* algorithms are suitable for predicting the output of systems for which the output is corrupted by additive white noise. For instance, the RTRL algorithm is a unidirected algorithm, whereas *a posteriori* (data-reusing) algorithms are unidirected–directed.

There is a need to relate terms coming from different communities, which refer to the same learning strategies. This correspondence for terms used in the signal processing, system identification, neural networks and adaptive systems communities is shown in Table 6.1.

6.12 Learning Algorithms and the Bias/Variance Dilemma

The optimal prediction performance would provide a compromise between the bias and the variance of the prediction error achieved by a chosen model. An analogy with

Table 6.1 Terms related to learning strategies used in different communities

Signal Processing	System ID	Neural Networks	Adaptive Systems
Output Error	Parallel	Supervised	Unidirected
Equation Error	Series–Parallel	Teacher Forcing	Directed

polynomials, which to a certain extent applies for trajectory-tracking problems, shows that if a number of points is approximated by a polynomial of an insufficient order, the polynomial fit has errors, i.e. the polynomial curve cannot pass through every point. On the other hand, if the order of the polynomial is greater that the number of points through which to fit, the polynomial will pass exactly through every point, but will oscillate in between, hence having a large variance.

If we consider a measure of the error as

$$E[(d(k) - y(k))^2], \tag{6.56}$$

then adding and subtracting a dummy variable $E[y(k)]$ within the square brackets of (6.56) yields (Principe *et al.* 2000)

$$E[(d(k) - y(k))^2] = E[E(y(k)) - d(k)]^2 + E[(y(k) - E[y(k)])^2]. \tag{6.57}$$

The first term on the right-hand side of (6.57) represents the squared bias term, whereas the second term on the right-hand side of (6.57) represents the variance. An inspection of (6.57) shows that teacher forcing (equation error) algorithms suffer from biased estimates, since $d(k) \neq E[y(k)]$ in the first term of (6.57), whereas the supervised (output error) algorithms might suffer from increased variance.

If the noise term $\epsilon(k)$ is included within the output, for the case when we desire to approximate mapping $f(x(k))$ by a neural network for which the output is $y(k)$, we want to minimise

$$E[y(k) - f(x)]. \tag{6.58}$$

If

$$\left.\begin{aligned} y(k) &= f^*(k) + \epsilon(k), & f^*(k) &= E[y|x], \\ E[\epsilon|x] &= 0, & \bar{f} &= E[f(x)], \end{aligned}\right\} \tag{6.59}$$

where, for convenience, the time index k is dropped, we have (Gemon 1992)

$$E[f(x)] = E[\epsilon^2] + E[f^* - \bar{f}]^2 + E[(f - \bar{f})^2] = \sigma^2 + B^2 + \mathrm{var}(f). \tag{6.60}$$

In Equation (6.60), σ^2 denotes the noise variance, $B^2 = E[f^* - \bar{f}]^2$ is the squared bias and $\mathrm{var}(f) = E[(f - \bar{f})^2]$ denotes the variance. The term σ^2 cannot be reduced, since it is due to the observation noise. The second and third term in (6.60) can be reduced by choosing an appropriate architecture and learning strategy, as shown before in this chapter. A thorough analysis of the bias/variance dilemma can be found in Gemon (1992) and Haykin (1994).

6.13 Recursive and Iterative Gradient Estimation Techniques

Since, for real-time applications, the coefficient update in the above algorithms is finished before the next input sample arrives, there is a possibility to reiterate learning algorithms about the current point in the state space of the network. Therefore, a relationship between the iteration index l and time index k can be established. The possibilities include the following.

- A purely recursive algorithm – one gradient iteration (coefficients update) per sampling interval k, i.e. $l = 1$. This is the most commonly used technique for recursive filtering algorithms.
- Several coefficients updates per sampling interval, i.e. $l > 1$ – this may improve the nonlinear filter performance, as for instance in *a posteriori* algorithms, where $l \geqslant 2$ for every time instant (Mandic and Chambers 2000e).
- P coefficients with greatest magnitude updated every sampling period – this helps to reduce computational complexity (Douglas 1997).
- Coefficients are only updated every K sampling periods – useful for processing signals with slowly varying statistics.

6.14 Exploiting Redundancy in Neural Network Design

Redundancy encountered in neural network design can improve the robustness and fault tolerance in the neural network approach to a specific signal processing task, as well as serve as a basis for network topology adaptation due to statistical changes in the input data. It gives more degrees of freedom than necessary for a particular task. In the case of neural networks for time series prediction, the possible sources of redundancy can be

- Redundancy in the global network architecture, which includes design with
  - more layers of neurons than necessary, which enables network topology adaptation while in use and improves robustness in the network,
  - more neurons within the layers than necessary, which improves robustness and fault-tolerance in the network,
  - memory neurons (Poddar and Unninkrishnan 1991), which are specialised neurons added to the network to store the past activity of the network. Memory neurons are time replicae of processing neurons, in all the network layers except the output layer.
- Redundancy among the internal connections within the network. According to the level of redundancy in the network we can define
  - fully recurrent networks (Connor 1994; McDonnell and Waagen 1994),
  - partially recurrent networks (Bengio 1995; Haykin and Li 1995),

  - time delays in the network as a source of redundancy (Baldi and Atiya 1994; Haykin and Li 1995; Nerrand *et al.* 1994).

- Data reusing, where we expect to make the trajectory along the error performance surface 'less stochastic'.[7]

6.15 Summary

It has been shown that a small neural network can represent high-order nonlinear systems, whereas a very large number of terms are required for an equivalent Volterra series representation. We have shown that when modelling an unknown dynamical system, or tracking unknown dynamics, it is important to concentrate on the embedding dimension of the network.

Architectures for neural networks as nonlinear adaptive filters have been introduced and learning strategies, such as *equation error* and *output error* strategy have been explained. Connection between the learning strategies from different engineering communities has been established. The online real-time recurrent learning algorithm for general recurrent neural networks has been derived inductively, starting from a linear adaptive IIR filter, via a recurrent perceptron, through to a general case. Finally, sources of redundancies in RNN architectures have been addressed.

[7] We wish to speed up the convergence of learning trajectory along the error performance surface. It seems that reusing of 'good' data can improve the convergence rate of a learning algorithm. This data reusing makes a learning algorithm *iterative* as well as recursive.

7

Stability Issues in RNN Architectures

7.1 Perspective

The focus of this chapter is on stability and convergence of relaxation realised through NARMA recurrent neural networks. Unlike other commonly used approaches, which mostly exploit Lyapunov stability theory, the main mathematical tool employed in this analysis is the contraction mapping theorem (CMT), together with the fixed point iteration (FPI) technique. This enables derivation of the asymptotic stability (AS) and global asymptotic stability (GAS) criteria for neural relaxive systems. For rigour, existence, uniqueness, convergence and convergence rate are considered and the analysis is provided for a range of activation functions and recurrent neural networks architectures.

7.2 Introduction

Stability and convergence are key issues in the analysis of dynamical adaptive systems, since the analysis of the dynamics of an adaptive system can boil down to the discovery of an attractor (a stable equilibrium) or some other kind of fixed point. In neural associative memories, for instance, the locally stable equilibrium states (attractors) store information and form neural memory. Neural dynamics in that case can be considered from two aspects, convergence of state variables (memory recall) and the number, position, local stability and domains of attraction of equilibrium states (memory capacity). Conveniently, LaSalle's invariance principle (LaSalle 1986) is used to analyse the state convergence, whereas stability of equilibria are analysed using some sort of linearisation (Jin and Gupta 1996). In addition, the dynamics and convergence of learning algorithms for most types of neural networks may be explained and analysed using fixed point theory.

Let us first briefly introduce some basic definitions. The full definitions and further details are given in Appendix I. Consider the following linear, finite dimensional, autonomous system[1] of order N

$$y(k) = \sum_{i=1}^{N} a_i(k)y(k-i) = \boldsymbol{a}^{\mathrm{T}}(k)\boldsymbol{y}(k-1). \tag{7.1}$$

Definition 7.2.1 (see Kailath (1980) and LaSalle (1986)). The system (7.1) is said to be *asymptotically stable* in $\Omega \subseteq \mathbb{R}^N$, if for any $\boldsymbol{y}(0)$, $\lim_{k\to\infty} y(k) = 0$, for $\boldsymbol{a}(k) \in \Omega$.

Definition 7.2.2 (see Kailath (1980) and LaSalle (1986)). The system (7.1) is *globally asymptotically stable* if for any initial condition and any sequence $\boldsymbol{a}(k)$, the response $y(k)$ tends to zero asymptotically.

For NARMA systems realised via neural networks, we have

$$y(k+1) = \boldsymbol{\Phi}(\boldsymbol{y}(k), \boldsymbol{w}(k)). \tag{7.2}$$

Let $\boldsymbol{\Phi}(k, k_0, \boldsymbol{Y}_0)$ denote the trajectory of the state change for all $k \geqslant k_0$, with $\boldsymbol{\Phi}(k_0, k_0, \boldsymbol{Y}_0) = \boldsymbol{Y}_0$. If $\boldsymbol{\Phi}(k, k_0, \boldsymbol{Y}^*) = \boldsymbol{Y}^*$ for all $k \geqslant 0$, then $\boldsymbol{Y}^*$ is called an *equilibrium* point. The largest set $\mathcal{D}(\boldsymbol{Y}^*)$ for which this is true is called the *domain of attraction* of the equilibrium $\boldsymbol{Y}^*$. If $\mathcal{D}(\boldsymbol{Y}^*) = \mathbb{R}^N$ and if $\boldsymbol{Y}^*$ is asymptotically stable, then $\boldsymbol{Y}^*$ is said to be asymptotically stable in large or *globally asymptotically stable*.

It is important to clarify the difference between asymptotic stability and absolute stability. Asymptotic stability may depend upon the input (initial conditions), whereas global asymptotic stability does not depend upon initial conditions. Therefore, for an absolutely stable neural network, the system state will converge to one of the asymptotically stable equilibrium states regardless of the initial state and the input signal. The equilibrium points include the isolated minima as well as the maxima and saddle points. The maxima and saddle points are not stable equilibrium points. Robust stability for the above discussed systems is still under investigation (Bauer *et al.* 1993; Jury 1978; Mandic and Chambers 2000c; Premaratne and Mansour 1995).

In conventional nonlinear systems, the system is said to be globally asymptotically stable, or asymptotically stable in large, if it has a unique equilibrium point which is globally asymptotically stable in the sense of Lyapunov. In this case, for an arbitrary initial state $\boldsymbol{x}(0) \in \mathbb{R}^N$, the state trajectory $\phi(k, \boldsymbol{x}(0), \boldsymbol{s})$ will converge to the unique equilibrium point $\boldsymbol{x}^*$, satisfying

$$\boldsymbol{x}^* = \lim_{k\to\infty} \phi[k, \boldsymbol{x}(0), \boldsymbol{s}]. \tag{7.3}$$

Stability in this context has been considered in terms of Lyapunov stability and M-matrices (Forti and Tesi 1994; Liang and Yamaguchi 1997). To apply the Lyapunov method to a dynamical system, a neural system has to be mapped onto a new system for which the origin is at an equilibrium point. If the network is stable, its 'energy' will decrease to a minimum as the system approaches and attains its equilibrium state. If a function that maps the objective function onto an 'energy function' can be found, then the network is guaranteed to converge to its equilibrium state (Hopfield and

[1] Stability of systems of this type is discussed in Appendix H.

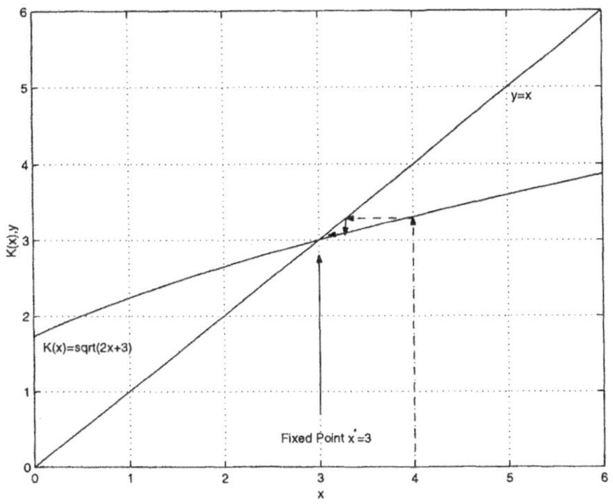

Figure 7.1 FPI solution for roots of $F(x) = x^2 - 2x - 3$

Tank 1985; Luh *et al.* 1998). The Lyapunov stability of neural networks is studied in detail in Han *et al.* (1989) and Jin and Gupta (1996).

The concept of fixed point will be central to much of what follows, for which the basic theorems and principles are introduced in Appendix G.

Point x^* is called a *fixed point* of a function K if it satisfies $K(x^*) = x^*$, i.e. the value x^* is unchanged under the application of function K. For instance, the roots of function $F(x) = x^2 - 2x - 3$ can be found by rearranging $x_{k+1} = K(x_k) = \sqrt{2x_k + 3}$ via fixed point iteration. The roots of the above function are -1 and 3. The FPI which started from $x_0 = 4$ converges to within 10^{-5} of the exact solution in nine steps, which is depicted in Figure 7.1. This example is explained in more detail in Appendix G.

One of the virtues of neural networks is their processing power, which rests upon their ability to converge to a set of fixed points in the state space. Stability analysis, therefore, is essential for the derivation of conditions that assure convergence to these fixed points. Stability, although necessary, is not sufficient for effective processing (see Appendix H), since in practical applications, it is desirable that a neural system converges to only a preselected set of fixed points. In the remainder of this chapter, two different aspects of equilibrium, i.e. the *static* aspect (existence and uniqueness of equilibrium states) and the *dynamic* aspect (global stability, rate of convergence), are studied. While analysing global asymptotic stability,[2] it is convenient to study the static problem of the existence and uniqueness of the equilibrium point first, which is the necessary condition for GAS.

[2] It is important to note that the iterates of random Lipschitz functions converge if the functions are contracting on the average (Diaconis and Freedman 1999). The theory of random operators is a probabilistic generalisation of operator theory. The study of probabilistic operator theory and its applications was initiated by the Prague school under the direction of Antonin Spacek, in the 1950s (Bharucha-Reid 1976). They recognised that it is necessary to take into consideration the fact that the operators used to describe the behaviour of systems may not be known exactly. The application of this theory in signal processing is still under consideration and can be used to analyse stochastic learning algorithms (Chambers *et al.* 2000).

7.3 Overview

The role of the nonlinear activation function in the global asymptotic convergence of recurrent neural networks is studied. For a fixed input and weights, a repeated application of the nonlinear difference equation which defines the output of a recurrent neural network is proven to be a relaxation, provided the activation function satisfies the conditions required for a contraction mapping. This relaxation is shown to exhibit linear asymptotic convergence. Nesting of modular recurrent neural networks is demonstrated to be a fixed point iteration in a spatial form.

7.4 A Fixed Point Interpretation of Convergence in Networks with a Sigmoid Nonlinearity

To solve many problems in the field of optimisation, neural control and signal processing, dynamic neural networks need to be designed to have only a unique equilibrium point. The equilibrium point ought to be globally stable to avoid the risk of spurious responses or the problem of local minima. Global asymptotic stability (GAS) has been analysed in the theory of both linear and nonlinear systems (Barnett and Storey 1970; Golub and Van Loan 1996; Haykin 1996a; Kailath 1980; LaSalle 1986; Priestley 1991). For nonlinear systems, it is expected that convergence in the GAS sense depends not only on the values of the parameter vector, but also on the parameters of the nonlinear function involved. As systems based upon sigmoid functions exhibit stability in the bounded input bounded output (BIBO) sense, due to the saturation type sigmoid nonlinearity, we investigate the characteristics of the nonlinear activation function to obtain GAS for a general RNN-based nonlinear system. In that case, both the external input vector to the system $\boldsymbol{x}(k)$ and the parameter vector $\boldsymbol{w}(k)$ are assumed to be a time-invariant part of the system under fixed point iteration.

7.4.1 Some Properties of the Logistic Function

To derive the conditions which the nonlinear activation function of a neuron should satisfy to enable convergence of real-time learning algorithms, activation functions of a neuron are analysed in the framework of contraction mappings and fixed point iteration.

Observation 7.4.1. *The logistic function*

$$\Phi(x) = \frac{1}{1+e^{-\beta x}} \tag{7.4}$$

is a contraction on $[a,b] \in \mathbb{R}$ for $0 < \beta < 4$ and the iteration

$$x_{i+1} = \Phi(x_i) \tag{7.5}$$

converges to a unique solution x^* from $\forall x_0 \in [a,b] \in \mathbb{R}$.

Proof. By the contraction mapping theorem (CMT) (Appendix G), function K is a contraction on $[a,b] \in \mathbb{R}$ if

Figure 7.2 The contraction mapping

(i) $x \in [a, b] \Rightarrow K(x) \in [a, b]$,

(ii) $\exists \gamma < 1 \in \mathbb{R}^+$ s.t. $|K(x) - K(y)| \leqslant \gamma |x - y| \quad \forall x, y \in [a, b]$.

The condition (i) is illustrated in Figure 7.2. The logistic function (7.4) is strictly monotonically increasing, since its first derivative is strictly greater than zero. Hence, in order to prove that Φ is a contraction on $[a, b] \in \mathbb{R}$, it is sufficient to prove that it contracts the upper and lower bound of interval $[a, b]$, i.e. a and b, which in turn gives

- $a - \Phi(a) \leqslant 0$,
- $b - \Phi(b) \geqslant 0$.

These conditions will be satisfied if the function Φ is smaller in magnitude than the curve $y = x$, i.e. if

$$|x| > \left| \frac{1}{1 + \mathrm{e}^{-\beta x}} \right|, \qquad \beta > 0. \tag{7.6}$$

Condition (ii) can be proven using the mean value theorem (MVT) (Luenberger 1969). Namely, as the logistic function Φ (7.4) is differentiable, for $\forall x, y \in [a, b]$, $\exists \xi \in (a, b)$ such that

$$|\Phi(x) - \Phi(y)| = |\Phi'(\xi)(x - y)| = |\Phi'(\xi)||x - y|. \tag{7.7}$$

The first derivative of the logistic function (7.4) is

$$\Phi'(x) = \left(\frac{1}{1 + \mathrm{e}^{-\beta x}} \right)' = \frac{\beta \mathrm{e}^{-\beta x}}{(1 + \mathrm{e}^{-\beta x})^2}, \tag{7.8}$$

which is strictly positive, and for which the maximum value is $\Phi'(0) = \beta/4$. Hence, for $\beta \leqslant 4$, the first derivative $\Phi' \leqslant 1$. Finally, for $\gamma < 1 \Leftrightarrow \beta < 4$, function Φ given in (7.4) is a contraction on $[a, b] \in \mathbb{R}$.

Convergence of FPI: if x^* is a zero of $x - \Phi(x) = 0$, or in other words the fixed point of function Φ, then for $\gamma < 1$ $(\beta < 4)$

$$|x_i - x^*| = |\Phi(x_{i-1}) - \Phi(x^*)| \leqslant \gamma |x_{i-1} - x^*|. \tag{7.9}$$

Thus, since for $\gamma < 1 \Rightarrow \{\gamma\}^i \xrightarrow{i} 0$

$$|x_i - x^*| \leqslant \gamma^i |x_0 - x^*| \;\Rightarrow\; \lim_{i \to \infty} x_i = x^* \tag{7.10}$$

and iteration $x_{i+1} = \Phi(x_i)$ converges to some $x^* \in [a, b]$. □

Convergence/divergence of the FPI clearly depends on the size of slope β in Φ. Considering the general nonlinear system Equation (7.2), this means that for a fixed input vector to the iterative process and fixed weights of the network, an FPI solution depends on the slope (first derivative) of the nonlinear activation function and some measure of the weight vector. If the solution exists, that is the only value to which

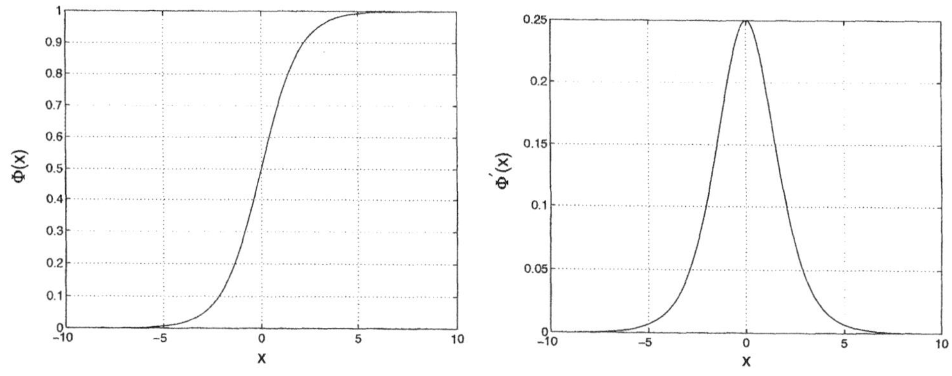

(a) The logistic nonlinear function

(b) The first derivative of the logistic function

Figure 7.3 The logistic function and its derivative

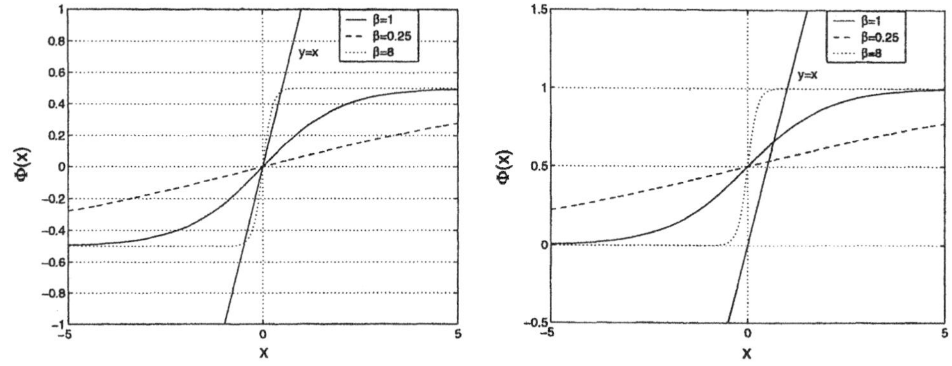

(a) Centred logistic functions

(b) Unipolar logistic functions

Figure 7.4 Various logistic functions

such a relaxation algorithm converges. Figure 7.3 shows the logistic function and its first derivative for $\beta = 1$. To depict Observation 7.4.1 further, we use a centred logistic function ($\Phi - \text{mean}(\Phi)$), as shown in Figure 7.4(a). For Φ a contraction, the condition (i) from CMT (Appendix G) must be satisfied. That is the case if the values of Φ are smaller in magnitude than the corresponding values of the function $y = x$. As shown in Figure 7.4(a), that condition is satisfied for a range of logistic functions with the slope $0 < \beta < 4$. Indeed, e.g. for $\beta = 8$, the logistic function has an intersection with the function $y = x$ (dotted curve in Figure 7.4(a)), which means that for $\beta > 4$, there are regions in Φ where $(a - \Phi(a)) \nleq 0$, which violates condition (i) of CMT and Observation 7.4.1.

7.4.2 Logistic Function, Rate of Convergence and Fixed Point Theory

The rate of convergence of a fixed point iteration can be judged by the closeness of x_{k+1} to x^* relative to the closeness of x_k to x^* (Dennis and Schnabel 1983; Gill *et al.* 1981).

Definition 7.4.2. A sequence $\{x_k\}$ is said to converge towards its fixed point x^* with order r if

$$0 \leqslant \lim_{k\to\infty} \frac{\|x_{k+1} - x^*\|}{\|x_k - x^*\|^r} < \infty, \tag{7.11}$$

where $r \in \mathbb{N}$ is the largest number such that the above inequality holds.

Since we are interested in the value of r that occurs in the limit, r is sometimes called the *asymptotic convergence rate*. If $r = 1$, the sequence is said to exhibit *linear* convergence, if $r = 2$, the sequence is said to exhibit *quadratic convergence.*

Definition 7.4.3. For a sequence $\{x_k\}$ which has an order of convergence r, the *asymptotic error constant* of the fixed point iteration is the value $\gamma \in \mathbb{R}^+$ which satisfies

$$\gamma = \lim_{k\to\infty} \frac{\|x_{k+1} - x^*\|}{\|x_k - x^*\|^r}. \tag{7.12}$$

When $r = 1$, i.e. for linear convergence, γ must be strictly less than unity in order for convergence to occur (Gill *et al.* 1981).

Example 7.4.4. Show that the convergent FPI process

$$x_{i+1} = \Phi(x_i) \tag{7.13}$$

exhibits a *linear asymptotic convergence* for which the error constant equals $|\Phi'(x^*)|$.

Solution. Consider the ratio $|e_{i+1}|/|e_i|$ of successive errors, where $e_i = x_i - x^*$

$$\frac{|e_{i+1}|}{|e_i|} = \frac{|x_{i+1} - x^*|}{|x_i - x^*|} = \frac{|\Phi(x_i) - \Phi(x^*)|}{|x_i - x^*|} \overset{\text{MVT}}{=} |\Phi'(\xi)| \tag{7.14}$$

for some $\xi \in (x_i, x^*)$. Having in mind that the iteration (7.13) converges to x^* when $i \to \infty$

$$\lim_{i\to\infty} \frac{|e_{i+1}|}{|e_i|} = \lim_{i\to\infty} |\Phi'(\xi)| = |\Phi'(x^*)|. \tag{7.15}$$

Therefore, iteration (7.13) exhibits linear asymptotic convergence with convergence rate $|\Phi'(x^*)|$. □

Example 7.4.5. Derive the error bound $e_i = |x_i - x^*|$ for the FPI process

$$x_{i+1} = \Phi(x_i). \tag{7.16}$$

Solution. Rewrite the error bound as

$$x_i - x^* = \Phi(x_{i-1}) - \Phi(x_i) + \Phi(x_i) - \Phi(x^*) \tag{7.17}$$

and therefore

$$|x_i - x^*| \leqslant \gamma|x_{i-1} - x_i| + \gamma|x_i - x^*|. \tag{7.18}$$

Table 7.1 Fixed point iterates for the logistic function

Starting value x_0	-10	10
First iterate	0.000 045	1
Second iterate	0.5	0.7311
Third iterate	0.6225	0.6750
Fourth iterate	0.6508	0.6626
Fifth iterate	0.6572	0.6598
Sixth iterate	0.6586	0.6592
Seventh iterate	0.6589	0.6591

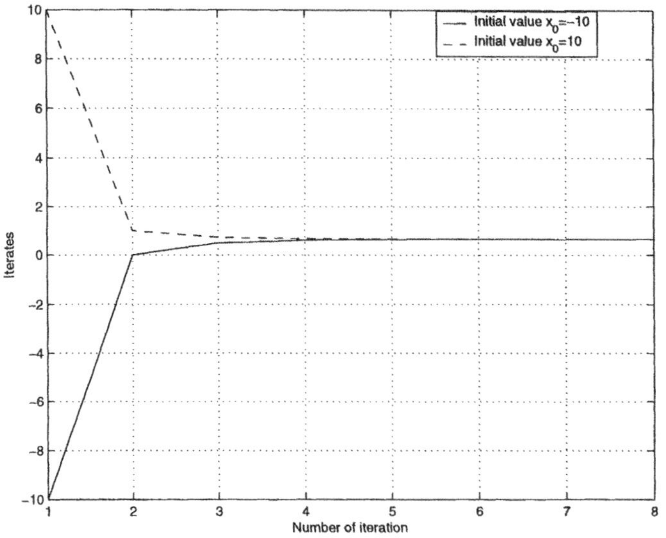

Figure 7.5 FPI for a logistic function and different initial values

Hence

$$|x_i - x^*| \leqslant \frac{\gamma}{1-\gamma}|x_{i-1} - x_i|. \tag{7.19}$$

□

Example 7.4.6. Show that when repeatedly applying logistic function Φ the interval $[-10, 10]$ degenerates towards a point $\zeta \in [-10, 10]$.

Solution. Observation 7.4.1 provides a general background for this example. Notice that $\beta = 1$. In order to show that a function converges in the FPI sense, it is sufficient to show that it contracts the bound points of the interval $[-10, 10]$, since it is a strictly monotonically increasing function. Let us therefore set up the iteration

$$x_{i+1} = \Phi(x_i), \qquad x_0 \in \{-10, 10\}. \tag{7.20}$$

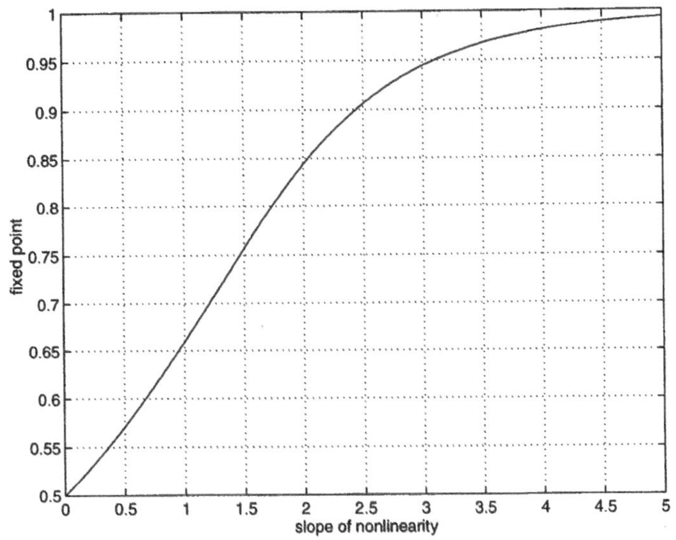

Figure 7.6 Fixed points for the *logistic* nonlinearity, as a function of slope β and starting point $x_0 = 10$

The results of the iteration are given in Table 7.1 and Figure 7.5. As seen from Table 7.1, for both initial values, function Φ provides a contraction of the underlying interval, i.e. it provides a set of mappings

$$\left.\begin{aligned} \Phi &: [-10, 10] \to [0.000\,045, 1], \\ \Phi &: [0.000\,045, 1] \to [0.5, 0.7311], \\ &\vdots \\ \Phi &: \zeta \to \zeta. \end{aligned}\right\} \tag{7.21}$$

Indeed, the iterates from either starting point $x_0 \in \{-10, 10\}$ converge to a value $\zeta \in [0.6589, 0.6591] \in [-10, 10]$. It can be shown that after 24 iterations, the fixed point ζ is

$$\Phi : [-10, 10] \xrightarrow{i} \zeta = 0.659\,046\,068\,407\,41, \tag{7.22}$$

which is shown in Figure 7.5. □

Example 7.4.7. Plot the fixed points of the logistic function

$$\Phi(x) = \frac{1}{1 + \mathrm{e}^{-\beta x}} \tag{7.23}$$

for a range of β.

Solution. The result of the experiment is shown in Figure 7.6. From Figure 7.6, the values of the fixed point increase with β and converge to unity when β increases. □

Example 7.4.8. Show that the logistic function from Example 7.4.6, exhibits a linear asymptotic convergence for which the convergence rate is $\gamma = 0.2247$.

Table 7.2 Error convergence for the FPI of the logistic function

	$x_0 = -10$	e_i	e_i/e_{i-1}	$x_0 = 10$	e_i	e_i/e_{i-1}
First iterate	0.000 045	0.659	—	1	0.341	—
Second iterate	0.5	0.159	0.2413	0.7311	0.0721	0.2114
Third iterate	0.6225	0.0365	0.2296	0.6750	0.016	0.2219
Fourth iterate	0.6508	0.0082	0.2247	0.6626	0.0036	0.2246
Fifth iterate	0.6572	0.0018	0.2247	0.6598	0.0008	0.2247
Sixth iterate	0.6586	0.0004	0.2247	0.6592	0.0002	0.2247
Seventh iterate	0.6589	0.0001	0.2247	0.6591	0.0001	0.2247

Solution. To show that the rate of convergence of the iterative process (7.13) is $|\Phi'(x^*)|$, let us calculate $\Phi'(x^*) \approx \Phi'(0.659) = 0.2247$. Let us now upgrade Table 7.1 in order to show the rate of convergence. The results are shown in Table 7.2. As $\Phi'(x^*) \approx 0.2247$, it is expected that, according to CMT, the ratio of successive errors converges to $\Phi'(x^*)$. Indeed, for either initial value in the FPI, the errors $e_i = x_i - x^*$ decrease with the order of iteration and the ratio of successive errors e_{i+1}/e_i converges to 0.2247 and reaches that value after as few iterations as $i = 4$ for $x_0 = -10$ and $i = 5$ for $x_0 = 10$. □

Properties of the tanh activation function in this context are given in Krcmar *et al.* (2000).

Remark 7.4.9. The function

$$\tanh(\beta x) = \frac{e^{\beta x} - e^{-\beta x}}{e^{\beta x} + e^{-\beta x}}$$

provides contraction mapping for $0 < \beta < 1$.

This is easy to show, following the analysis for the logistic function and noting that $\tanh'(\beta x) = 4\beta/(e^{-\beta x} + e^{\beta x})^2$, which is strictly positive and for which the maximum value is $\beta = 1$ for $x = 0$. Convergence of FPI for $\beta = 1$ and $\beta = 1.2$ for a tanh activation function is shown in Figure 7.7. The graphs show convergence from two different starting values, $y = -10$ and $y = 10$. For $\beta = 1$, relaxations from both starting values converge towards zero, whereas for $\beta = 1.2$, which is greater than the bound given in Remark 7.4.9, we have two different fixed points. For convergence of learning algorithms for adaptive filters based upon neural networks, we desire only one stable fixed point, and the further emphasis will be on bounds on the weights and nonlinearity which preserve this condition.

7.5 Convergence of Nonlinear Relaxation Equations Realised Through a Recurrent Perceptron

We next analyse convergence towards an equilibrium based upon a recurrent perceptron using contraction mapping and corresponding fixed point iteration. Unlike in the linear case, the external input data to (7.2) do not need to be a zero vector, but simply kept constant.

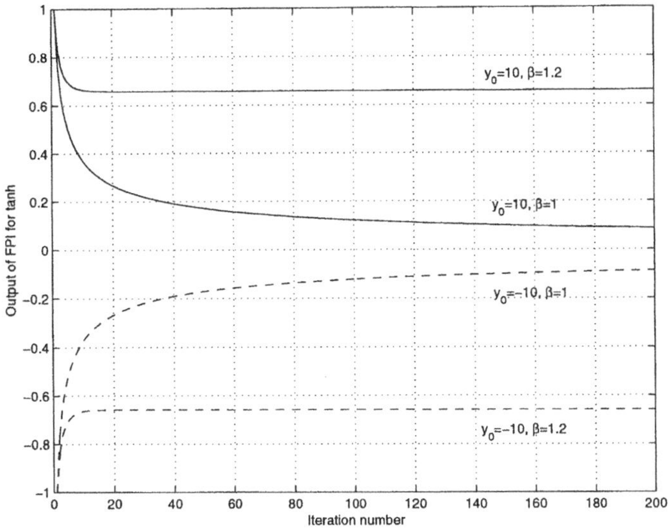

Figure 7.7 Fixed points for the tanh activation function

Proposition 7.5.1 (see Mandic and Chambers 1999b). *GAS relaxation for a recurrent perceptron given by*

$$y(k+1) = \Phi(\boldsymbol{u}(k)^{\mathrm{T}}\boldsymbol{w}(k)), \tag{7.24}$$

where $\boldsymbol{u}_k^{\mathrm{T}} = [y(k-1), \dots, y(k-N), 1, x(k-1), \dots, x(k-M)]$, *is a contraction mapping and converges to some value* $y^* \in (0,1)$ *for* $\beta \sum_{j=1}^{N} |w_j(k)| < 4$.

Proof. Equation (7.24) can be written as

$$y(k+1) = \Phi\biggl(\sum_{j=1}^{N+M+1} w_j z_j(k)\biggr), \tag{7.25}$$

where $z_j(k)$ is the jth element of input $\boldsymbol{u}(k)$. The iteration (7.25) is biased and can be expressed as

$$y(k+1) = \Phi(y(k), \dots, y(k-N+1), \text{const.}). \tag{7.26}$$

The existence, uniqueness and convergence features of mapping (7.24), follow from properties of the logistic function. Iteration (7.24), for a contractive Φ converges to a fixed point $y^* = \Phi(y^* + \text{const.})$, where the constant is given by

$$\text{const.} = \biggl(\sum_{j=N+1}^{N+M+1} w_j z_j(k)\biggr).$$

It is assumed that the weights are not time-variant. Since the condition for convergence of the logistic function to a fixed point is $0 < \beta < 4$, it follows that the slope in the logistic function β and the weights $w_1, \dots, w_N$ in the weight vector $\boldsymbol{w}$ are not

Table 7.3 Fixed point iterates for the NARMA perceptron

Starting value y_0	-10	10
First iterate	0.006 68	0.795 71
Second iterate	0.445 10	0.520 35
Third iterate	0.487 69	0.494 82
Fourth iterate	0.491 73	0.492 41
Fifth iterate	0.492 11	0.492 18
Sixth iterate	0.492 15	0.492 16

independent and that the effective slope in the logistic function now becomes the product $\beta \sum_{j=1}^{N} w_j$. Therefore

$$\left|\beta \sum_{j=1}^{N} w_j\right| \leqslant \beta \sum_{j=1}^{N} |w_j| < 4 \quad \Leftrightarrow \quad \|\boldsymbol{w}\|_1 < \frac{4}{\beta} \tag{7.27}$$

is the condition of GAS convergence of (7.2) realised through a recurrent NARMA perceptron. □

A comparison of the nonlinear GAS result (7.27) with its linear counterpart shows that they are both based upon the $\|\cdot\|_1$ norm of the corresponding coefficient vector. In the nonlinear case, however, the measure of nonlinearity is also included.

Example 7.5.2. Show that the iteration

$$y(k) = \Phi(y(k-1)) = \frac{1}{1 + \mathrm{e}^{-0.25y(k-1)+0.5}} \tag{7.28}$$

with initial values $y_0 = -10$ and $y_0 = 10$ converges towards a point $y^* \in [-10, 10]$.

Solution. Note that $\beta = 0.25$ and $w = 1$. The numerical values for iteration (7.28) are given in Table 7.3. Indeed, the iterates from either starting point converge to a value $y^* \in [0.492\,15, 0.492\,16] \subset [-10, 10]$. It can be shown that after 15 iterations for $y_0 = -10$ and 16 iterations for $y_0 = 10$, the fixed point to which the FPI (7.28) converges is $y^* = 0.492\,159\,969\,021\,68$. □

Corollary 7.5.3 (see Mandic and Chambers 1999b). *In the case of the realisation of (7.2) by a NARMA recurrent perceptron, convergence towards a point in the FPI sense* does not *depend on the number of external input signals, nor on their values, as long as they are finite.*

The convergence rate is the ratio of the distances between the current and previous iterate of an FPI and a fixed point y^*, i.e. $(y(k) - y^*)/(y(k-1) - y^*)$. This reveals how quickly an FPI process converges towards a point.

Observation 7.5.4 (see Mandic and Chambers 1999b). *A realisation of an iterative process (7.2) by a recurrent perceptron converges towards a fixed point y^* exhibiting linear convergence with convergence rate $\Phi'(y^*)$ (Example 7.4.8).*

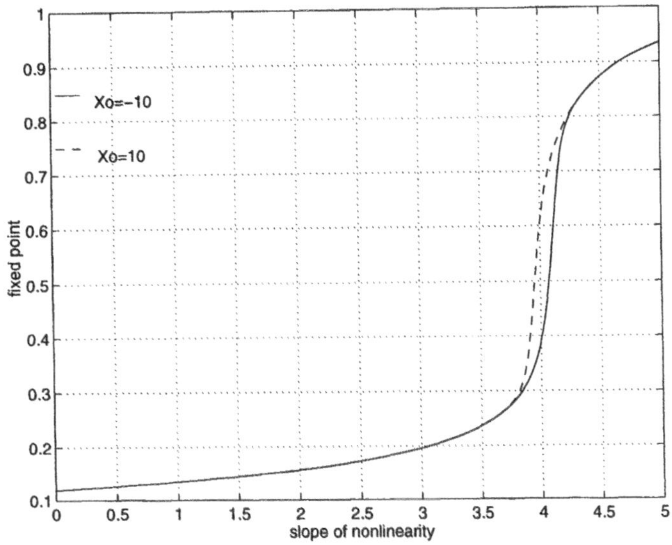

Figure 7.8 Fixed points for the biased *logistic* nonlinearity

Example 7.5.5. Plot the fixed points of the biased logistic function

$$\Phi(x) = \frac{1}{1 + \mathrm{e}^{-\beta x + \mathrm{bias}}} \tag{7.29}$$

for a range of β and bias $= 2$.

Solution. To depict the effects of varying β, noise was added to the system. From Figure 7.8, the values of fixed points increase with β and converge to unity when β increases. However, for β large enough, the fixed points to which the iteration $x_{i+1} = \Phi(x_i)$ converges might not be unique. Indeed, the broken line in Figure 7.8 represents the iteration whose starting value was $x_0 = 10$, while the solid line in Figure 7.8 represents the case with $x_0 = -10$. For a range of β around $\beta = 4$, the iterations from different starting points do not converge to the same value. The values of fixed points for the biased logistic function differ from the corresponding values for the pure logistic function. Moreover, the fixed points differ for various values of the bias in the biased logistic function. □

Remark 7.5.6. For stability of FPI for a tanh activation function replace the bound $\beta < 4$ by $\beta < 1$, i.e. $\|\boldsymbol{w}\|_1 < 1/\beta$.

7.6 Relaxation in Nonlinear Systems Realised by an RNN

Let $\boldsymbol{Y}_i = [y_1^i, \dots, y_N^i]^{\mathrm{T}}$ be a vector comprising the outputs of a general RNN at iteration i of the FPI. The input vector to a network is $\boldsymbol{u_i} = [y_1^i, \dots, y_N^i, 1, x_{N+1}, \dots, x_{N+M+1}]^{\mathrm{T}}$. The weight matrix $\boldsymbol{W}$ consists of N rows and $N + M + 1$ columns. Then, by a CMT in $\mathbb{R}^N$, the iterative process applied on the general RNN converges, if $\boldsymbol{M} = [a, b]^N$ is a closed subset of $\mathbb{R}^N$ such that

(i) $\boldsymbol{\Phi}: \boldsymbol{M} \rightarrow \boldsymbol{M}$;

(ii) if for some norm $\|\cdot\|$, $\exists\gamma < 1$ such that $\|\boldsymbol{\Phi}(\boldsymbol{x}) - \boldsymbol{\Phi}(\boldsymbol{y})\| \leqslant \gamma\|\boldsymbol{x} - \boldsymbol{y}\|$, $\forall \boldsymbol{x}, \boldsymbol{y} \in \boldsymbol{M}$ the equation

$$\boldsymbol{x} = \boldsymbol{\Phi}(\boldsymbol{x}) \tag{7.30}$$

has a unique solution $\boldsymbol{x}^* \in \boldsymbol{M}$, and the iteration

$$\boldsymbol{x}_{i+1} = \boldsymbol{\Phi}(\boldsymbol{x}_i) \tag{7.31}$$

converges to $\boldsymbol{x}^*$ for any starting value $\boldsymbol{x}_0 \in \boldsymbol{M}$.

Actually, since the function $\boldsymbol{\Phi}$ in this case is a multivariate function, $\boldsymbol{\Phi} = [\Phi_1, \dots, \Phi_N]^{\mathrm{T}}$, where N is the number of neurons of the RNN, we have a set of mappings

$$\left.\begin{aligned} y_1^i &= \Phi_1(\boldsymbol{u}_{i-1}^{\mathrm{T}}\boldsymbol{W}_1), \\ \vdots \quad & \quad \vdots \\ y_N^i &= \Phi_N(\boldsymbol{u}_{i-1}^{\mathrm{T}}\boldsymbol{W}_N), \end{aligned}\right\} \tag{7.32}$$

where $\{\boldsymbol{W}_i\}$ are the appropriate columns in $\boldsymbol{W}$. An obvious problem is that the convergence is norm dependent. Therefore, that condition should be replaced by some condition based upon the features of $\boldsymbol{\Phi}$.

Let us denote the Jacobian of $\boldsymbol{\Phi}$ by $\boldsymbol{J}$. If $\boldsymbol{M} \in \mathbb{R}^N$ is a convex set and $\boldsymbol{\Phi}$ is continuously differentiable on $\boldsymbol{M} = [a, b]^N \subset \mathbb{R}^N$ and satisfies the conditions of the CMT, then

$$\max_{\boldsymbol{z} \in \boldsymbol{M}} \|\boldsymbol{J}(\boldsymbol{z})\| \leqslant \gamma. \tag{7.33}$$

For convergence, the FPI at every neuron should be convergent. The following analysis gives the bound for the elements of the weight matrix $\boldsymbol{W}$ of the RNN with respect to the derivatives of the components of $\Phi = [\Phi_1, \dots, \Phi_N]$. Recall that for the case of a single recurrent perceptron, the condition for GAS was

$$\sum_{j=1}^{N} |w_j| < \frac{4}{\beta} \quad \Leftrightarrow \quad \|\boldsymbol{w}\|_1 < \frac{4}{\beta} = \frac{1}{\Phi'_{\max}}.$$

However, for a network of N neurons, it is possible to have a convergent FPI, even if some of the neurons violate the previous conditions. When it comes to the monotonic convergence, it is important that the process at every neuron converges uniformly. This is straightforward to show, since for any $\boldsymbol{x}, \boldsymbol{y} \subset \mathbb{R}^N$, which are processed by a neural network, we have

$$\begin{aligned} |\boldsymbol{\Phi}(\boldsymbol{x}) - \boldsymbol{\Phi}(\boldsymbol{y})| &= \sum_{i=1}^{N} \left| \sum_{j=1}^{N} w_{i,j}\Phi_j(x_j) - \sum_{j=1}^{N} w_{i,j}\Phi_j(y_j) \right| \\ &\leqslant \sum_{i=1}^{N} \sum_{j=1}^{N} |w_{i,j}||\Phi_j(x_j) - \Phi_j(y_j)| \\ &\leqslant \sum_{j=1}^{N} |\Phi'_{\max}||x_j - y_j| \sum_{i=1}^{N} |w_{i,j}|. \end{aligned} \tag{7.34}$$

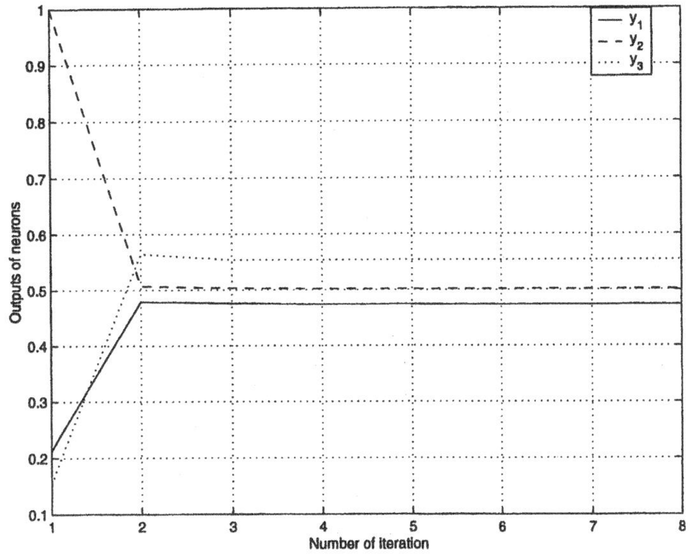

Figure 7.9 FPI for a general RNN

For uniform convergence at every particular neuron, it is the diagonal weights of the weight matrix (self-feedback) which together with the slope β_i have an influence on the convergence in the FPI sense. As in the case of a recurrent NARMA perceptron, the feedback of a general RNN may consist of a number n of delayed versions of its output, in addition to the state feedback from the remaining neurons in the network. In that case, the number of feedback inputs to the network becomes $N + n - 1$ and the condition for GAS becomes

$$\max_{1 \leqslant k \leqslant N} \{|w_{k,k}|, |w_{k,N+1}|, \ldots, |w_{k,N+n-1}|\} < \frac{4}{(N + n - 1) \max_{1 \leqslant i \leqslant N} \beta_i}. \tag{7.35}$$

Observation 7.6.1. *The rate of convergence of relaxation in RNNs does not depend on the length of the tap delay input line.*

Proof. It is already shown that all the variables related to the MA part of the underlying NARMA process form a constant during the FPI iteration, while the feedback variables are updated in every iteration. Hence, no matter how many external input signals, their contribution to the FPI relaxation is embodied in a constant. Therefore, the iteration

$$\boldsymbol{Y}_{i+1} = \Phi(\boldsymbol{Y}_i, \boldsymbol{X}, \boldsymbol{W}) \tag{7.36}$$

does not depend on the number of external input samples. □

Example 7.6.2. Analyse the convergence of the iteration process for a general RNN with three neurons and six external input signals and a logistic activation function.

Solution. Let us choose the initial values $\boldsymbol{X_0} = \text{rand}(10,1)*1$, $\boldsymbol{W} = \text{rand}(10,3)*2-1$, using the notation of MATLAB, and start the iteration process. Here $\text{rand}(M, N)$

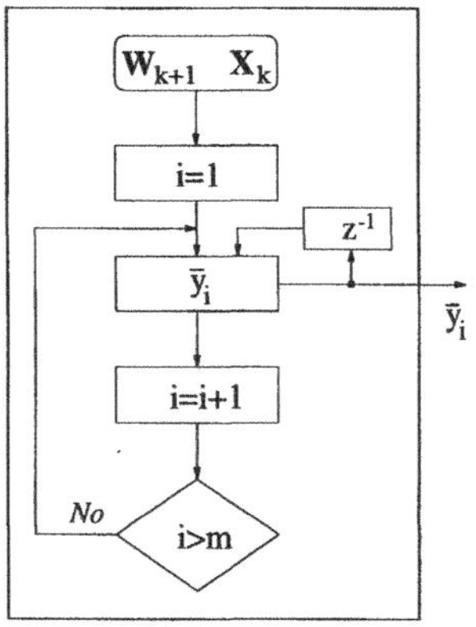

(a) Iterative process

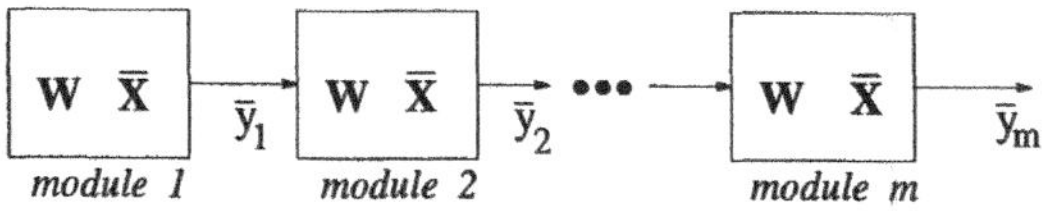

(b) Iterative process realised spatially

Figure 7.10 Spatial realisation of an iterative process

denotes an $(M \times N)$-dimensional matrix of uniformly distributed random numbers $\in [0,1]$. The convergence of the outputs of neurons in the FPI sense is depicted in Figure 7.9. For every neuron, the iteration process converges or, if in vector form, the output vector of the RNN converges to a fixed vector of the iteration. □

7.7 The Iterative Approach and Nesting

Nesting corresponds to the procedure of reducing the interval size in set theory. In signal processing, however, nesting is essentially a nonlinear spatial structure which corresponds to the cascaded structure in linear signal processing (Baltersee and Chambers 1998; Haykin and Li 1995; Mandic and Chambers 1998b; Mandic *et al.* 1998). The RNN-based nested sigmoid scheme can be written as (Haykin 1994; Poggio and Girosi 1990)

$$F(W,X) = \Phi\biggl(\sum_n w_n \Phi\biggl(\sum_i v_i \Phi\biggl(\cdots \Phi\biggl(\sum_j u_j X_j\biggr)\cdots\biggr)\biggr)\biggr), \tag{7.37}$$

where Φ is a sigmoidal function. This corresponds to a multilayer network of units that sum their inputs with 'weights' $W = \{w_n, v_i, \dots, u_j, \dots\}$ and then perform a sigmoidal transformation of this sum. Our aim is to show that nesting can exhibit contraction mapping and that repeatedly applied nesting can lead to convergence in the FPI sense. Therefore, instead of having a spatial, nested, pipelined structure, nesting can be obtained through a temporal, iterative, relaxive structure (Mandic and Chambers 2000c), as shown in Figure 7.10. Quantities that change under iteration in Figure 7.10 have a bar above the symbol.

Observation 7.7.1. *The compound nested logistic functions*

$$\begin{aligned}\hat{x} &= \Phi(x_N)\\ &= \Phi(\Phi(x_{N-1}))\\ &\ \vdots\\ &= \Phi\Bigl(\underbrace{\Phi(\Phi(\cdots(\Phi(x_1))\cdots))}_{N}\Bigr)\end{aligned} \tag{7.38}$$

provide a contraction mapping for $\beta < 4$ and the FPI converges towards a point $x^* \in [a, b]$.

Proof. Notice that the nesting process (7.38) represents an implicitly written fixed point iteration process

$$x_{i+1} = \Phi(x_i) \quad \Leftrightarrow \quad x_{i+1} = \Phi(\Phi(x_{i-1})) = \Phi\Bigl(\underbrace{\Phi(\Phi(\cdots(\Phi(x_1))\cdots))}_{N}\Bigr). \tag{7.39}$$

Hence, nesting (7.38) and fixed point iteration (7.13) are a realisation of the same process and have already been considered. Let us therefore just show the diagram of the effects of the nesting process for the logistic function with slope $\beta = 1$, depicted in Figure 7.11. From Figure 7.11, it is apparent that nesting (7.38) provides contraction mapping of its argument. Hence, it is expected that the nesting process (7.38) with N stages converges towards the point $x^* \in [|\Phi'(x^*)|^N a, |\Phi'(x^*)|^N b]$. For N small, the fixed point iteration achieved through a nesting process (7.38) may not reach its fixed point. However, from Tables 7.1 and 7.2 and Figure 7.11, even with $N = 4$, the error $|x_4 - x^*| < 0.01$, which suffices for practical applications. □

To summarise:

- for the nesting process to be a contraction mapping, the range of slopes β for the logistic function Φ should be bounded, with $0 < \beta < 4$;
- the nesting process of a sufficient order applied to an interval $[a, b] \in \mathbb{R}$ converges to a point $x^* \in [a, b]$, which is a fixed point of the fixed point iteration $x_{i+1} = \Phi(x_i)$;
- the nesting process (7.38) exhibits a *linear asymptotic convergence* whose rate is $|\Phi'(x^*)|$, where x^* is the fixed point of mapping Φ.

The nesting process (7.38) provides the iteration spatially, rather than temporally. Such a strategy is known as *pipelining* and is widely used in advanced computer architectures (Hwang and Briggs 1986). Using the *pipelining* strategy, a task is divided in subtasks, each of them being represented by a module. Pipelining corresponds to unfolding the finite iterative process into a spatial structure of the same length as the number of iterations in the former. Now, from (7.37), the pipelined structure represents indeed a spatial realisation of an essentially temporal iterative process, and converges under the same conditions as the nesting process (7.38). A realisation of

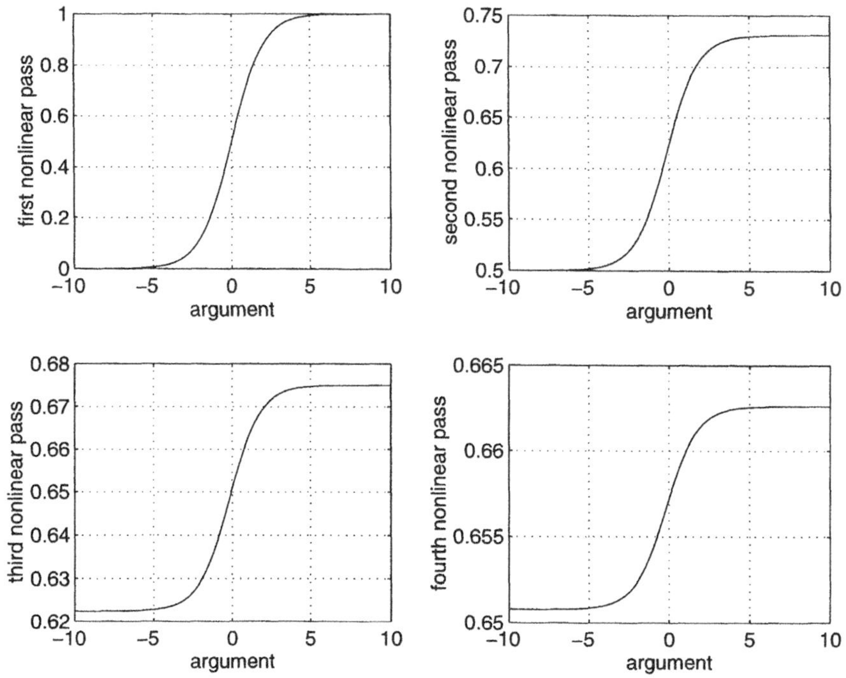

Figure 7.11 Nested *logistic* nonlinearity

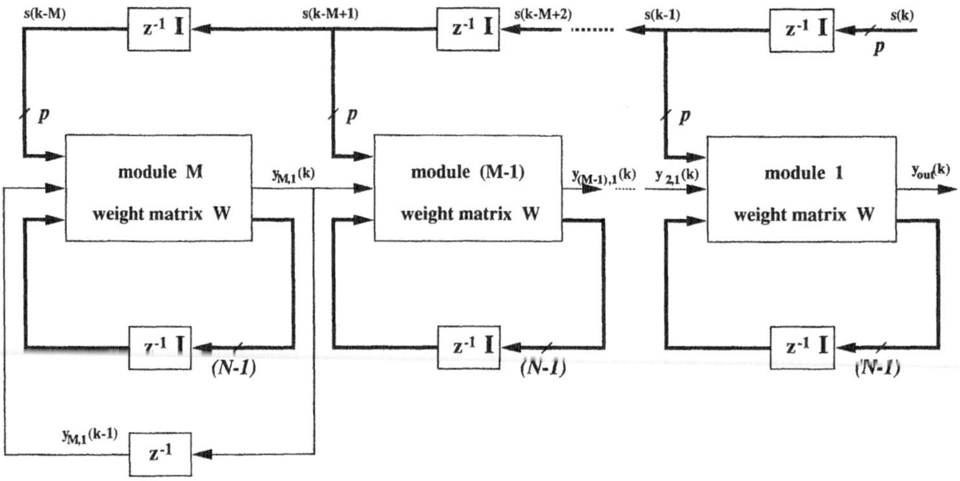

Figure 7.12 Pipelined Recurrent Neural Network

process (7.38) is the so-called pipelined recurrent neural network (PRNN) (Haykin and Li 1995), shown in Figure 7.12, which provides a spatial form of the iteration (7.37). Therefore, for instance, instead of having a temporal FPI on a recurrent perceptron (Figure 6.2), it suffices, for a finite-length FPI, to consider a spatial PRNN structure.

7.8 Upper Bounds for GAS Relaxation within FCRNNs

Neural systems of the form

$$\boldsymbol{x}(k+1) = \boldsymbol{A}\boldsymbol{x}(k) + \boldsymbol{B}\boldsymbol{\sigma}[\boldsymbol{W}\boldsymbol{x}(k) + \boldsymbol{s}] \tag{7.40}$$

have been widely considered (Jin *et al.* 1994). Here, $\boldsymbol{x}$ is the state vector of the network and $\boldsymbol{\sigma}(\,\cdot\,)$ is a vector of nonlinear activation functions. On the other hand, the weight matrix $\boldsymbol{W}$ of a recurrent neural network can be split up into the feedback part (index a) and the feedforward part (index b), which gives $\boldsymbol{Y}(k+1) = \boldsymbol{\Phi}(\boldsymbol{W}_a\boldsymbol{Y}(k) + \boldsymbol{W}_b\boldsymbol{x}(k))$, which can degenerate into the form (7.40). Namely, for a contractive activation function Φ, we have (Mandic and Chambers 2000e)

$$\Phi(a+b) < \Phi(a) + \Phi(b) < a + \Phi(b), \tag{7.41}$$

and results for system (7.40) provide the upper bound for stability of the fully connected RNN system described above (Mandic *et al.* 2000).

7.9 Summary

The relationships between the number of neurons in the RNN, the slope in the activation function β and a measure of $\boldsymbol{W}$ have been provided, which guarantee convergence of a relaxation process realised by fully connected recurrent neural networks. Based upon the fixed point iteration (FPI), it has been shown that these conditions rest entirely upon the slope of the activation function β and a measure of the $\|\cdot\|_1$ norm of the weight vector of a recurrent perceptron. A connection between nesting and FPI, which is the basis of the GAS convergence, has been established, and a pipelined recurrent neural network (PRNN) has been shown to be a spatial realisation of the FPI process. The results obtained can be applied when recurrent neural networks are used as computational models, in particular, as optimisation models. The results can also be used as stability analysis tools for some classes of nonlinear control systems.

8

Data-Reusing Adaptive Learning Algorithms

8.1 Perspective

In this chapter, a class of data-reusing learning algorithms for recurrent neural networks is analysed. This is achieved starting from a case of feedforward neurons, through to the case of networks with feedback, trained with gradient descent learning algorithms. It is shown that the class of data-reusing algorithms outperforms the standard (*a priori*) algorithms for nonlinear adaptive filtering in terms of the instantaneous prediction error. The relationships between the *a priori* and *a posteriori* errors, learning rate and the norm of the input vector are derived in this context.

8.2 Introduction

The so-called *a posteriori* error estimates provide us with, roughly speaking, some information after computation. From a practical point of view, they are valuable and useful, since real-life problems are often nonlinear, large, ill-conditioned, unstable or have multiple solutions and singularities (Hlavacek and Krizek 1998). The *a posteriori* error estimators are local in a computational sense, and the computational complexity of *a posteriori* error estimators should be far less expensive than the computation of an exact numerical solution of the problem. An account of the essence of *a posteriori* techniques is given in Appendix F.

In the area of linear adaptive filters, the most comprehensive overviews of *a posteriori* techniques can be found in Treichler (1987) and Ljung and Soderstrom (1983). These techniques are also known as data-reusing techniques (Douglas and Rupp 1997; Roy and Shynk 1989; Schnaufer and Jenkins 1993; Sheu *et al.* 1992). The quality of an *a posteriori* error estimator is often measured by its *efficiency index*, i.e. the ratio of the estimated error to the true error. It has been shown that the *a posteriori* approach in the neural network framework introduces a kind of normalisation of the employed learning algorithm (Mandic and Chambers 1998c). Consequently, it is expected that the instantaneous *a posteriori* output error $\bar{e}(k)$ is smaller in magnitude than the corresponding *a priori* error $e(k)$ for a non-expansive nonlinearity Φ (Mandic and Chambers 1998c; Treichler 1987).

8.2.1 *Towards an A Posteriori Nonlinear Predictor*

To obtain an *a posteriori* RNN-based nonlinear predictor, let us, for simplicity, consider a NARMA recurrent perceptron, the output of which can be expressed as

$$y(k) = \Phi(\boldsymbol{u}^{\mathrm{T}}(k)\boldsymbol{w}(k)), \tag{8.1}$$

where the *information vector*

$$\boldsymbol{u}(k) = [x(k-1), \ldots, x(k-M), 1, y(k-1), \ldots, y(k-N)]^{\mathrm{T}} \tag{8.2}$$

comprises both the external input and feedback signals. As the updated weight vector $\boldsymbol{w}(k+1)$ is available before the arrival of the next input vector $\boldsymbol{u}(k+1)$, an *a posteriori* output estimate $\bar{y}(k)$ can be formed as

$$\bar{y}(k) = \Phi(\boldsymbol{u}^{\mathrm{T}}(k)\boldsymbol{w}(k+1)). \tag{8.3}$$

The corresponding instantaneous *a priori* and *a posteriori* errors at the output neuron of a neural network are given, respectively, as

$$e(k) = d(k) - y(k) \qquad \textit{a priori} \text{ error}, \tag{8.4}$$

$$\bar{e}(k) = d(k) - \bar{y}(k) \qquad \textit{a posteriori} \text{ error}, \tag{8.5}$$

where $d(k)$ is some teaching signal. The *a posteriori* outputs (8.3) can be used to form an *a posteriori information vector*

$$\bar{\boldsymbol{u}}(k) = [x(k-1), \ldots, x(k-M), 1, \bar{y}(k-1), \ldots, \bar{y}(k-N)]^{\mathrm{T}}, \tag{8.6}$$

which can replace the *a priori* information vector (8.2) in the output (8.3) and weight update calculations (6.43)–(6.45). This also results in greater accuracy (Ljung and Soderstrom 1983). An alternate representation of such an algorithm is the so-called *a posteriori* error gradient descent algorithm (Ljung and Soderstrom 1983; Treichler 1987), explained later in this chapter.

A simple data-reusing algorithm for linear adaptive filters

The procedure of calculating the instantaneous error, output and weight update may be repeated for a number of times, keeping the same external input vector $\boldsymbol{x}(k)$ and teaching signal $d(k)$, which results in improved error estimation. Let us consider such a *data-reusing* LMS algorithm for FIR adaptive filters, described by (Mandic and Chambers 2000e)

$$\left.\begin{aligned} e_i(k) &= d(k) - \boldsymbol{x}^{\mathrm{T}}(k)\boldsymbol{w}_i(k), \\ \boldsymbol{w}_{i+1}(k) &= \boldsymbol{w}_i(k) + \eta e_i(k)\boldsymbol{x}(k), \\ \text{subject to} \quad |e_{i+1}(k)| &\leqslant \gamma|e_i(k)|, \quad 0 < \gamma < 1, \quad i = 1, \ldots, L. \end{aligned}\right\} \tag{8.7}$$

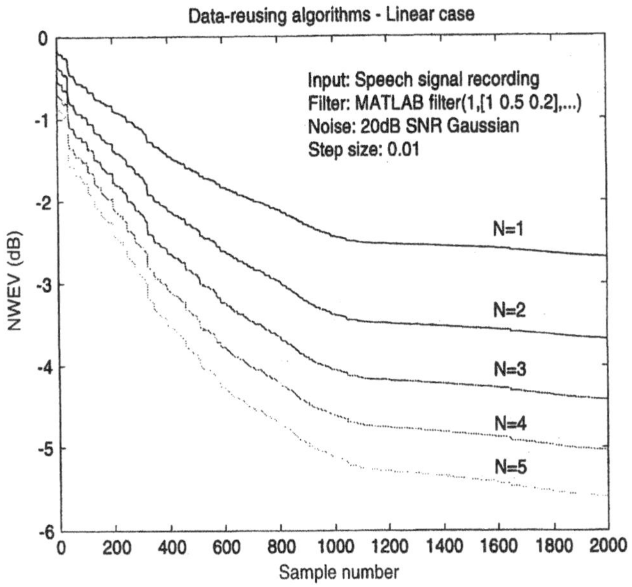

Figure 8.1 Convergence curves for a repeatedly applied data-reusing algorithm

From (8.7), $\boldsymbol{w}(k+1)$ is associated with the index $(L+1)$, i.e. $\boldsymbol{w}(k+1) = \boldsymbol{w}_{L+1}(k)$, whereas for $L = 1$, the problem reduces to the standard *a priori* algorithm, i.e. $\boldsymbol{w}_1(k) = \boldsymbol{w}(k)$, $\boldsymbol{w}_2(k) = \boldsymbol{w}(k+1)$. Convergence curves for such a reiterated LMS algorithm for a data-reusing FIR filter applied to echo cancellation are shown in Figure 8.1. The averaged squared prediction error becomes smaller with the number of iterations, N. For $N \to \infty$, the prediction error becomes the one of the NLMS[1] algorithm. A geometrical perspective of the procedure (8.7) is given in Appendix F and Figures F.2 and F.3. This provides advantageous stabilising features as compared to standard algorithms. This is further elaborated in Section F.2.2 of Appendix F. In practice, however, the advantage of the *a posteriori* algorithms is not always significant, and depends on the physics of the problem and the chosen filter.

8.2.2 Note on the Computational Complexity

It has been shown that the computational complexity of the *a priori* RTRL algorithm is $\mathcal{O}(N^4)$ (Haykin 1994; Williams and Zipser 1995), with N denoting the number of neurons in the RNN. If, in order to improve the performance, the number of neurons in the network is increased from N to $(N+1)$, the time required for the new adaptation process to finish can be dramatically increased. To depict that problem, the relative change in the computational load when the number of neurons increases, i.e. the ratio $(N+1)^4/N^4$, is shown in Figure 8.2. In other words, that means that the *a posteriori*

[1] In fact, for the linear case, the NLMS algorithm is approached by repeating this kind of data-reusing for an infinite number of times (Nitzberg 1985; Roy and Shynk 1989; Schnaufer and Jenkins 1993). For further details, see Appendix F.

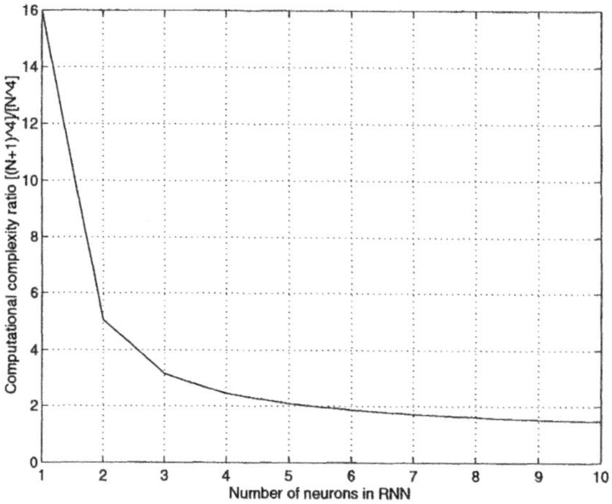

Figure 8.2 Ratio of the increase of computational burden with N

procedure applied to the network with N neurons should have the computational load $\mathcal{C}_L$ given by

$$\mathcal{C}_L(N^4) \leqslant \mathcal{C}_{L\ a\ posteriori} < \mathcal{C}_L((N+1)^4). \tag{8.8}$$

8.2.3 Chapter Summary

A detailed account of various data-reusing techniques for nonlinear adaptive filters realised as neural networks is provided. The relationships between the *a priori* and *a posteriori* errors are derived and the corresponding bounds on learning rates are analysed. This class of algorithms performs better than standard algorithms, does not introduce a significant additional computational burden, and for a class of data-reusing algorithms, when iterated for an infinite number of times, converges to a class of normalised algorithms.

8.3 A Class of Simple *A Posteriori* Algorithms

Consider a simple computational model of a feedforward neural adaptive filter shown in Figure 6.4. The aim is to preserve

$$|\bar{e}(k)| \leqslant \gamma |e(k)|, \qquad 0 \leqslant \gamma < 1 \tag{8.9}$$

at each iteration, for both feedforward and recurrent neural networks acting as a nonlinear predictor. The problem of obtaining the *a posteriori* error can be represented in the gradient descent setting as (Mandic and Chambers 2000e)

$$\left.\begin{aligned} \boldsymbol{w}(k+1) &= \boldsymbol{w}(k) - \eta \nabla_{\boldsymbol{w}} E(k), \\ \bar{e}(k) &= d(k) - \Phi(\boldsymbol{x}^{\mathrm{T}}(k)\boldsymbol{w}(k+1)), \\ \text{subject to} \quad |\bar{e}(k)| &\leqslant \gamma |e(k)|, \qquad 0 < \gamma < 1. \end{aligned}\right\} \tag{8.10}$$

From (8.10), the actual learning is performed in the standard manner, i.e. *a priori* using $e(k)$, whereas an improved *a posteriori* error $\bar{e}(k)$ is calculated at every discrete time interval using the updated weight vector $\boldsymbol{w}(k+1)$. The gradient descent algorithm for this computational model, with the cost function in the form of $E(k) = \frac{1}{2}e^2(k)$, is given by

$$\left.\begin{aligned} e(k) &= d(k) - \Phi(\boldsymbol{x}^{\mathrm{T}}(k)\boldsymbol{w}(k)), \\ \boldsymbol{w}(k+1) &= \boldsymbol{w}(k) + \eta(k)e(k)\Phi'(\boldsymbol{x}^{\mathrm{T}}(k)\boldsymbol{w}(k))\boldsymbol{x}(k), \\ \bar{e}(k) &= d(k) - \Phi(\boldsymbol{x}^{\mathrm{T}}(k)\boldsymbol{w}(k+1)). \end{aligned}\right\} \tag{8.11}$$

This case represents a generalisation of the LMS algorithm for FIR adaptive linear filters. Let us express the *a posteriori* error term from above as

$$\bar{e}(k) = d(k) - \Phi(\boldsymbol{x}^{\mathrm{T}}(k)\boldsymbol{w}(k)) - [\Phi(\boldsymbol{x}^{\mathrm{T}}(k)\boldsymbol{w}(k+1)) - \Phi(\boldsymbol{x}^{\mathrm{T}}(k)\boldsymbol{w}(k))]. \tag{8.12}$$

Using the CMT, for a contractive, monotonically increasing Φ and positive $e(k)$ and $\bar{e}(k)$, we have

$$\Phi(\boldsymbol{x}^{\mathrm{T}}(k)\boldsymbol{w}(k+1)) - \Phi(\boldsymbol{x}^{\mathrm{T}}(k)\boldsymbol{w}(k)) = \alpha(k)\boldsymbol{x}^{\mathrm{T}}(k)\Delta\boldsymbol{w}(k), \tag{8.13}$$

where $\alpha(k) = \Phi'(\xi) < 1$, $\xi \in (\boldsymbol{x}^{\mathrm{T}}(k)\boldsymbol{w}(k), \boldsymbol{x}^{\mathrm{T}}(k)\boldsymbol{w}(k+1))$. Using (8.11)–(8.13) yields

$$\bar{e}(k) = [1 - \eta(k)\alpha(k)\Phi'(k)\|\boldsymbol{x}(k)\|_2^2]e(k), \tag{8.14}$$

where $\Phi'(k) = \Phi'(\boldsymbol{x}^{\mathrm{T}}(k)\boldsymbol{w}(k))$. The learning rate

$$\eta(k) = \frac{1}{\alpha(k)\Phi'(k)\|\boldsymbol{x}(k)\|_2^2},$$

which minimises (8.14), is approximately that of a normalised nonlinear gradient descent algorithm (9.15), given in Chapter 9.

To obtain the bounds of such an *a posteriori* error, premultiplying the weight update equation in (8.11) by $\boldsymbol{x}^{\mathrm{T}}(k)$ and applying the nonlinear activation function Φ on either side yields (Mandic and Chambers 2000e)

$$\Phi(\boldsymbol{x}^{\mathrm{T}}(k)\boldsymbol{w}(k+1)) = \Phi(\boldsymbol{x}^{\mathrm{T}}(k)\boldsymbol{w}(k) + \eta(k)e(k)\Phi'(k)\|\boldsymbol{x}(k)\|_2^2). \tag{8.15}$$

Further analysis depends on the function Φ, which can exhibit either contractive or expansive behaviour. For simplicity, let us consider a class of contractive functions Φ, which satisfy[2]

$$\Phi(a+b) \leqslant \Phi(a) + \Phi(b). \tag{8.16}$$

With $a = \boldsymbol{x}^{\mathrm{T}}(k)\boldsymbol{w}(k)$ and $b = \eta(k)e(k)\Phi'(k)\|\boldsymbol{x}(k)\|_2^2$, applying (8.16) to (8.15) and subtracting $d(k)$ from both sides of the resulting equation, due to contractivity of Φ, we obtain

$$\bar{e}(k) \geqslant e(k) - \Phi(\eta(k)e(k)\Phi'(k)\|\boldsymbol{x}(k)\|_2^2). \tag{8.17}$$

[2] This is the case, for instance, for many sigmoid functions. For many other functions this is satisfied in a certain range of interest. For instance, for $a = 0$, positive b and a saturating, mononically increasing, positive sigmoid, $\Phi(a+b) < \Phi(b) < b$. The condition $\Phi(a+b) \leqslant \Phi(a) + \Phi(b)$ is satisfied for the logistic function on all of its range and for the positive range of the tanh activation function. For many other functions, $|\Phi(a+b)| \leqslant |\Phi(a) + \Phi(b)|$.

For Φ a contraction, $|\Phi(\xi)| < |\xi|$, $\forall \xi \in \mathbb{R}$, and (8.17) finally becomes

$$\bar{e}(k) > [1 - \eta(k)\Phi'(k)\|\boldsymbol{x}(k)\|_2^2]e(k), \tag{8.18}$$

which is the lower bound for the *a posteriori* error for a contractive nonlinear activation function. In this case, the range allowed for the learning rate $\eta(k)$ in (8.18) with constraint (8.9) is[3]

$$0 < \eta(k) < \frac{1}{\Phi'(k)\|\boldsymbol{x}(k)\|_2^2}. \tag{8.19}$$

For Φ a linear function,

$$0 < \eta(k) < \frac{1}{\|\boldsymbol{x}(k)\|_2^2}, \tag{8.20}$$

which boils down to the learning rate of the NLMS algorithm. Therefore, the *a posteriori* algorithm in this context introduces a kind of normalisation of the corresponding learning algorithm.

8.3.1 The Case of a Recurrent Neural Filter

In this case, the gradient updating equation regarding a recurrent perceptron can be symbolically expressed as (Haykin 1994) (see Appendix D)

$$\frac{\partial y(k)}{\partial \boldsymbol{w}(k)} = \boldsymbol{\Pi}(k+1) = \Phi'(\boldsymbol{u}^{\mathrm{T}}(k)\boldsymbol{w}(k))[\boldsymbol{u}(k) + \boldsymbol{w}_a(k)\boldsymbol{\Pi}(k)], \tag{8.21}$$

where the vector $\boldsymbol{\Pi}$ denotes the set of corresponding gradients of the output neuron and the vector $\boldsymbol{u}(k)$ encompasses both the external and feedback inputs to the recurrent perceptron. The correction to the weight vector at the time instant k becomes

$$\Delta \boldsymbol{w}(k) = \eta(k)e(k)\boldsymbol{\Pi}(k). \tag{8.22}$$

Following the same principle as for feedforward networks, the lower bound for the *a posteriori* error algorithm in single-node recurrent neural networks with a contractive activation function is obtained as

$$\bar{e}(k) > [1 - \eta(k)\boldsymbol{u}^{\mathrm{T}}(k)\boldsymbol{\Pi}(k)]e(k), \tag{8.23}$$

whereas the corresponding range allowed for the learning rate $\eta(k)$ is given by

$$0 < \eta(k) < \frac{1}{|\boldsymbol{u}^{\mathrm{T}}(k)\boldsymbol{\Pi}(k)|}. \tag{8.24}$$

---

[3] Condition (8.18) is satisfied for any $\eta > 0$. However, we want to preserve $|\bar{e}(k)| < |e(k)|$ (8.10), with the constraint that both $\bar{e}(k)$ and $e(k)$ have the same sign, and hence the learning rate η has to satisfy (8.19).

8.3.2 The Case of a General Recurrent Neural Network

For recurrent neural networks of the Williams–Zipser type (Williams and Zipser 1989a), with N neurons, one of which is the output neuron, the weight matrix update for an RTRL training algorithm can be expressed as

$$\Delta \boldsymbol{W}(k) = \eta(k)e(k)\frac{\partial y_1(k)}{\partial \boldsymbol{W}(k)} = \eta(k)e(k)\boldsymbol{\Pi}_1(k), \tag{8.25}$$

where $\boldsymbol{W}(k)$ represents the weight matrix and

$$\boldsymbol{\Pi}_1(k) = \frac{\partial y_1(k)}{\partial \boldsymbol{W}(k)}$$

is the matrix of gradients at the output neuron

$$\pi^1_{n,l}(k) = \frac{\partial y_1(k)}{\partial w_{n,l}},$$

where the index n runs along the N neurons in the network and the index l runs along the inputs to the network. This equation is similar to the one for a recurrent perceptron, with the only difference being that weight matrix $\boldsymbol{W}$ replaces weight vector $\boldsymbol{w}$ and gradient matrix $\boldsymbol{\Pi} = [\boldsymbol{\Pi}_1, \dots, \boldsymbol{\Pi}_N]$ replaces gradient vector $\boldsymbol{\Pi}$. Notice that in order to update matrix $\boldsymbol{\Pi}_1$, a modified version of (8.21) has to update gradient matrices $\boldsymbol{\Pi}_i$, $i = 2, \dots, N$. More details about this procedure can be found in Williams and Zipser (1989a) and Haykin (1994).

The lower bound for the *a posteriori* error obtained by an *a priori* learning – *a posteriori* error RTRL algorithm (8.25) with constraint (8.9), and a contractive nonlinear activation function Φ – is therefore

$$\bar{e}(k) > [1 - \eta(k)\boldsymbol{u}^{\mathrm{T}}(k)\boldsymbol{\Pi}_1(k)]e(k), \tag{8.26}$$

whereas the range of allowable learning rates $\eta(k)$ is

$$0 < \eta(k) < \frac{1}{|\boldsymbol{u}^{\mathrm{T}}(k)\boldsymbol{\Pi}_1(k)|}. \tag{8.27}$$

8.3.3 Example for the Logistic Activation Function

It is shown in Chapter 7 that the condition for the logistic activation function to be a contraction is $\beta < 4$. As such a function is monotone and ascending, the bound on its first derivative is $\Phi'(\xi) \leqslant \beta/4$, $\forall \xi \in \mathbb{R}$. That being the case, the bounds on the *a posteriori* error and learning rate for the feedforward case become, respectively,

$$\bar{e}(k) > \tfrac{1}{4}[4 - \eta(k)\beta\|\boldsymbol{x}(k)\|_2^2]e(k) \tag{8.28}$$

and

$$0 < \eta(k) < \frac{4}{\beta\|\boldsymbol{x}(k)\|_2^2}. \tag{8.29}$$

Similar conditions can be derived for the recurrent case. Further relationships between η, β and $\boldsymbol{w}$ are given in Chapter 12.

8.4 An Iterated Data-Reusing Learning Algorithm

This class of algorithms employs L reuses of the weight update per sample and is a nonlinear version of algorithm (8.7). A data-reusing gradient descent algorithm for a nonlinear FIR filter is given by (Douglas and Rupp 1997; Mandic and Chambers 1998c)

$$\left.\begin{aligned} e_i(k) &= d(k) - \Phi(\boldsymbol{x}^{\mathrm{T}}(k)\boldsymbol{w}_i(k)), \qquad i = 1, \dots, L, \\ \boldsymbol{w}_{i+1}(k) &= \boldsymbol{w}_i(k) + \eta(k)e_i(k)\Phi'(\boldsymbol{x}^{\mathrm{T}}(k)\boldsymbol{w}_i(k))\boldsymbol{x}(k), \\ \text{subject to} \quad |e_{i+1}(k)| &\leqslant \gamma|e_i(k)|, \qquad 0 < \gamma < 1, \; i = 1, \dots, L, \end{aligned}\right\} \tag{8.30}$$

where $\boldsymbol{w}_i(k)$ is the weight vector at the ith iteration of (8.30), $\boldsymbol{x}(k)$ is the input vector, $d(k)$ is some teaching signal and $e_i(k)$ is the prediction error from the ith iteration of (8.30). For $L = 1$, the problem reduces to the standard *a priori* algorithm, whereas $\boldsymbol{w}(k+1)$ is associated with the index $(L+1)$, i.e.

$$\left.\begin{aligned} \boldsymbol{w}_1(k) &= \boldsymbol{w}(k), \\ \boldsymbol{w}_{L+1}(k) &= \boldsymbol{w}(k+1). \end{aligned}\right\} \tag{8.31}$$

Starting from the last iteration in (8.30), i.e. for $i = L$, we obtain

$$\begin{aligned} \boldsymbol{w}(k+1) = \boldsymbol{w}_{L+1}(k) &= \boldsymbol{w}_L(k) + \eta(k)e_L(k)\Phi'(\boldsymbol{x}^{\mathrm{T}}(k)\boldsymbol{w}_L(k))\boldsymbol{x}(k) \\ &= \boldsymbol{w}_{L-1}(k) + \eta(k)e_{L-1}(k)\Phi'(\boldsymbol{x}^{\mathrm{T}}(k)\boldsymbol{w}_{L-1}(k))\boldsymbol{x}(k) \\ &\quad + \eta(k)e_L(k)\Phi'(\boldsymbol{x}^{\mathrm{T}}(k)\boldsymbol{w}_L(k))\boldsymbol{x}(k) \\ &= \boldsymbol{w}(k) + \sum_{i=1}^{L} \eta(k)e_i(k)\Phi'(\boldsymbol{x}^{\mathrm{T}}(k)\boldsymbol{w}_i(k))\boldsymbol{x}(k). \end{aligned} \tag{8.32}$$

Consider the expression for the instantaneous error from the $(i+1)$th iteration at the output neuron

$$\begin{aligned} e_{i+1}(k) &= d(k) - \Phi(\boldsymbol{x}^{\mathrm{T}}(k)\boldsymbol{w}_{i+1}(k)) \\ &= [d(k) - \Phi(\boldsymbol{x}^{\mathrm{T}}(k)\boldsymbol{w}_i(k))] - [\Phi(\boldsymbol{x}^{\mathrm{T}}(k)\boldsymbol{w}_{i+1}(k)) - \Phi(\boldsymbol{x}^{\mathrm{T}}(k)\boldsymbol{w}_i(k))]. \end{aligned} \tag{8.33}$$

The second term on the right-hand side of (8.33) depends on the function Φ, which can exhibit either contractive or expansive behaviour (Appendix G). For a contractive Φ, assuming positive quantities, $\exists\alpha(k) = \Phi'(\xi)$, $\xi \in (\boldsymbol{x}^{\mathrm{T}}(k)\boldsymbol{w}_i(k), \boldsymbol{x}^{\mathrm{T}}(k)\boldsymbol{w}_{i+1}(k))$ such that the right-hand term in square brackets from (8.33) can be replaced by $\alpha(k)\boldsymbol{x}^{\mathrm{T}}(k)\Delta\boldsymbol{w}_i(k)$, which yields

$$e_{i+1}(k) = e_i(k)[1 - \eta(k)\alpha(k)\Phi'(\boldsymbol{x}^{\mathrm{T}}(k)\boldsymbol{w}_i(k))\|\boldsymbol{x}(k)\|_2^2]. \tag{8.34}$$

To calculate the bound on such an error, premultiplying the first equation in (8.30) by $\boldsymbol{x}^{\mathrm{T}}(k)$ and applying the nonlinear activation function Φ on either side yields

$$\Phi(\boldsymbol{x}^{\mathrm{T}}(k)\boldsymbol{w}_{i+1}(k)) = \Phi(\boldsymbol{x}^{\mathrm{T}}(k)\boldsymbol{w}_i(k) + \eta(k)e_i(k)\Phi'(\boldsymbol{x}^{\mathrm{T}}(k)\boldsymbol{w}_i(k))\|\boldsymbol{x}(k)\|_2^2). \tag{8.35}$$

Further analysis depends on whether Φ is a contraction or an expansion. It is convenient to assume that $e_i(k)$, $i = 1, \dots, L$, have the same sign during iterations (Appendix F, Figure F.3). From (8.15)–(8.18), we have

$$e_{i+1}(k) > [1 - \eta(k)\Phi'(\boldsymbol{x}^{\mathrm{T}}(k)\boldsymbol{w}_i(k))\|\boldsymbol{x}(k)\|_2^2]e_i(k) \tag{8.36}$$

from iteration to iteration of (8.30). Assume that

$$\Phi'(k) \approx \Phi'(\boldsymbol{x}^{\mathrm{T}}(k)\boldsymbol{w}_1(k)) \approx \cdots \approx \Phi'(\boldsymbol{x}^{\mathrm{T}}(k)\boldsymbol{w}_L(k)),$$

then after L iterations[4] of (8.36), we have

$$e(k+1) > [1 - \eta(k)\Phi'(k)\|\boldsymbol{x}(k)\|_2^2]^L e(k). \tag{8.37}$$

The term in the square brackets from above has its modulus less than unity. In that case, the whole procedure is a fixed point iteration, whose convergence is given in Appendix G.

From (8.37) and the condition $|\bar{e}(k)| < |e(k)|$, the range allowed for the learning rate $\eta(k)$ in the data-reusing adaptation (8.30) is

$$0 < \eta(k) < \frac{1}{\Phi'(k)\|\boldsymbol{x}(k)\|_2^2}. \tag{8.38}$$

8.4.1 The Case of a Recurrent Predictor

The correction to the weight vector of the jth neuron, at the time instant k becomes

$$\Delta\boldsymbol{w}_j(k) = \eta(k)e(k)\boldsymbol{\Pi}_1^{(j)}(k), \tag{8.39}$$

where $\boldsymbol{\Pi}_1^{(j)}(k)$ represents the jth row of the gradient matrix $\boldsymbol{\Pi}_1(k)$. From the above analysis

$$0 < \eta(k) < \max_j \frac{1}{|\boldsymbol{u}^{\mathrm{T}}(k)\boldsymbol{\Pi}_1^{(j)}(k)|}. \tag{8.40}$$

8.5 Convergence of the *A Posteriori* Approach

In the case of nonlinear adaptive filters, there is generally no Wiener solution, and hence the convergence is mainly considered through Lyapunov stability (DeRusso *et al.* 1998; Zurada and Shen 1990), or through contraction mapping (Mandic and Chambers 1999b). Here, due to the assumption that for this class of data-reusing algorithms, the *a priori* and the *a posteriori* errors have the same sign through the data-reusing fixed point iteration, and $|\bar{e}(k)| < |e(k)|$, convergence of the *a posteriori* (data-reusing) error algorithm is defined by convergence of the underlying *a priori* error learning algorithm, which is detailed in Chapter 10. The limit behaviour of the above class of algorithms can be achieved for the infinite number of data-reuse iterations, i.e. when

[4] The term in the square brackets from (8.37) is strictly less than unity and becomes smaller with L. Also, $e(k) = e_1(k)$, $e(k+1) = e_{L+1}(k)$. In fact, the relation (8.36) represents a fixed point iteration, which, due to CMT, converges for $|1 - \eta(k)\Phi'(\boldsymbol{x}^{\mathrm{T}}(k)\boldsymbol{w}_i(k))\|\boldsymbol{x}(k)\|_2^2| < 1$.

$L \to \infty$. In that case, for instance, $e_i(k) > [1 - \eta(k)\Phi'(k)\|\boldsymbol{x}(k)\|_2^2]^{i-1}e(k)$, which from (8.36) forms a geometric series, which converges to a normalised nonlinear gradient descent algorithm (Figure F.3), and consequently the ratio $e_{i+1}(k)/e_i(k) \to 0$.

8.6 *A Posteriori* Error Gradient Descent Algorithm

The *a posteriori* outputs (8.3) can be used to form an updated *a posteriori information vector*

$$\bar{\boldsymbol{u}}(k) = [x(k-1), \dots, x(k-M), 1, \bar{y}(k-1), \dots, \bar{y}(k-N)]^{\mathrm{T}}, \tag{8.41}$$

which can replace the *a priori* information vector (8.2) in the output (8.3) and weight update calculations (6.43)–(6.45). An alternate representation of such an algorithm is the so-called *a posteriori* error gradient descent algorithm (Ljung and Soderstrom 1983; Treichler 1987), which is the topic of this section. Since the updated weight vector $\boldsymbol{w}(k+1)$ is available before the new input vector $\boldsymbol{x}(k+1)$ arrives, an *a posteriori* error gradient can be expressed as (Douglas and Rupp 1997; Ljung and Soderstrom 1983; Treichler 1987)

$$\bar{\nabla}_{\boldsymbol{w}}(\tfrac{1}{2}\bar{e}^2(k)) = \frac{\partial(\frac{1}{2}\bar{e}^2(k))}{\partial \boldsymbol{w}(k+1)}. \tag{8.42}$$

Using the above expression and, for simplicity, constraining the *a posteriori* information vector $\bar{\boldsymbol{u}}(k)$ to the case of a nonlinear dynamical neuron without feedback yields (Ljung and Soderstrom 1983; Treichler 1987)

$$\frac{\partial(\frac{1}{2}\bar{e}^2(k))}{\partial \boldsymbol{w}(k+1)} = -\Phi'(\boldsymbol{x}^{\mathrm{T}}(k)\boldsymbol{w}(k+1))\bar{e}(k)\boldsymbol{x}(k). \tag{8.43}$$

The *a posteriori* error can be now expressed as (Mandic and Chambers 1998b,c)

$$\begin{aligned}
\bar{e}(k) &= d(k) - \Phi(\boldsymbol{x}^{\mathrm{T}}(k)\boldsymbol{w}(k+1)) \\
&= d(k) - \Phi(\boldsymbol{x}^{\mathrm{T}}(k)\boldsymbol{w}(k)) + \Phi(\boldsymbol{x}^{\mathrm{T}}(k)\boldsymbol{w}(k)) - \Phi(\boldsymbol{x}^{\mathrm{T}}(k)\boldsymbol{w}(k+1)) \\
&= e(k) - [\Phi(\boldsymbol{x}^{\mathrm{T}}(k)\boldsymbol{w}(k+1)) - \Phi(\boldsymbol{x}^{\mathrm{T}}(k)\boldsymbol{w}(k))],
\end{aligned} \tag{8.44}$$

which contains terms with the time index $(k+1)$. Let us therefore express the term[5]

$$\Phi(\boldsymbol{x}^{\mathrm{T}}(k)\boldsymbol{w}(k+1)) = \Phi(\boldsymbol{x}^{\mathrm{T}}(k)\boldsymbol{w}(k) + \boldsymbol{x}^{\mathrm{T}}(k)\Delta\boldsymbol{w}(k)) \tag{8.45}$$

via its first-order Taylor expansion about the point $\boldsymbol{x}^{\mathrm{T}}(k)\boldsymbol{w}(k)$ as

$$\begin{aligned}
\Phi(\boldsymbol{x}^{\mathrm{T}}(k)\boldsymbol{w}(k+1)) &\approx \Phi(\boldsymbol{x}^{\mathrm{T}}(k)\boldsymbol{w}(k)) + \frac{\partial\Phi(\boldsymbol{x}^{\mathrm{T}}(k)\boldsymbol{w}(k))}{\partial\boldsymbol{w}(k)}\Delta\boldsymbol{w}(k) \\
&= \Phi(\boldsymbol{x}^{\mathrm{T}}(k)\boldsymbol{w}(k)) + \eta\bar{e}(k)\Phi'^2(k)\boldsymbol{x}^{\mathrm{T}}(k)\boldsymbol{x}(k),
\end{aligned} \tag{8.46}$$

---

[5] Notice that using Lipschitz continuity of Φ, the modulus of the term on the right-hand side of (8.44), i.e. $[\Phi(\boldsymbol{x}^{\mathrm{T}}(k)\boldsymbol{w}(k+1)) - \Phi(\boldsymbol{x}^{\mathrm{T}}(k)\boldsymbol{w}(k))]$ is bounded from above by

$$|\eta\bar{e}(k)\Phi'(\boldsymbol{x}^{\mathrm{T}}(k)\boldsymbol{w}(k+1))\boldsymbol{x}^{\mathrm{T}}(k)\boldsymbol{x}(k)|.$$

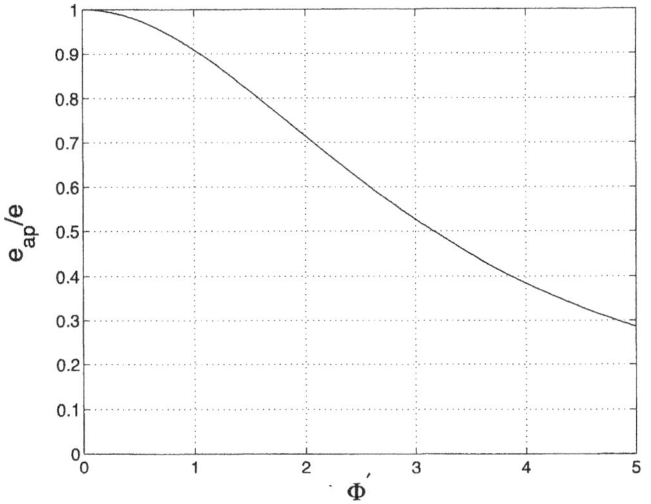

Figure 8.3 Ratio between the *a posteriori* and *a priori* errors for various slopes of the activation function

where $\Phi'(k) = \Phi'(\boldsymbol{x}^{\mathrm{T}}(k)\boldsymbol{w}(k)) \approx \Phi'(\boldsymbol{x}^{\mathrm{T}}(k)\boldsymbol{w}(k+1))$. Now, combining (8.44) and (8.46) yields the *a posteriori* error

$$\bar{e}(k) = \frac{e(k)}{1 + \eta\Phi'^2(k)\boldsymbol{x}^{\mathrm{T}}(k)\boldsymbol{x}(k)} = \frac{e(k)}{1 + \eta\Phi'^2(k)\|\boldsymbol{x}(k)\|_2^2}. \tag{8.47}$$

The *a posteriori* error $\bar{e}(k)$ (8.47) is smaller in magnitude than the corresponding *a priori* error $e(k)$, since the denominator of (8.47) is strictly greater than unity. For expansive activation functions, $\Phi' > 1$ and the effect described by (8.47) is even more emphasised. However, the *a priori* error in this case can be quite large in magnitude. The effective learning rate for this algorithm becomes[6]

$$\eta(k) = \frac{1}{1 + \eta\Phi'^2(k)\|\boldsymbol{x}(k)\|_2^2}, \tag{8.48}$$

which does not exhibit the problem of unboundedness of η for small $\|\boldsymbol{x}(k)\|_2^2$, as experienced with the class of normalised algorithms. The ratio between $\bar{e}$ and e as defined by (8.47) is shown in Figure 8.3. For the simulation, the parameters were $\eta = 0.1$ and $\|\boldsymbol{x}\| = 1$. The *a posteriori* error is clearly smaller than the *a priori* error for any Φ' of interest. For expansive functions, the ratio is smaller than for contractive functions, which means that data-reusing has the effect of stabilisation of the algorithm in this case.

[6] For a linear Φ, using the matrix inversion lemma, the learning rate (8.48) is equivalent to the learning rate of a general NLMS algorithm $\mu_0(k)/\|\boldsymbol{x}(k)\|_2^2$ (Douglas and Rupp 1997).

8.6.1 *A Posteriori Error Gradient Algorithm for Recurrent Neural Networks*

Recall that when using the *a posteriori* information vector (8.41), the output of a recurrent perceptron becomes

$$\bar{y}(k) = \Phi\left(\sum_{j=1}^{M} w_j(k+1)x(k-j)+w_{M+1}(k+1)+\sum_{m=1}^{N} w_{m+M+1}(k+1)\bar{y}(k-m)\right), \quad (8.49)$$

or compactly as

$$\bar{y}(k) = \Phi(\bar{\boldsymbol{u}}^{\mathrm{T}}(k)\boldsymbol{w}(k+1)). \quad (8.50)$$

Following the approach from Treichler (1987) and the above analysis, the *a posteriori* error gradient adaptation regarding the recurrent perceptron can be symbolically expressed as

$$\bar{\boldsymbol{\Pi}}(k+1) = \Phi'(\bar{\boldsymbol{u}}^{\mathrm{T}}(k)\boldsymbol{w}(k+1))[\bar{\boldsymbol{u}}(k) + \boldsymbol{w}(k+1)\bar{\boldsymbol{\Pi}}(k)]. \quad (8.51)$$

Using the *a posteriori* error gradient technique from Ljung and Soderstrom (1983) and Treichler (1987), the weight update of this algorithm becomes

$$\eta\bar{e}(k)\bar{\boldsymbol{\Pi}}(k), \quad (8.52)$$

where $\bar{e}(k) = d(k) - \Phi(\bar{\boldsymbol{u}}^{\mathrm{T}}(k)\boldsymbol{w}(k+1))$. Hence (Mandic and Chambers 1998c, 1999d),

$$\bar{e}(k) = \frac{d(k) - \Phi(\bar{\boldsymbol{u}}^{\mathrm{T}}(k)\boldsymbol{w}(k))}{1 + \eta\|\bar{\boldsymbol{\Pi}}(k)\|_2^2}. \quad (8.53)$$

The denominator in (8.53) is strictly greater than unity, which makes the *a posteriori* error $\bar{e}(k)$ smaller in magnitude than the *a priori* error (Mandic and Chambers 1998a). The analysis for a full recurrent neural network is straightforward.

8.7 Experimental Results

The simulations on two speech signals, denoted by s1 and s2, which come from two different individuals, were undertaken in order to support the derived algorithms. The amplitudes of the signals were adjusted to lie within the range of the function Φ, i.e. within $(0, 1)$ for the logistic function. The measure that was used to assess the performance of the predictors was the forward prediction gain R_{p} given by

$$R_{\mathrm{p}} \triangleq 10\log_{10}\left(\frac{\hat{\sigma}_{\mathrm{s}}^2}{\hat{\sigma}_{\mathrm{e}}^2}\right) \text{ dB}, \quad (8.54)$$

where $\hat{\sigma}_{\mathrm{s}}^2$ denotes the estimated variance of the speech signal $\{s(k)\}$, whereas $\hat{\sigma}_{\mathrm{e}}^2$ denotes the estimated variance of the forward prediction error signal $\{e(k)\}$. In the experiments, the initialisation procedure used the same strategy, namely epochwise, with 200 epochs ran over 300 samples, as described in Mandic *et al.* (1998) and Baltersee and Chambers (1998). The network chosen for the analysis was with $N = 2$ neurons and one external input signal to the network. Such a network was tested on both the *a priori* and *a posteriori* algorithms and the simulation results are shown

Table 8.1 Comparison between prediction gains *Rp* between an *a priori* scheme and various *a posteriori* schemes

Prediction gain	s1	s2
R_p (dB) *a priori* scheme $\boldsymbol{W}(k)$, feedback nonupdated, $N = 2$	7.78	8.48
R_p (dB) *a posteriori* scheme $\boldsymbol{W}(k+1)$, feedback nonupdated, $N = 2$	7.84	8.55
R_p (dB) *a posteriori scheme* $\boldsymbol{W}(k+1)$, feedback updated, *a posteriori* $\bar{\boldsymbol{\Pi}}$, $N = 2$	9.23	8.78
R_p (dB) *a priori* scheme $\boldsymbol{W}(k)$, feedback nonupdated, $N = 3$	7.79	8.49

in Table 8.1. In order to show the merit of the *a posteriori* approach on the network with $N = 2$ neurons, the results achieved were further compared to the performance obtained using the *a priori* network with $N = 3$ neurons. The results show that the performance of an RNN predictor improves with the amount of *a posteriori* information involved. The best results were achieved for the scheme with the *a posteriori* gradients, which were used in the recursive weights adaptation algorithm, and the corresponding feedback values were updated, i.e. the *a posteriori* information vector was used. The improvement in the performance in that case, as compared with the *a priori* scheme with $N = 3$, was, for example, 1.44 dB for the signal s1.

In the second experiment, an *a posteriori* algorithm with an updated *a posteriori* information vector was applied to modular nested recurrent architectures (PRNN) as described in Mandic *et al.* (1998) and Baltersee and Chambers (1998). The experiments were undertaken on speech. For modular networks with typically $M = 5$ modules and $N = 2$ neurons per recurrent module, there was not much improvement in prediction gain between the *a posteriori* gradient formulation and the standard algorithm. However, for an ERLS learning algorithm (Appendix D), there was an improvement of several percent in the prediction gain when using the algorithm with an updated *a posteriori* information vector (8.41).

8.8 Summary

Relationships between the *a priori* and *a posteriori* prediction error, learning rate and slope of the nonlinear activation function of a nonlinear adaptive filter realised by a neural network have been derived. This has been undertaken for learning algorithms based upon gradient descent for both the feedforward and recurrent case. A general data-reusing (*a posteriori* adaptation) adaptive algorithm for nonlinear adaptive filters realised as neural networks was then analysed. These algorithms use L weight updates per fixed external input vector and teaching signal. Therefore, their performance is in between the standard gradient descent and normalised algorithm. Relationships between the errors from consecutive iterations of the algorithm are derived based upon the corresponding *a priori* prediction error and gradients of the nonlinear activation function of a neuron, as well as the $\mathcal{L}_2$ norm of the input data vector and the learning rate η. However, in practice, the benefits of data-reusing techniques may not be significant. Relationships between η, β and $\|\boldsymbol{x}\|_2^2$ deserve more attention and will be addressed in Chapter 12.

9

A Class of Normalised Algorithms for Online Training of Recurrent Neural Networks

9.1 Perspective

A normalised version of the real-time recurrent learning (RTRL) algorithm is introduced. This has been achieved via local linearisation of the RTRL around the current point in the state space of the network. Such an algorithm provides an adaptive learning rate normalised by the $\mathcal{L}_2$ norm of the gradient vector at the output neuron. The analysis is general and also covers simpler cases of feedforward networks and linear FIR filters.

9.2 Introduction

Gradient-descent-based algorithms for training neural networks, such as the backpropagation, backpropagation through time, recurrent backpropagation (RBP) and real-time recurrent learning (RTRL) algorithm, typically suffer from slow convergence when dealing with statistically nonstationary inputs. In the area of linear adaptive filters, similar problems with the LMS algorithm have been addressed by utilising normalised algorithms, such as NLMS. We therefore introduce a normalised RTRL-based learning algorithm with the idea to impose similar stabilisation and convergence effects on training of RNNs, as normalisation imposes on the LMS algorithm.

In the area of linear FIR adaptive filters, it is shown (Soria-Olivas *et al.* 1998) that a normalised gradient-descent-based learning algorithm can be derived starting from the Taylor series expansion of the instantaneous output error of an adaptive FIR filter, given by

$$e(k+1) = e(k) + \sum_{i=1}^{N} \frac{\partial e(k)}{\partial w_i(k)} \Delta w_i(k) + \frac{1}{2!} \sum_{i=1}^{N} \sum_{j=1}^{N} \frac{\partial^2 e(k)}{\partial w_i(k) \partial w_j(k)} \Delta w_i(k) \Delta w_j(k) + \cdots . \tag{9.1}$$

From the mathematical description of LMS[1] from Chapter 2, we have

$$\frac{\partial e(k)}{\partial w_i(k)} = -x(k-i+1), \qquad i = 1, 2, \dots, N, \tag{9.2}$$

and

$$\Delta w_i(k) = \mu(k)e(k)x(k-i+1), \qquad i = 1, 2, \dots, N. \tag{9.3}$$

Due to the linearity of the FIR filter, the second- and higher-order partial derivatives in (9.1) vanish.

Combining (9.1)–(9.3) yields

$$e(k+1) = e(k) - \mu(k)e(k)\|\boldsymbol{x}(k)\|_2^2 \tag{9.4}$$

for which the nontrivial solution gives the learning rate of a normalised LMS algorithm

$$\mu_{\text{NLMS}}(k) = \frac{1}{\|\boldsymbol{x}(k)\|_2^2}. \tag{9.5}$$

The stability analysis of adaptive algorithms can be undertaken using contractive operators and fixed point iteration. For the contractive operator $\boldsymbol{T}$, it follows that

$$\|\boldsymbol{T}\boldsymbol{z}_1 - \boldsymbol{T}\boldsymbol{z}_2\| \leqslant \gamma\|\boldsymbol{z}_1 - \boldsymbol{z}_2\|, \qquad 0 \leqslant \gamma < 1, \quad \boldsymbol{z}_1, \boldsymbol{z}_2 \in \mathbb{R}^N. \tag{9.6}$$

The convergence analysis of LMS, for instance, can be undertaken starting from the misalignment[2] vector $\boldsymbol{v}(k) = \boldsymbol{w}(k) - \tilde{\boldsymbol{w}}(k)$ by setting $\boldsymbol{z}_1 = \boldsymbol{v}(k+1)$, $\boldsymbol{z}_2 = \boldsymbol{v}(0)$ and $\boldsymbol{T} = [\boldsymbol{I} - \mu(k)\boldsymbol{x}(k)\boldsymbol{x}^{\mathrm{T}}(k)]$ (Gholkar 1990). Detailed convergence analysis for a class of gradient-based learning algorithms for recurrent neural networks is given in Chapter 10.

9.3 Overview

A class of normalised gradient-based algorithms is derived starting from the LMS algorithm for linear adaptive filters through to a normalised algorithm for training recurrent neural networks. For each case the adaptive learning rate has been derived. Stability of such algorithms is addressed in Chapter 10. The normalised algorithms are shown to outperform standard algorithms with fixed learning rate.

---

[1] The two core equations for adaptation of the LMS algorithm are

$$e(k) = d(k) - \boldsymbol{x}^{\mathrm{T}}(k)\boldsymbol{w}(k),$$
$$\boldsymbol{w}(k+1) = \boldsymbol{w}(k) + \mu(k)e(k)\boldsymbol{x}(k).$$

[2] The misalignment vector is defined as $\boldsymbol{v}(k) = \boldsymbol{w}(k) - \tilde{\boldsymbol{w}}(k)$, where $\tilde{\boldsymbol{w}}(k)$ is the set of optimal weights of the system.

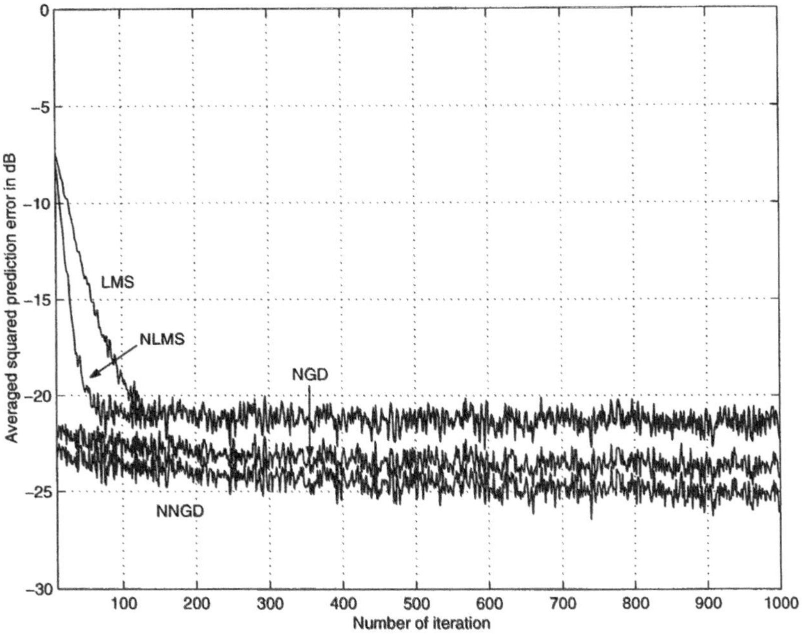

Figure 9.1 Comparison of convergence of the averaged squared prediction error with the LMS, NLMS, NGD and NNGD algorithms, with logistic activation function, for a coloured input

9.4 Derivation of the Normalised Adaptive Learning Rate for a Simple Feedforward Nonlinear Filter

The equations that define the adaptation for a neural adaptive filter with one neuron (Figure 2.6), trained by a nonlinear gradient descent (NGD) algorithm, are

$$e(k) = d(k) - \Phi(\boldsymbol{x}^{\mathrm{T}}(k)\boldsymbol{w}(k)), \tag{9.7}$$

$$\boldsymbol{w}(k+1) = \boldsymbol{w}(k) + \eta(k)\Phi'(\boldsymbol{x}^{\mathrm{T}}(k)\boldsymbol{w}(k))e(k)\boldsymbol{x}(k), \tag{9.8}$$

where $e(k)$ is the instantaneous error at the output neuron, $d(k)$ is some training (desired) signal, $\boldsymbol{x}(k) = [x_1(k), \ldots, x_N(k)]^{\mathrm{T}}$ is the input vector, $\boldsymbol{w}(k) = [w_1(k), \ldots, w_N(k)]^{\mathrm{T}}$ is the weight vector, $\Phi(\,\cdot\,)$ is a nonlinear activation function of a neuron and $(\,\cdot\,)^{\mathrm{T}}$ denotes the vector transpose. The learning rate η is supposed to be a small positive real number. Following the approach from Mandic (2000a), if the output error (9.7) is expanded using a Taylor series expansion, we have

$$e(k+1) = e(k) + \sum_{i=1}^{N} \frac{\partial e(k)}{\partial w_i(k)} \Delta w_i(k) + \frac{1}{2!} \sum_{i=1}^{N} \sum_{j=1}^{N} \frac{\partial^2 e(k)}{\partial w_i(k) \partial w_j(k)} \Delta w_i(k) \Delta w_j(k) + \cdots. \tag{9.9}$$

From (9.7) and (9.8), the elements of (9.9) are

$$\frac{\partial e(k)}{\partial w_i(k)} = -\Phi'(\boldsymbol{x}^{\mathrm{T}}(k)\boldsymbol{w}(k))x_i(k), \qquad i = 1, 2, \ldots, N, \tag{9.10}$$

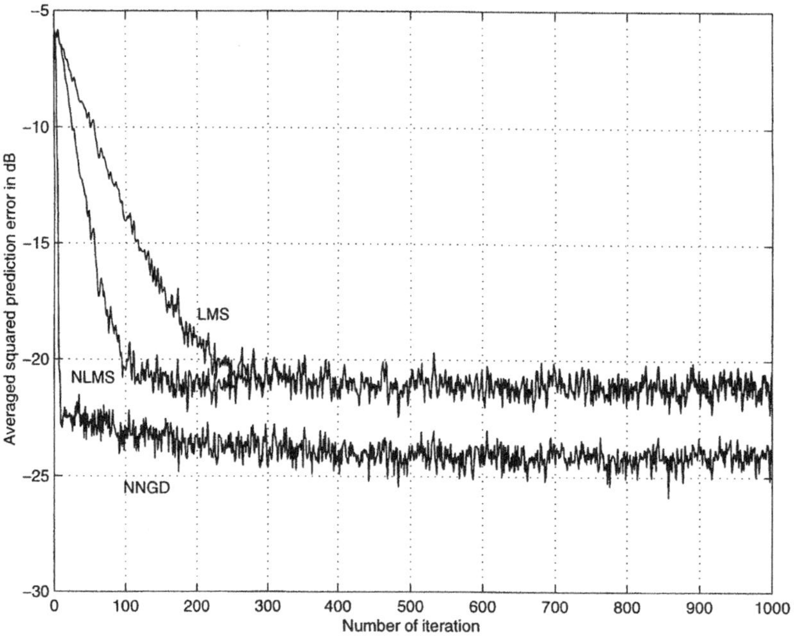

Figure 9.2 Comparison of convergence of the averaged squared prediction error of the LMS, NLMS and NNGD algorithms for a coloured input and tanh activation function with $\beta = 1$

and

$$\Delta w_i(k) = w_i(k+1) - w_i(k) = \eta(k)\Phi'(\boldsymbol{x}^{\mathrm{T}}(k)\boldsymbol{w}(k))e(k)x_i(k), \qquad i = 1, 2, \dots, N. \tag{9.11}$$

The second partial derivatives are

$$\frac{\partial^2 e(k)}{\partial w_i(k)\partial w_j(k)} = -\Phi''(\boldsymbol{x}^{\mathrm{T}}(k)\boldsymbol{w}(k))x_i(k)x_j(k), \qquad i, j = 1, 2, \dots, N. \tag{9.12}$$

Let us denote $\mathrm{net}(k) = \boldsymbol{x}^{\mathrm{T}}(k)\boldsymbol{w}(k)$. Combining (9.9)–(9.12) yields

$$\begin{aligned} e(k+1) = {} & e(k) - \eta(k)[\Phi'(\mathrm{net}(k))]^2 e(k) \sum_{i=1}^{N} x_i^2(k) \\ & - \frac{1}{2!}\eta^2(k)e^2(k)[\Phi'(\mathrm{net}(k))]^2 \Phi''(\mathrm{net}(k)) \sum_{i=1}^{N}\sum_{j=1}^{N} x_i^2(k)x_j^2(k) + \cdots . \end{aligned} \tag{9.13}$$

A truncated Taylor series expansion of (9.13) gives

$$e(k+1) = e(k)[1 - \eta(k)[\Phi'(\mathrm{net}(k))]^2 \|\boldsymbol{x}(k)\|_2^2]. \tag{9.14}$$

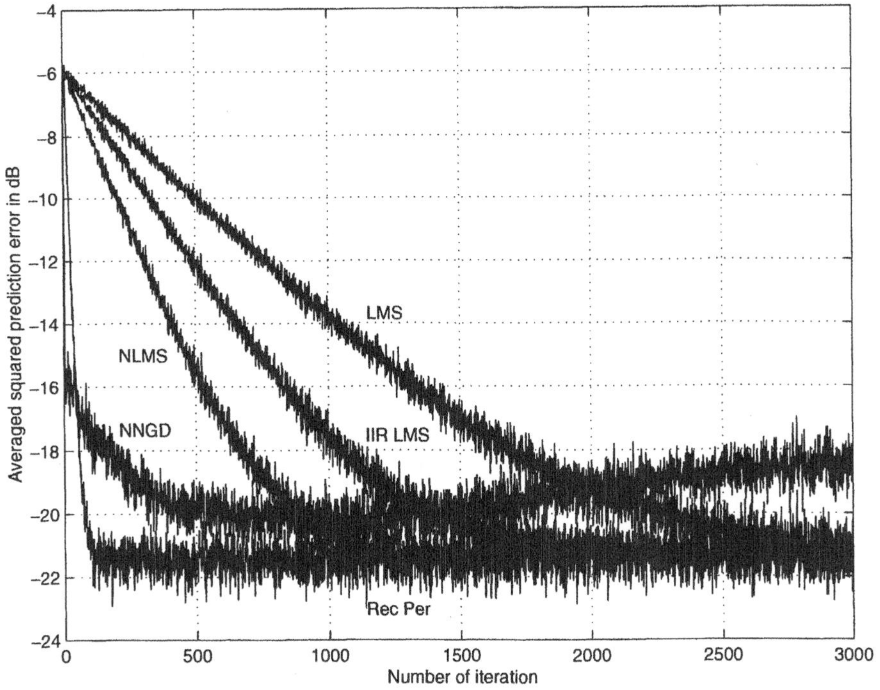

Figure 9.3 Convergence comparison of averaged squared prediction error for feedforward and recurrent structures, tanh activation function with $\beta = 4$ and coloured input

The aim is for the error $e(k+1)$ in (9.14) to vanish, which is the case for the nontrivial solution

$$\eta_{\mathrm{OPT}}(k) = \frac{1}{[\Phi'(\mathrm{net}(k))]^2 \|\boldsymbol{x}(k)\|_2^2}, \tag{9.15}$$

which is the step size of a normalised gradient descent (NNGD) algorithm for a nonlinear FIR filter. Taking into account the bounds[3] on the values of higher derivatives of Φ, for a contractive activation function we may adjust the derived learning rate with a positive constant C, as

$$\eta_{\mathrm{OPT}}(k) = \frac{1}{C + [\Phi'(\mathrm{net}(k))]^2 \|\boldsymbol{x}(k)\|_2^2}. \tag{9.16}$$

The magnitude of the learning rate varies in time with the tap input power and the first derivative of the activation function, which provides a normalisation of the algorithm. Further discussion on the size and role of constant C in (9.16) can be found in Mandic and Krcmar (2001) and Krcmar and Mandic (2001). The adaptive learning rate from (9.15) degenerates into the learning rate of the NLMS algorithm for a linear activation function. A normalised backpropagation algorithm for a general feedforward neural network is given in Mandic and Chambers (2000f). Although the

[3] For the logistic function, for instance, the second-order term in the Taylor series expansion is positive.

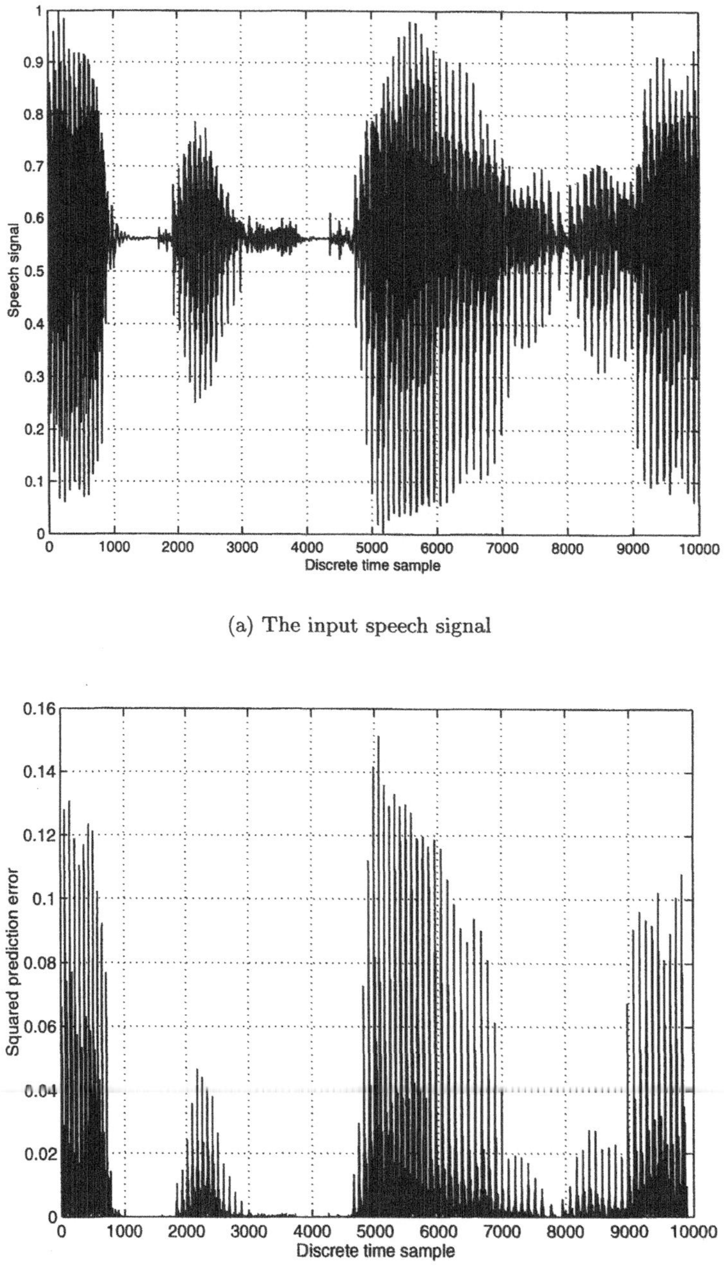

(a) The input speech signal

(b) Standard RTRL algorithm

Figure 9.4 Squared instantaneous prediction error for the RTRL and NRTRL algorithms with speech inputs

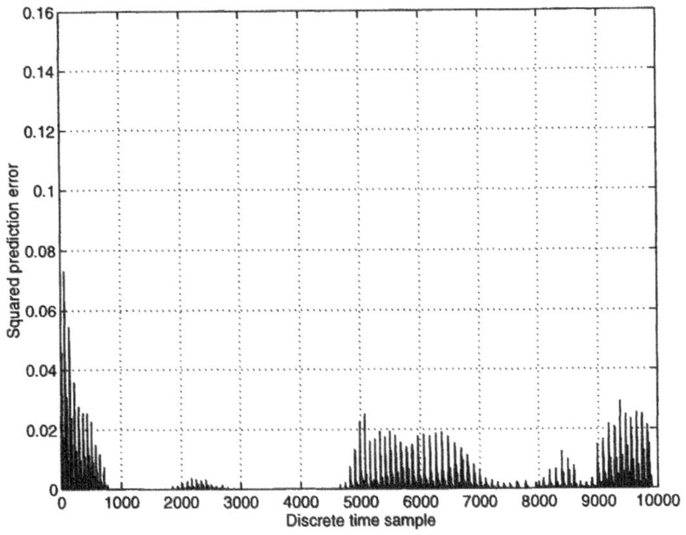

(c) Normalised RTRL algorithm

Figure 9.4 *Cont.*

derivation of the normalised algorithm is simple, it assumes statistical independence between the weights, input vector, teaching signal and learning rate, which is often not the case in practical applications. Therefore, the optimal learning rate for practical applications should be chosen to be smaller than the one derived above. This is one of the reasons why there is a need to add a positive constant C to the denominator of (9.15).

In Mandic (2000a), a simulation was undertaken on speech, a nonlinear and nonstationary signal, for a nonlinear FIR filter with tap length $N = 10$, with $\eta = 0.3$, $C = 1$ and $\beta = 4$. The quantitative performance measure was the standard prediction gain, a logarithmic ratio between the expected signal and error variances $R_p = 10\log(\hat{\sigma}_s^2/\hat{\sigma}_e^2)$. For this setting, the prediction gain for the LMS was 7.24 dB, 8.26 dB for the NLMS, 7.67 dB for a nonlinear GD and 9.28 dB for the NNGD algorithm, confirming the analysis from the previous section.

We next compare the performances of FIR filters trained by LMS and NLMS, IIR filters trained by LMS, nonlinear FIR filters trained by NGD and NNGD and a NARMA recurrent perceptron trained by the RTRL. The order of FIR filters was $N = 10$. The input was a white noise sequence passed through an AR channel given by

$$y(k) = 1.79y(k-1) - 1.85y(k-2) + 1.27y(k-3) - 0.41y(k-4) + \nu(k), \qquad (9.17)$$

where $\nu(k)$ denotes the white input noise. The resulting input signal was rescaled so as to fit within the range of the logistic and tanh activation function. A Monte Carlo simulation with 200 trials was undertaken for all the experiments.

Figure 9.1 shows a comparison between convergence curves for the LMS, NLMS,[4] NGD (a standard nonlinear gradient descent) and NNGD algorithms for a coloured input from AR channel (9.17). The slope of the logistic function was $\beta = 4$, which partly coincides with the linear curve $y = x$. The NNGD algorithm for a feedforward dynamical neuron clearly outperforms the other employed algorithms. The NGD algorithm also outperformed the LMS and NLMS algorithms. Figure 9.2 shows the convergence curves for a tanh activation function and the input from the same AR channel. The NNGD algorithm has consistently improved convergence performance over the LMS and NLMS algorithms.

Convergence curves for LMS, NLMS, NNGD, IIR LMS and a NARMA(6,1) recurrent perceptron for a correlated input (AR channel) and tanh activation function with $\beta = 4$ are shown in Figure 9.3. A NARMA recurrent perceptron outperformed all the other algorithms in simulations. This does not mean, however, that recurrent structures perform best in all practical applications.

9.5 A Normalised Algorithm for Online Adaptation of Recurrent Neural Networks

An output error of a fully connected recurrent neural network can be expanded via a Taylor series expansion as (Mandic and Chambers 2000b)

$$\begin{aligned} e(k+1) = e(k) &+ \sum_{i=1}^{N} \sum_{j=1}^{M+N+1} \frac{\partial e(k)}{\partial w_{i,j}(k)} \Delta w_{i,j}(k) \\ &+ \frac{1}{2!} \sum_{i=1}^{N} \sum_{m=1}^{M+N+1} \sum_{j=1}^{N} \sum_{n=1}^{M+N+1} \frac{\partial^2 e(k)}{\partial w_{i,m}(k) \partial w_{j,n}(k)} \Delta w_{i,m}(k) \Delta w_{j,n}(k) + \cdots, \end{aligned} \tag{9.18}$$

where M is the order of the input signal tap delay line and N is the number of neurons. This is a complicated expression and only the first two terms of (9.18) will be considered. Due to the internal feedback in RNNs, the partial derivatives $\partial e(k)/\partial w_{i,j}(k)$ are not straightforward to calculate (Appendix D). From (9.18), using an approach similar to the one explained for a simple feedforward neural filter and neglecting the higher-order terms in the Taylor series expansion gives

$$\begin{aligned} e(k+1) &= e(k) - \eta(k) e(k) \sum_{i=1}^{N} \sum_{j=1}^{M+N+1} \left[\frac{\partial y_1(k)}{\partial w_{i,j}(k)} \right]^2 \\ &= e(k) - \eta(k) e(k) \sum_{i=1}^{N} \| \boldsymbol{\Pi}_1^{(i)}(k) \|_2^2, \end{aligned} \tag{9.19}$$

[4] For numerical stability, the learning rate for NLMS was chosen as $\mu(k) = \mu_0/(\epsilon + \|\boldsymbol{x}\|_2^2)$, where $\mu_0 < 1$ is a positive constant and ϵ is some small positive constant that prevents divergence for small $\|\boldsymbol{x}\|_2$. This explains the better performance of NNGD over NLMS for an input coming from a linear AR channel.

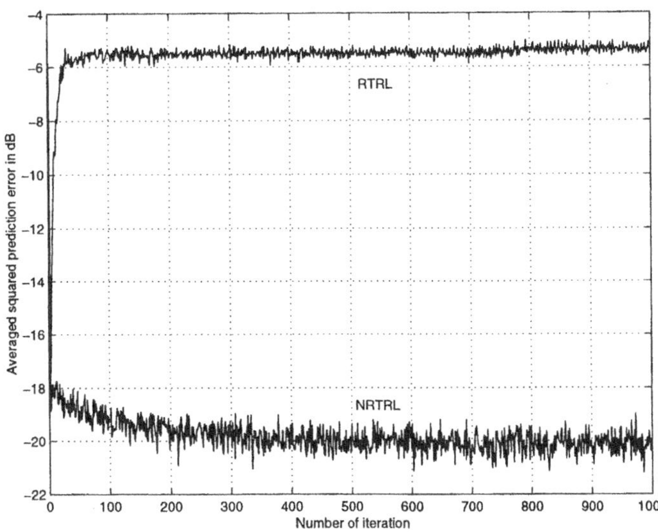

(a) Convergence comparison between RTRL and NRTRL

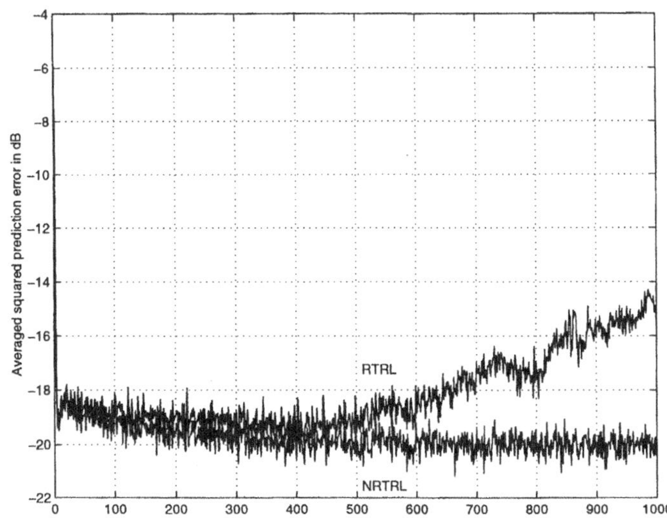

(b) Convergence comparison between RTRL and NRTRL when RTRL fails

Figure 9.5 Convergence comparison of averaged squared prediction error for a RTRL and NRTRL trained recurrent structure, tanh activation function with $\beta = 2$ and coloured input

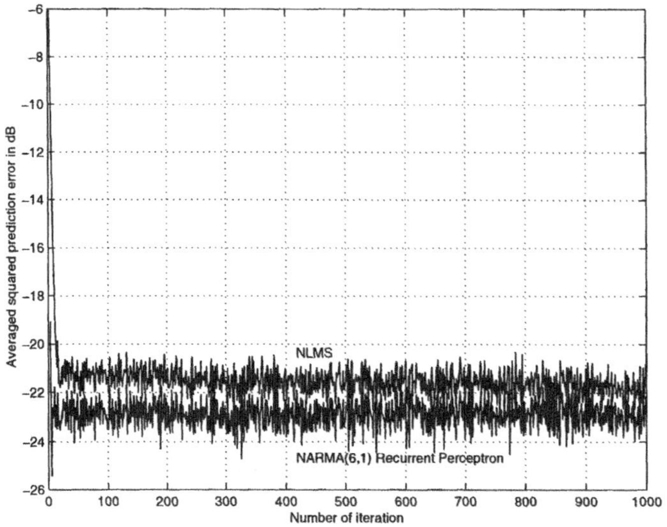

(a) Convergence curves for NLMS for $N = 10$ and RTRL for a NARMA(4,1) recurrent perceptron for a nonlinear input (9.22), *logistic* activation function with $\beta = 4$

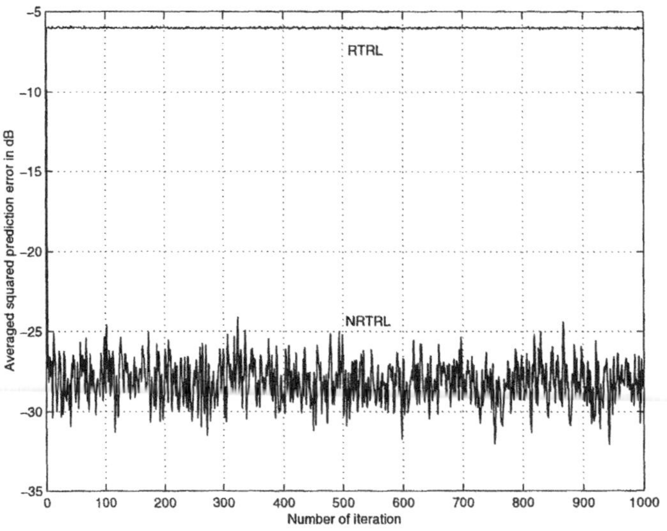

(b) Convergence curves for RTRL and NRTRL, for a NARMA(10,2) recurrent perceptron, tanh activation function with $\beta = 8$ for a nonlinear input (9.23)

Figure 9.6 Convergence of RTRL and NRTRL for nonlinear inputs where $\boldsymbol{\Pi}_1^{(i)}$ denotes the gradients at the output neuron y_1 with respect to the weights from the ith neuron. Hence, the optimal value of learning rate $\eta_{\mathrm{OPT}}(k)$ for an RTRL trained RNN is

$$\eta_{\mathrm{OPT}}(k) = \frac{1}{\sum_{i=1}^{N} \|\boldsymbol{\Pi}_1^{(i)}(k)\|_2^2}. \tag{9.20}$$

The normalisation factor is the tap input power to an RNN multiplied by the derivative of the nonlinear activation function and augmented by the product of gradients and feedback weights. Hence, we will refer to the result from (9.20) as the normalised real-time recurrent learning (NRTRL) algorithm. For a normalised algorithm for a recurrent perceptron, we have

$$\eta_{\mathrm{OPT}}(k) = \frac{1}{\|\boldsymbol{\Pi}(k)\|_2^2}. \tag{9.21}$$

Due to the derivation of η_{OPT} from a truncated Taylor series expansion, a positive constant C should be added to the term in the denominator of (9.20) and (9.21).

Figure 9.4 shows the comparison of instantaneous squared prediction errors between the RTRL and NRTRL for a nonstationary (speech) signal. The NRTRL algorithm from Figure 9.4(c), clearly achieves significantly better performance than the RTRL algorithm (Figure 9.4(b)). To quantify this, if the measure of performance is the standard prediction gain, the NRTRL achieved approximately 7 dB better performance than the RTRL algorithm. Convergence comparison between the RTRL and NRTRL algorithms for the cases where both algorithms converge (Figure 9.5(a)) and when RTRL diverges (Figure 9.5(b)) is shown in Figure 9.5. A small constant was added to the denominator of the optimal learning rate η_{OPT}. The input was a coloured signal from an AR channel and the slope of the tanh activation function was $\beta = 2$ (notice that the contractivity might have been violated). In both cases depicted in Figure 9.5, the NRTRL comprehensively outperformed the RTRL algorithm. In Figure 9.6, a comparison between convergence curves for benchmark nonlinear inputs defined as (Narendra and Parthasarathy 1990)

$$y(k+1) = \frac{y(k)y(k-1)y(k-2)x(k-1)[y(k-2)-1] + x(k)}{1 + y^2(k-1) + y^2(k-2)}, \tag{9.22}$$

$$y(k+1) = \frac{y(k)}{1 + y^2(k)} + x^3(k), \tag{9.23}$$

is given. In Figure 9.6(a), a NARMA(4,1) recurrent perceptron trained by RTRL outperformed a FIR filter with $N = 10$ trained by NLMS for input (9.22).

In Figure 9.6(b), comparison between convergence curves for RTRL and NRTRL on a benchmark nonlinear input (9.23) is given. The employed tanh activation function was expansive with $\beta = 8$ and the simulations were undertaken for a NARMA(10,2) recurrent perceptron. The NRTRL outperformed RTRL for this case.

Simulations show that the performance of the NRTRL is highly dependent on the choice of the constant C in the denominator of the optimal learning rate. Dependent on the choice of C, the NRTRL can have worse, similar or better performance than RTRL. However, in most practical cases, $C < 1$ is a sufficiently good range for the NRTRL to outperform the RTRL. To further depict the dependence of performance on the

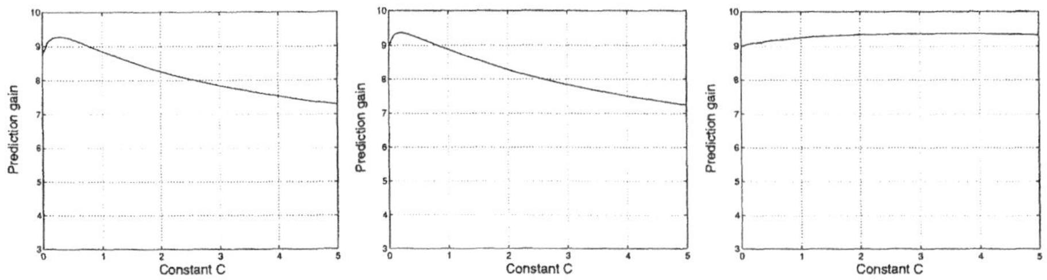

(a) NARMA(4,2), tanh function, $\beta = 1$

(b) NARMA(6,1), tanh function, $\beta = 1$

(c) NARMA(6,1), tanh function, $\beta = 4$

Figure 9.7 Prediction gain versus the Taylor series remainder C for a speech signal and NARMA recurrent perceptrons

value of C, three experiments were undertaken on a real speech signal. The prediction gain was calculated for various values of parameter C. The filter used was a NARMA recurrent perceptron. In Figure 9.7(a), prediction gain for a NARMA(4,2) perceptron with a tanh activation function with $\beta = 1$ had its maximum for $C = 0.3$. The experiment was repeated for a NARMA(6,1) recurrent perceptron, and the maximum of the prediction gain was obtained for $C = 0.22$, which is shown in Figure 9.7(b). Finally, for the same network, an expansive tanh activation function was used, with $\beta = 4$. As expected, in this case, the best performance was achieved for $C > 1$, which is shown in Figure 9.7(c).

9.6 Summary

An optimal adaptive learning rate has been derived for the RTRL algorithm for continually running fully connected recurrent neural networks. The learning rate is optimal in the sense that it minimises the instantaneous squared prediction error at the output neuron for every time instant while the network is running. This algorithm normalises the learning rate of the RTRL and is hence referred to as the normalised RTRL (NRTRL) algorithm. The NRTRL is stabilised by the $\mathcal{L}_2$ norm of the input data vector and local gradients at the output neuron of the network. The additional computational complexity involved is not significant, when compared to the entire computational complexity of the RTRL algorithm. Simulations show that normalised algorithms outperform the standard algorithms in both the feedforward and recurrent case.

10

Convergence of Online Learning Algorithms in Neural Networks

10.1 Perspective

An analysis of convergence of real-time algorithms for online learning in recurrent neural networks is presented. For convenience, the analysis is focused on the real-time recurrent learning (RTRL) algorithm for a recurrent perceptron. Using the assumption of contractivity of the activation function of a neuron and relaxing the rigid assumptions of the fixed optimal weights of the system, the analysis presented is general and is applicable to a wide range of existing algorithms. It is shown that some of the results obtained for stochastic gradient algorithms for linear systems can be considered as a bound for stability of RNN-based algorithms, as long as the contractivity condition holds.

10.2 Introduction

The following criteria (Bershad *et al.* 1990) are most commonly used to assess the performance of adaptive algorithms.

1. Convergence (consistency of the statistics).
2. Transient behaviour (how quickly the algorithm reacts to changes in the statistics of the input).
3. Convergence rate (how quickly the algorithm approaches the optimal solution), which can be linear, quadratic or superlinear.

The standard approach for the analysis of convergence of learning algorithms for linear adaptive filters is to look at convergence of the mean weight error vector, convergence in the mean square and at the steady-state misadjustment (Gholkar 1990; Haykin 1996a; Kuan and Hornik 1991; Widrow and Stearns 1985). The analysis of convergence of steepest-descent-based algorithms has been ongoing ever since their introduction (Guo and Ljung 1995; Ljung 1984; Slock 1993; Tarrab and Feuer 1988). Some of the recent results consider the exact expectation analysis of the LMS algorithm for linear adaptive filters (Douglas and Pan 1995) and the analysis of LMS with Gaussian inputs (Bershad 1986). For neural networks as nonlinear adaptive filters, the analysis is far more difficult, and researchers have often resorted to numerical experiments (Ahmad *et al.* 1990). Convergence of neural networks has been considered in Shynk and Roy (1990), Bershad *et al.* (1993a) and Bershad *et al.* (1993b), where the authors used the Gaussian model for input data and a Rosenblatt perceptron learning algorithm. These analyses, however, were undertaken for a hard limiter nonlinearity, which is not convenient for nonlinear adaptive filters. Convergence of RTRL was addressed in Mandic and Chambers (2000b) and Chambers *et al.* (2000).

An error equation for online training of a recurrent perceptron can be expressed as

$$e(k) = s(k) - \Phi(\boldsymbol{u}^{\mathrm{T}}(k)\boldsymbol{w}(k)), \tag{10.1}$$

where $s(k)$ is the teaching (desired) signal, $\boldsymbol{w}(k) = [w_1(k), \dots, w_N(k)]^{\mathrm{T}}$ is the weight vector and $\boldsymbol{u}(k) = [u_1(k), \dots, u_N(k)]^{\mathrm{T}}$ is an input vector. A weight update equation for a general class of stochastic gradient-based nonlinear neural algorithms can be expressed as

$$\boldsymbol{w}(k+1) = \boldsymbol{w}(k) + \eta(k)F(\boldsymbol{u}(k))g(\boldsymbol{u}(k), \boldsymbol{w}(k)), \tag{10.2}$$

where $\eta(k)$ is the learning rate, $F : \mathbb{R}^N \to \mathbb{R}^N$ usually consists of N copies of the scalar function f and $g(\,\cdot\,)$ is a scalar function related to the error $e(k)$. The function F is related to data nonlinearities, which have an influence on the convergence of the algorithm. The function g is related to error nonlinearities, and it affects the cost function to be minimised. Error nonlinearities are mostly chosen to be sign-preserving (Sethares 1992).

Let us assume additive noise $q(k) \sim \mathcal{N}(0, \sigma_q^2)$ in the output of the system, which can be expressed as

$$s(k) = \Phi(\boldsymbol{u}^{\mathrm{T}}(k)\tilde{\boldsymbol{w}}(k)) + q(k), \tag{10.3}$$

where $\tilde{\boldsymbol{w}}(k)$ are optimal filter weights and $q(k)$ is an i.i.d. sequence. The error equation (10.1) now becomes

$$e(k) = \Phi(\boldsymbol{u}^{\mathrm{T}}(k)\tilde{\boldsymbol{w}}(k)) - \Phi(\boldsymbol{u}^{\mathrm{T}}(k)\boldsymbol{w}(k)) + q(k). \tag{10.4}$$

To examine the stability of algorithm (10.2), researchers often resort to linearisation. For RTRL, F is an identity matrix and g is some nonlinear, sign-preserving function of the output error. A further assumption is that the learning rate η is sufficiently small to allow the algorithm to be linearised around its current point in the state space. From Lyapunov stability theory, the system

$$\boldsymbol{z}(k+1) = F(k, \boldsymbol{z}(k)) \tag{10.5}$$

can be analysed via its linearised version

$$\boldsymbol{z}(k+1) = \boldsymbol{A}(k)\boldsymbol{z}(k), \tag{10.6}$$

where $\boldsymbol{A}$ is the Jacobian of F. This is the Lyapunov indirect method and assumes that $\boldsymbol{A}(k)$ is bounded in the neighbourhood of the current point in the state space and that

$$\lim_{\|z\| \to 0} \max_{k} \frac{\|F(k, z) - A(k)z\|}{\|z\|} = 0, \tag{10.7}$$

which guarantees that time variation in the nonlinear terms of the Taylor series expansion of (10.5) does not become arbitrarily large in time (Chambers *et al.* 2000). Results on Lyapunov stability for a class of nonlinear systems can be found in Wang and Michel (1994) and Tanaka (1996).

Averaging methods for the analysis of stability and convergence of adaptive algorithms, for instance, use a linearised version of the system matrix of (10.2)

$$v(k) = [I - \eta u(k)u^{\mathrm{T}}(k)]\tilde{w}(k), \tag{10.8}$$

which is then replaced by the ensemble average (Anderson *et al.* 1986; Kushner 1984; Solo and Kong 1994)

$$E[I - \eta u(k)u^{\mathrm{T}}(k)] = I - \eta R_{u,u}, \tag{10.9}$$

where $v(k)$ is the misalignment vector which will be defined later and $R_{u,u}$ is the autocorrelation matrix of the tap-input vector $u(k)$.

It is also often assumed that the filter coefficients are statistically independent of the input data currently in the filter memory, which is convenient, but essentially incorrect. This assumption is one of the *independence assumptions*, which are (Haykin 1996a)

1. the sequence of tap input vectors are statistically independent;

2. the tap input vector is statistically independent of all the previous samples of the desired response;

3. the desired response is statistically independent of all the previous samples of the desired response; and

4. the tap input vector and the desired response consist of *mutually Gaussian-distributed* random variables.

The weight error vector hence depends on the previous sample input vectors, the previous samples of the desired response and the initial value of the tap weight vector. Convergence analysis of stochastic gradient algorithms is still ongoing, mainly to relax the independence assumptions (Douglas and Pan 1995; Guo *et al.* 1997; Solo and Kong 1994).

The following are the most frequently used convergence criteria in the analysis of adaptive algorithms:

1. convergence of the weight fluctuation in the mean $\|E[v(k)]\| \to 0$, as $k \to \infty$, where $v(k) = w(k) - \tilde{w}(k)$;

2. mean squared error convergence calculated from $E[v(k)v^{\mathrm{T}}(k)]$; and

3. steady-state mean squared error, which is obtained from mean squared error convergence (misadjustment).

To allow for time-varying input signal statistics, in the following analysis we use a fairly general condition that the optimal filter weights $\tilde{\boldsymbol{w}}(k)$ are governed by the modified first-order Markov model as (Bershad *et al.* 1990),

$$\tilde{\boldsymbol{w}}(k+1) = \lambda\tilde{\boldsymbol{w}}(k) + \sqrt{1-\lambda^2}\boldsymbol{n}(k), \tag{10.10}$$

where $\lambda \in [0,1]$ is the parameter which defines the time variation of $\tilde{\boldsymbol{w}}(k)$ and $\boldsymbol{n}(k)$ is an i.i.d. Gaussian noise vector. A zero-mean initialisation of model (10.10) is assumed ($E[\tilde{\boldsymbol{w}}(k)] = \boldsymbol{0}$). This model covers most of the learning algorithms employed, be they linear or nonlinear. For instance, the momentum algorithm models the weight update as an AR process. In addition, learning algorithms based upon the Kalman filter model weight fluctuations as a white noise sequence (random walk), which is in fact a first-order Markov process (Appendix D). The standard case of a single optimal solution to the stochastic gradient optimisation process (non time-varying) can be obtained by setting $\lambda = 1$.

10.3 Overview

Based upon the stability results introduced in Chapter 7, the analysis of convergence for stochastic gradient algorithms for nonlinear adaptive filters is provided. The analysis is mathematically strict and covers most of the previously introduced algorithms. This approach can be extended to more complicated architectures and learning algorithms.

10.4 Convergence Analysis of Online Gradient Descent Algorithms for Recurrent Neural Adaptive Filters

The problem of optimal nonlinear gradient-descent-based training can be presented in a similar fashion to the linear case (Douglas 1994), as

$$\begin{aligned} &\text{minimise} && \|\boldsymbol{w}(k+1) - \boldsymbol{w}(k)\| && (10.11)\\ &\text{subject to} && s(k) - \Phi(\boldsymbol{u}^{\mathrm{T}}(k)\boldsymbol{w}(k+1)) = 0, && (10.12)\end{aligned}$$

where $\|\cdot\|$ denotes some norm (most commonly the 2-norm). The equation that defines the adaptation of a recurrent neural network is

$$\boldsymbol{w}(k+1) = \boldsymbol{w}(k) - \eta(k)\nabla_{\boldsymbol{w}(k)}E(k), \tag{10.13}$$

where $E(k) = \frac{1}{2}e^2(k)$ is the cost function to be minimised. The correction to the weight vector for a recurrent perceptron at time instant k becomes (Williams and Zipser 1989a)

$$\Delta\boldsymbol{w}(k) = \eta(k)e(k)\boldsymbol{\Pi}(k), \tag{10.14}$$

where

$$\boldsymbol{\Pi}(k) = \left[\frac{\partial y(k)}{\partial w_1(k)}, \dots, \frac{\partial y(k)}{\partial w_N(k)}\right]^{\mathrm{T}}$$

represents the gradient vector at the output of the neuron. Consider the weight update equation for a general RTRL trained RNN

$$\boldsymbol{w}(k+1) = \boldsymbol{w}(k) + \eta(k)e(k)\boldsymbol{\Pi}(k). \tag{10.15}$$

Following the approach from Chambers *et al.* (2000) and using (10.4) and (10.15), we have

$$\boldsymbol{w}(k+1) = \boldsymbol{w}(k)+\eta(k)q(k)\boldsymbol{\Pi}(k)+\eta(k)\Phi(\boldsymbol{u}^{\mathrm{T}}(k)\tilde{\boldsymbol{w}}(k))\boldsymbol{\Pi}(k)-\eta(k)\Phi(\boldsymbol{u}^{\mathrm{T}}(k)\boldsymbol{w}(k))\boldsymbol{\Pi}(k). \tag{10.16}$$

The misalignment vector $\boldsymbol{v}$ can be expressed as

$$\boldsymbol{v}(k) = \boldsymbol{w}(k) - \tilde{\boldsymbol{w}}(k). \tag{10.17}$$

Let us now subtract $\tilde{\boldsymbol{w}}(k+1)$ from both sides of (10.16), which yields

$$\begin{aligned}\boldsymbol{v}(k+1) = \boldsymbol{w}(k) - \tilde{\boldsymbol{w}}(k+1) + \eta(k)q(k)\boldsymbol{\Pi}(k)& \\ - \eta(k)[\Phi(\boldsymbol{u}^{\mathrm{T}}(k)\boldsymbol{w}(k)) - \Phi(\boldsymbol{u}^{\mathrm{T}}(k)\tilde{\boldsymbol{w}}(k))]\boldsymbol{\Pi}(k).&\end{aligned}$$

Using (10.10), we have

$$\begin{aligned}\boldsymbol{v}(k+1) = \boldsymbol{w}(k) - \tilde{\boldsymbol{w}}(k) + \tilde{\boldsymbol{w}}(k) - \lambda\tilde{\boldsymbol{w}}(k) - \sqrt{1-\lambda^2}\boldsymbol{n}(k) + \eta(k)q(k)\boldsymbol{\Pi}(k)& \\ - \eta(k)[\Phi(\boldsymbol{u}^{\mathrm{T}}(k)\boldsymbol{w}(k)) - \Phi(\boldsymbol{u}^{\mathrm{T}}(k)\tilde{\boldsymbol{w}}(k))]\boldsymbol{\Pi}(k).&\end{aligned} \tag{10.18}$$

It then follows that $\boldsymbol{v}(k+1)$ becomes

$$\begin{aligned}\boldsymbol{v}(k+1) = \boldsymbol{v}(k) + \eta(k)q(k)\boldsymbol{\Pi}(k) - \eta(k)[\Phi(\boldsymbol{u}^{\mathrm{T}}(k)\boldsymbol{w}(k)) - \Phi(\boldsymbol{u}^{\mathrm{T}}(k)\tilde{\boldsymbol{w}}(k))]\boldsymbol{\Pi}(k)& \\ + (1-\lambda)\tilde{\boldsymbol{w}}(k) - \sqrt{1-\lambda^2}\boldsymbol{n}(k).&\end{aligned} \tag{10.19}$$

For $\Phi(k)$ a sign-preserving[1] contraction mapping (as in the case of the logistic function), the term in the square brackets from (10.19) is bounded from above by $\Theta|\boldsymbol{u}^{\mathrm{T}}(k)\boldsymbol{v}(k)|$, $0 < \Theta < 1$ (Mandic and Chambers 2000e). Further analysis towards the weight convergence becomes rather involved because of the nature of $\boldsymbol{\Pi}(k)$. Let us denote $\boldsymbol{u}^{\mathrm{T}}(k)\boldsymbol{w}(k) = \mathrm{net}(k)$. Since the gradient vector $\boldsymbol{\Pi}$ is a vector of partial derivatives of the output $y(k)$,

$$\boldsymbol{\Pi}(k) = \frac{\partial y(k)}{\partial \boldsymbol{w}(k)} = \Phi'(\mathrm{net}(k))[\boldsymbol{u}(k) + \boldsymbol{w}_a(k)\boldsymbol{\Pi}_a(k)], \tag{10.20}$$

where the subscript 'a' denotes the elements which are due to the feedback of the system, we restrict ourselves to an approximation,

$$\boldsymbol{\Pi}(k) \longrightarrow \Phi'(\mathrm{net}(k))\boldsymbol{u}(k).$$

[1] For the sake of simplicity, we assume Φ sign preserving, i.e. for positive a, b, $b > a$, $\Phi(b) - \Phi(a) < b - a$. For other contractive activation functions, $|\Phi(a) - \Phi(b)| < |a - b|$, and norms of the corresponding expressions from the further analysis should be taken into account. The activation functions most commonly used in neural networks are sigmoidal, monotonically increasing, contractive, with a positive first derivative, so that this assumption holds.

This should not affect the generality of the result, since it is possible to return to the $\boldsymbol{\Pi}$ terms after the convergence results are obtained. In some cases, due to the problem of vanishing gradient, this approximation is quite satisfactory (Bengio *et al.* 1994). In fact, after approximating $\boldsymbol{\Pi}$, the structure degenerates into a single-layer, single neuron feedforward neural network (Mandic and Chambers 2000f). For Φ a monotonic ascending contractive activation function, $\exists\alpha(k) \in (0, \Theta]$, such that the term $[\Phi(\boldsymbol{u}^{\mathrm{T}}(k)\boldsymbol{w}(k)) - \Phi(\boldsymbol{u}^{\mathrm{T}}(k)\tilde{\boldsymbol{w}}(k))]$ from (10.19) can be replaced[2] by $\alpha(k)\boldsymbol{u}^{\mathrm{T}}(k)\boldsymbol{v}(k)$. Now, analysing (10.19) with the newly introduced parameter $\alpha(k)$, we have

$$\boldsymbol{v}(k+1) = \boldsymbol{v}(k) + \eta(k)q(k)\Phi'(\mathrm{net}(k))\boldsymbol{u}(k) - \alpha(k)\eta(k)\boldsymbol{u}^{\mathrm{T}}(k)\boldsymbol{v}(k)\Phi'(\mathrm{net}(k))\boldsymbol{u}(k) + (1-\lambda)\tilde{\boldsymbol{w}}(k) - \sqrt{1-\lambda^2}\boldsymbol{n}(k). \quad (10.21)$$

For a contractive activation function $0 < \Phi'(\mathrm{net}(k)) < 1$ (Mandic and Chambers 1999b) and can be replaced[3] by $\gamma(k)$. Equation (10.21) now becomes

$$\boldsymbol{v}(k+1) = \boldsymbol{v}(k) + \gamma(k)\eta(k)q(k)\boldsymbol{u}(k) - \alpha(k)\gamma(k)\eta(k)\boldsymbol{u}(k)\boldsymbol{u}^{\mathrm{T}}(k)\boldsymbol{v}(k) + (1-\lambda)\tilde{\boldsymbol{w}}(k) - \sqrt{1-\lambda^2}\boldsymbol{n}(k). \quad (10.22)$$

After including the zero-mean assumption for the driving noise, $\boldsymbol{n}(k)$ and the mutual statistical independence assumption between $\eta(k)$, $\boldsymbol{u}(k)$, $\boldsymbol{n}(k)$, $\tilde{\boldsymbol{w}}(k)$, $\alpha(k)$, $\gamma(k)$ and $\boldsymbol{v}(k)$, we have

$$E[\boldsymbol{v}(k+1)] = E[\boldsymbol{I} - \alpha\gamma\eta(k)\boldsymbol{u}(k)\boldsymbol{u}^{\mathrm{T}}(k)]E[\boldsymbol{v}(k)], \quad (10.23)$$

where $\gamma = E[\gamma(k)]$ and $\alpha = E[\alpha(k)]$, which are also in the range $(0, 1)$. For convergence,

$$0 < \|E[\boldsymbol{I} - \alpha\gamma\eta(k)\boldsymbol{u}(k)\boldsymbol{u}^{\mathrm{T}}(k)]\| < 1$$

as both α and γ are positive scalars for monotonic ascending contractive activation functions. For stability of the algorithm, the limits on $\eta(k)$ are thus[4]

$$0 < \eta(k) < E\left[\frac{2}{\alpha\gamma\boldsymbol{u}^{\mathrm{T}}(k)\boldsymbol{u}(k)}\right]. \quad (10.24)$$

Equation (10.24) tells us that the stability limit for the NLMS algorithm is the bound for the simplified recurrent perceptron algorithm. By continuity, the NLMS algorithm for IIR adaptive filters is the bound for the stability analysis of a single-neuron RTRL algorithm. The mean square and steady-state convergence analysis follow the same form and are presented below.

---

[2] In fact, by the CMT, $\exists\xi \in (\boldsymbol{u}^{\mathrm{T}}(k)\boldsymbol{w}(k), \boldsymbol{u}^{\mathrm{T}}(k)\tilde{\boldsymbol{w}}(k))$ such that

$$|\Phi(\boldsymbol{u}^{\mathrm{T}}(k)\boldsymbol{w}(k)) - \Phi(\boldsymbol{u}^{\mathrm{T}}(k)\tilde{\boldsymbol{w}}(k))| = |\Phi'(\xi)||\boldsymbol{u}^{\mathrm{T}}(k)\boldsymbol{w}(k) - \boldsymbol{u}^{\mathrm{T}}(k)\tilde{\boldsymbol{w}}(k)| = |\Phi'(\xi)||\boldsymbol{u}^{\mathrm{T}}(k)\boldsymbol{v}(k)|.$$

Hence, for a sigmoidal monotonic ascending, contractive Φ (logistic, tanh), the first derivative is strictly positive and $\alpha(k) = \Phi'(\xi)$. Assume positive a, b, $b > a$, then $\Phi(b) - \Phi(a) = \alpha(k)(b - a)$.

[3] From (10.20), there is a finite $\gamma(k)$ such that $\|\boldsymbol{\Pi}(k)\| = \gamma(k)\|\boldsymbol{u}(k)\|$. For simplicity, we approximate $\boldsymbol{\Pi}(k)$ as above and use $\gamma(k)$ as defined by the CMT. The derived results, however, are valid for any finite $\gamma(k)$, i.e. are directly applicable for both the recurrent and feedforward architectures.

[4] Using the independence assumption, $E[\boldsymbol{u}(k)\boldsymbol{u}^{\mathrm{T}}(k)]$ is a diagonal matrix and its norm can be replaced by $E[\boldsymbol{u}^{\mathrm{T}}(k)\boldsymbol{u}(k)]$.

10.5 Mean-Squared and Steady-State Mean-Squared Error Convergence

To investigate the mean squared convergence properties of stochastic gradient descent algorithms for recurrent neural networks, we need to analyse $\boldsymbol{R}_{\boldsymbol{v},\boldsymbol{v}}(k+1)$ which is defined as $\boldsymbol{R}_{\boldsymbol{v},\boldsymbol{v}}(k+1) = E[\boldsymbol{v}(k+1)\boldsymbol{v}^{\mathrm{T}}(k+1)]$. From (10.22), cross-multiplying and applying the expectation operator to both sides and using the definition of $\boldsymbol{R}_{\boldsymbol{v},\boldsymbol{v}}(k+1)$, α and γ and the previous assumptions, we obtain[5]

$$\begin{aligned}\boldsymbol{R}_{\boldsymbol{v},\boldsymbol{v}}(k+1) = \boldsymbol{R}_{\boldsymbol{v},\boldsymbol{v}}(k) &- \alpha\gamma E[\eta(k)\boldsymbol{u}(k)\boldsymbol{u}^{\mathrm{T}}(k)]\boldsymbol{R}_{\boldsymbol{v},\boldsymbol{v}}(k) - \boldsymbol{R}_{\boldsymbol{v},\boldsymbol{v}}(k)E[\boldsymbol{u}(k)\boldsymbol{u}^{\mathrm{T}}(k)\eta(k)]\gamma\alpha \\ &+ \alpha^2\gamma^2 E[\eta(k)\boldsymbol{u}(k)\boldsymbol{u}^{\mathrm{T}}(k)\boldsymbol{v}(k)\boldsymbol{v}^{\mathrm{T}}(k)\boldsymbol{u}(k)\boldsymbol{u}^{\mathrm{T}}(k)\eta(k)] \\ &+ \gamma^2 E[\eta(k)\boldsymbol{u}(k)\boldsymbol{u}^{\mathrm{T}}(k)\eta(k)]\sigma_q^2 \\ &+ (1-\lambda)^2 E[\tilde{\boldsymbol{w}}(k)\tilde{\boldsymbol{w}}^{\mathrm{T}}(k)] + (1-\lambda^2)E[\boldsymbol{n}(k)\boldsymbol{n}^{\mathrm{T}}(k)], \qquad (10.25)\end{aligned}$$

where σ_q^2 is the variance of the noise signal $q(k)$. The expectation terms are now evaluated using $\eta = E[\eta(k)]$ and σ_u^2 as the variance of the i.i.d. input signal $\boldsymbol{u}(k)$, which implies

$$E[\eta(k)\boldsymbol{u}(k)\boldsymbol{u}^{\mathrm{T}}(k)]\boldsymbol{R}_{\boldsymbol{v},\boldsymbol{v}}(k) = \boldsymbol{R}_{\boldsymbol{v},\boldsymbol{v}}(k)E[\boldsymbol{u}(k)\boldsymbol{u}^{\mathrm{T}}(k)\eta(k)] = \eta\sigma_u^2\boldsymbol{R}_{\boldsymbol{v},\boldsymbol{v}}(k), \qquad (10.26)$$

$$E[\eta(k)\boldsymbol{u}(k)\boldsymbol{u}^{\mathrm{T}}(k)\eta(k)] = \eta^2\sigma_u^2\boldsymbol{I} \qquad (10.27)$$

and by the fourth-order standard factorisation property of zero mean Gaussian variables[6] (Papoulis 1984)

$$E[\eta(k)\boldsymbol{u}(k)\boldsymbol{u}^{\mathrm{T}}(k)\boldsymbol{v}(k)\boldsymbol{v}^{\mathrm{T}}(k)\boldsymbol{u}(k)\boldsymbol{u}^{\mathrm{T}}(k)\eta(k)] = \eta^2\sigma_u^4[2\boldsymbol{R}_{\boldsymbol{v},\boldsymbol{v}}(k) + \boldsymbol{I}\,\mathrm{tr}\{\boldsymbol{R}_{\boldsymbol{v},\boldsymbol{v}}(k)\}]. \qquad (10.28)$$

---

[5] For small quantities $E[x^2(k)] \approx (E[x(k)])^2$, so that $E[\alpha^2(k)] \approx \alpha^2$, $E[\gamma^2(k)] \approx \gamma^2$ and $E[\eta^2(k)] \approx \eta^2$. Experiments show that this is a realistic assumption for the range of allowed $\alpha(k)$, $\gamma(k)$ and $\eta(k)$. Moreover, if η is fixed, $\eta(k) = \eta$ and $E[\eta^2] = \eta^2$.

[6] $E[\boldsymbol{x}_n\boldsymbol{x}_n^{\mathrm{T}}\boldsymbol{x}_n\boldsymbol{x}_n^{\mathrm{T}}]_{kl} = E[x(n-k)\sum_{i=1}^{N} x^2(n-i)x(n-l)]$, which by the standard factorisation property of real, zero mean Gaussian variables becomes

$$\begin{aligned}E[\boldsymbol{x}_1\boldsymbol{x}_2^{\mathrm{T}}\boldsymbol{x}_3\boldsymbol{x}_4^{\mathrm{T}}]_{kl} &= E[\boldsymbol{x}_1\boldsymbol{x}_2]E[\boldsymbol{x}_3\boldsymbol{x}_4] + E[\boldsymbol{x}_1\boldsymbol{x}_3]E[\boldsymbol{x}_2\boldsymbol{x}_4] + E[\boldsymbol{x}_1\boldsymbol{x}_4]E[\boldsymbol{x}_2\boldsymbol{x}_3] \\ &= 2\sum_{i=1}^{N} E[x(n-k)x(n-i)]E[x(n-l)x(n-i)] \\ &\quad + E[x(n-k)x(n-l)]\sum_{i=1}^{N} E[x^2(n-i)],\end{aligned}$$

which, in turn, implies

$$E[\boldsymbol{x}_n\boldsymbol{x}_n^{\mathrm{T}}\boldsymbol{x}_n\boldsymbol{x}_n^{\mathrm{T}}] = 2\boldsymbol{R}^2 + \boldsymbol{R}\,\mathrm{tr}\{\boldsymbol{R}\},$$

where $\mathrm{tr}\{\cdot\}$ is the trace operator. Now for i.i.d. Gaussian input signals $\boldsymbol{x}_n$, we have

$$E[x(n-i)x(n-j)] = \begin{cases} 0, & \text{if } i \neq j, \\ \sigma_x^2, & \text{if } i = j, \end{cases}$$

so that

$$E[\boldsymbol{x}_n\boldsymbol{x}_n^{\mathrm{T}}\boldsymbol{x}_n\boldsymbol{x}_n^{\mathrm{T}}]_{kl} = \begin{cases} 0, & \text{if } l \neq k, \\ (N+2)\sigma_x^4, & \text{if } l = k, \end{cases} \quad \text{and} \quad E[\boldsymbol{x}_n\boldsymbol{x}_n^{\mathrm{T}}\boldsymbol{x}_n\boldsymbol{x}_n^{\mathrm{T}}] = (N+2)\sigma_x^4\boldsymbol{I},$$

as required.

The first-order Markov model (10.10) used as the time-varying optimal weight system implies[7] that

$$E[\tilde{\boldsymbol{w}}(k)\tilde{\boldsymbol{w}}^{\mathrm{T}}(k)] = \sigma_n^2 \boldsymbol{I}, \tag{10.29}$$

$$E[\boldsymbol{n}(k)\boldsymbol{n}^{\mathrm{T}}(k)] = \sigma_n^2 \boldsymbol{I}, \tag{10.30}$$

where σ_n^2 is the variance of the signal $\boldsymbol{n}(k)$. Combining (10.25)–(10.30), we have

$$\begin{aligned} \boldsymbol{R}_{\boldsymbol{v},\boldsymbol{v}}(k+1) = \boldsymbol{R}_{\boldsymbol{v},\boldsymbol{v}}(k) - 2\alpha\gamma\eta\sigma_u^2 \boldsymbol{R}_{\boldsymbol{v},\boldsymbol{v}}(k) + \alpha^2\gamma^2\eta^2\sigma_u^4[2\boldsymbol{R}_{\boldsymbol{v},\boldsymbol{v}}(k) + \boldsymbol{I}\,\mathrm{tr}\{\boldsymbol{R}_{\boldsymbol{v},\boldsymbol{v}}(k)\}] \\ + \gamma^2\eta^2\sigma_u^2\sigma_q^2 \boldsymbol{I} + 2(1-\lambda)\sigma_n^2 \boldsymbol{I}. \end{aligned} \tag{10.31}$$

The mean squared misalignment ξ, which is a commonly used quantity in the assessment of the performance of an algorithm, can be now defined as

$$\xi(k+1) = E[\boldsymbol{v}^{\mathrm{T}}(k+1)\boldsymbol{v}(k+1)], \tag{10.32}$$

which can be obtained from $\boldsymbol{R}_{\boldsymbol{v},\boldsymbol{v}}(k+1)$ by taking its trace. Thus, we have

$$\begin{aligned} \xi(k+1) = [1 - 2\alpha\gamma\eta\sigma_u^2 + \alpha^2\gamma^2\eta^2\sigma_u^4(N+2)]\xi(k) \\ + \gamma^2\eta^2\sigma_u^2\sigma_q^2 N + 2(1-\lambda)N\sigma_n^2, \end{aligned} \tag{10.33}$$

where N is the length of vector $\boldsymbol{u}(k)$.

10.5.1 Convergence in the Mean Square

In order to guarantee convergence of the mean-square error (MSE), which is given under the above assumptions as

$$\mathrm{MSE}(k) = \sigma_u^2 \xi(k),$$

the update of the MSE has to be governed by a contraction mapping, i.e. from (10.33)

$$0 < |\alpha\gamma\eta\sigma_u^2[2 - \alpha\gamma\eta\sigma_u^2(N+2)]| < 2.$$

For convergence, the bounds on the learning rate η become[8]

$$0 < \eta < \frac{2}{\alpha\gamma\sigma_u^2(N+2)}. \tag{10.34}$$

The derived result is the upper bound for the learning rate which preserves the mean square convergence of the RTRL algorithm for a recurrent perceptron. Depending on the choice of γ, this is directly applicable for learning algorithms for both feedforward and recurrent neural networks. For a highly contractive Φ, α is small and η can be larger. For a linear activation function, $\alpha = \gamma = 1$, and the result (10.34) degenerates into the result for the LMS for linear FIR filters.

---

[7] Vectors $\tilde{\boldsymbol{w}}$ and $\boldsymbol{n}$ are drawn from the same statistical distribution $\mathcal{N}(\boldsymbol{0}, \sigma_n^2 \boldsymbol{I})$.

[8] Compare (10.34) with (10.24). From (10.24), for an i.i.d. input,

$$E\left[\frac{2}{\alpha\gamma\boldsymbol{u}^{\mathrm{T}}(k)\boldsymbol{u}(k)}\right] \approx \frac{2}{\alpha\gamma N\sigma_u^2},$$

which means that the MSE stability condition (10.34) is more stringent than the mean weight error stability condition (10.24).

10.5.2 Steady-State Mean-Squared Error

Let us first derive the steady-state misalignment. Normally, this is obtained by setting $\xi = \xi(k) = \xi(k+1)$ in (10.33) and solving for ξ, and thus

$$\begin{aligned} \xi &= \frac{\gamma^2\eta^2\sigma_u^2\sigma_q^2 N + 2(1-\lambda)N\sigma_n^2}{\alpha\gamma\eta\sigma_u^2[2-\alpha\gamma\eta\sigma_u^2(N+2)]} \\ &= \frac{\gamma\eta\sigma_q^2 N}{\alpha[2-\alpha\gamma\eta\sigma_u^2(N+2)]} + \frac{2(1-\lambda)N\sigma_n^2}{\alpha\gamma\eta\sigma_u^2[2-\alpha\gamma\eta\sigma_u^2(N+2)]}. \end{aligned} \tag{10.35}$$

The steady-state MSE is then

$$\text{MSE} = \sigma_u^2\xi. \tag{10.36}$$

The results for systems with a single fixed optimal weight solution can be obtained from the above by setting $\lambda = 1$.

10.6 Summary

Techniques for convergence analysis for an online stochastic gradient descent algorithm for neural adaptive filters have been provided. These are based upon the previously addressed contraction mapping properties of nonlinear neurons. The analysis has been undertaken for a general case of time-varying behaviour of the optimal weight vector. The learning algorithms for linear filters have been shown to be the bounds for the algorithms employed for neural networks. The analysis is applicable to both recurrent and feedforward architectures and can be straightforwardly extended to more complicated structures and learning algorithms.

11

Some Practical Considerations of Predictability and Learning Algorithms for Various Signals

11.1 Perspective

In this chapter, predictability, detecting nonlinearity and performance with respect to the prediction horizon are considered. Methods for detecting nonlinearity of signals are first discussed. Then, different algorithms are compared for the prediction of nonlinear and nonstationary signals, such as real NO_2 air pollutant and heart rate variability signals, together with a synthetic chaotic signal. Finally, bifurcations and attractors generated by a recurrent perceptron are analysed to demonstrate the ability of recurrent neural networks to model complex physical phenomena.

11.2 Introduction

When modelling a signal, an initial linear analysis is first performed on the signal, as linear models are relatively quick and easy to implement. The performance of these models can then determine whether more flexible nonlinear models are necessary to capture the underlying structure of the signal. One such standard model of linear time series, the auto-regressive integrated moving average, or ARIMA(p, d, q) model popularised by Box and Jenkins (1976), assumes that the time series x_k is generated by a succession of 'random shocks' ϵ_k, drawn from a distribution with zero mean and variance σ_ϵ^2. If x_k is non-stationary, then successive differencing of x_k via the differencing operator, $\nabla x_k = x_k - x_{k-1}$ can provide a stationary process. A stationary process $z_k = \nabla^d x_k$ can be modelled as an autoregressive moving average

$$z_k = \sum_{i=1}^{p} a_i z_{k-i} + \sum_{i=1}^{q} b_i \epsilon_{k-i} + \epsilon_k. \tag{11.1}$$

Of particular interest are pure autoregressive (AR) models, which have an easily understood relationship to the nonlinearity detection technique of DVS (deterministic

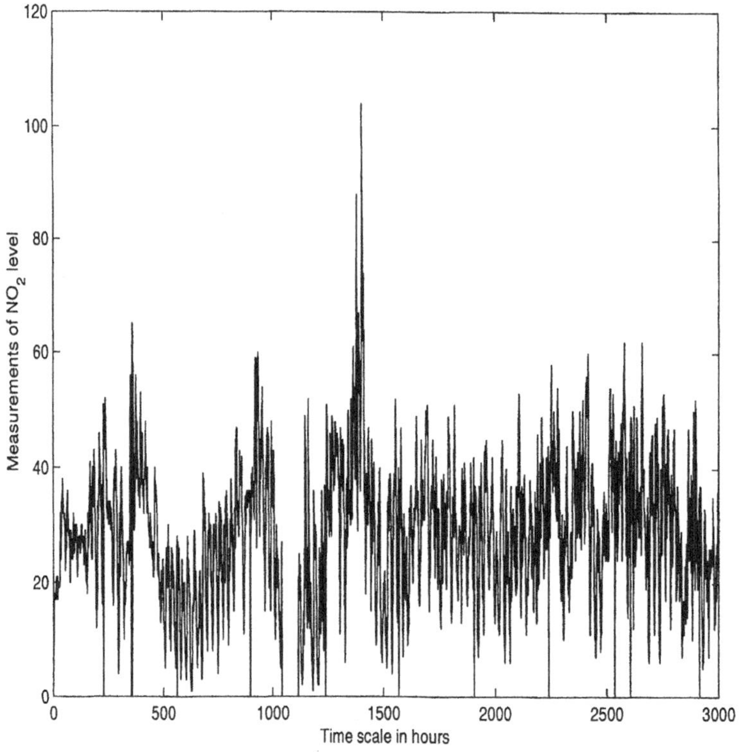

(a) The raw NO_2 time series

Figure 11.1 The NO_2 time series and its autocorrelation function

versus stochastic) plots. Also, an ARMA(p, q) process can be accurately represented as a pure AR(p') process, where $p' \gg p + d$ (Brockwell and Davis 1991). Penalised likelihood methods such as AIC or BIC (Box and Jenkins 1976) exist for choosing the order of the autoregressive model to be fitted to the data; or the point where the *autocorrelation function* (ACF) essentially vanishes for all subsequent lags can also be used. The autocorrelation function for a wide-sense stationary time series x_k at lag h gives the correlation between x_k and x_{k+h}; clearly, a non-zero value for the ACF at a lag h suggests that for modelling purposes at least the previous h lags should be used ($p \geqslant h$).

For instance, Figure 11.1 shows a raw NO_2 signal and its autocorrelation function (ACF) for lags of up to 40; the ACF does not vanish with lag and hence a high-order AR model is necessary to model the signal. Note the peak in the ACF at a lag of 24 hours and the rise to a smaller peak at a lag of 48 hours. This is evidence of *seasonal* behaviour, that is, the measurement at a given time of day is likely to be related to the measurement taken at the same time on a different day. The issue of seasonal time series is dealt with in Appendix J.

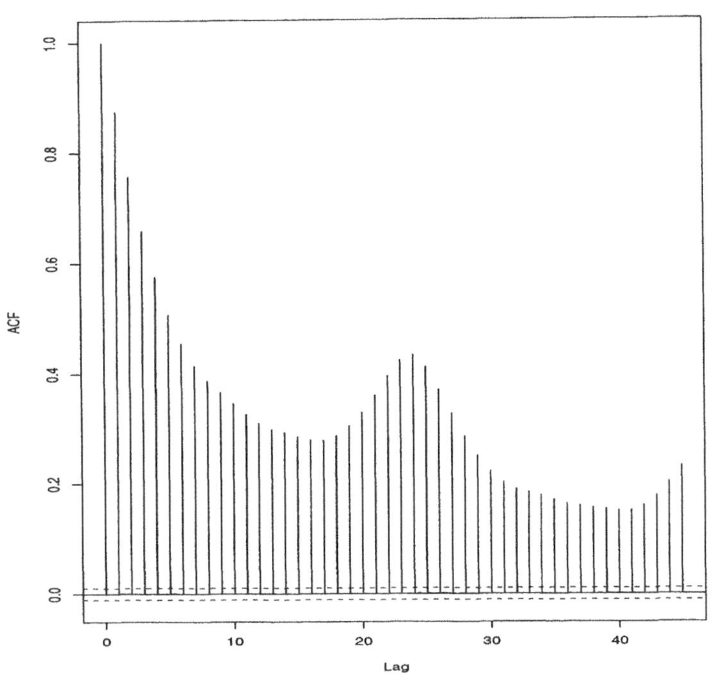

(b) The ACF of the NO_2 series

Figure 11.1 *Cont.*

11.2.1 Detecting Nonlinearity in Signals

Before deciding whether to use a linear or nonlinear model of a process, it is important to check whether the signal itself is linear or nonlinear. Various techniques exist for detecting nonlinearity in time series. Detecting nonlinearity is important because the existence of nonlinear structure in the series opens the possibility of highly accurate short-term predictions. This is not true for series which are largely stochastic in nature. Following the approach from Theiler *et al.* (1993), to gauge the efficacy of the techniques for detecting nonlinearity, a surrogate dataset is simulated from a high-order autoregressive model fit to the original series. Two main methods to achieve this exist, the first involves fitting a finite-order ARMA(p, q) model (we use a high-order AR(p) model to fit the data). The model coefficients are then used to generate the surrogate series, with the surrogate residuals ϵ_k taken as random permutations of the residuals from the original series. The second method involves taking a Fourier transform of the series. The phases at each frequency are replaced randomly from the uniform $(0, 2\pi)$ distribution while keeping the magnitude of each frequency the same as for the original series. The surrogate series is then obtained by taking the inverse Fourier transform. This series will have approximately the same autocorrelation function as the original series, with the approximation becoming exact in the limit as $N \to \infty$. A discussion of the respective merits of the two methods of generating surrogate data is given in Theiler *et al.* (1993), the method used here is the former. Evidence of nonlinearity from any method of detection is negated if the method gives a similar result when applied to the surrogate series, which is known to be linear (Theiler *et al.* 1993).

11.3 Overview

This chapter deals with some practical issues when performing prediction of nonlinear and nonstationary signals. Techniques for detecting nonlinearity and chaotic behaviour of signals are first introduced and a detailed analysis is provided for the NO_2 air pollutant measurements taken at hourly intervals from the Leeds meteo station, UK. Various linear and nonlinear algorithms are compared for prediction of air pollutants, heart rate variability and chaotic signals. The chapter concludes with an insight into the capability of recurrent neural networks to generate and model complex nonlinear behaviour such as chaos.

11.4 Measuring the Quality of Prediction and Detecting Nonlinearity within a Signal

Existence and/or discovery of an attractor in the phase space demonstrates whether the system is deterministic, purely stochastic or contains elements of both. To reconstruct the attractor examine plots in the m-dimensional space of $[x_k, x_{k-\tau}, \dots, x_{k-(m-1)\tau}]^{\mathrm{T}}$. It is critically important for the dimension of the space, m, in which the attractor resides, to be large enough to 'untangle' the attractor. This is known as the *embedding dimension* (Takens 1981). The value of τ, the *lag time* or *lag spacing*, is also important, particularly with noise present. The first inflection point of the autocorrelation function is a possible starting value for τ (Beule *et al.* 1999). Alternatively, if the series is known to be sampled coarsely, the value of τ can be taken as unity (Casdagli and Weigend 1993). A famous example of an attractor is given by the *Lorenz equations* (Lorenz 1963)

$$\left.\begin{aligned} \dot{x} &= \sigma(y - x), \\ \dot{y} &= rx - y - xz, \\ \dot{z} &= xy - bz, \end{aligned}\right\} \tag{11.2}$$

where σ, r and $b > 0$ are parameters of the system of equations. In Lorenz (1963) these equations were studied for the case $\sigma = 10$, $b = \frac{8}{3}$ and $r = 28$. A Lorenz attractor is shown in Figure 11.13(a). The discovery of an attractor for an air pollution time series would demonstrate chaotic behaviour; unfortunately, the presence of noise makes such a discovery unlikely. More robust techniques are necessary to detect the existence of deterministic structure in the presence of substantial noise.

11.4.1 Deterministic Versus Stochastic Plots

Deterministic versus stochastic (DVS) plots (Casdagli and Weigend 1993) display the (robust) prediction error $E(n)$ for local linear models against the number of nearest neighbours, n, used to fit the model, for a range of embedding dimensions m. The data are separated into a *test* set and a *training* set, where the test set is the last M elements of the series. For each element in the test set x_k, its corresponding delay vector in m-dimensional space

$$\boldsymbol{x}(k) = [x_{k-\tau}, x_{k-2\tau}, \dots, x_{k-m\tau}]^{\mathrm{T}} \tag{11.3}$$

is constructed. This delay vector is then examined against the set of all the delay vectors constructed from the training set. From this set the n nearest neighbours are defined to be the n delay vectors $\boldsymbol{x}(k')$ which have the shortest Euclidean distance to $\boldsymbol{x}(k)$. These n nearest neighbours $\boldsymbol{x}(k')$ along with their corresponding target values $x_{k'}$ are used as the variables to fit a simple linear model. This model is then given $\boldsymbol{x}(k)$ as an input which provides a prediction $\hat{x}_k$ for the target value x_k, with a robust prediction error of

$$|x_k - \hat{x}_k|. \tag{11.4}$$

This procedure is repeated for all the test set, enabling calculation of the mean robust prediction error,

$$E(n) = \frac{1}{M} \sum_{x_k \in T} |x_k - \hat{x}_k|, \tag{11.5}$$

where T is the test set. If the optimal number of nearest neighbours n, taken to be the value giving the lowest prediction error $E(n)$, is at, or close to, the maximum possible n, then globally linear models perform best and there is no indication of nonlinearity in the signal. As this global linear model uses all possible length m vectors of the series, it is equivalent to an AR model of order m when $\tau = 1$. Small optimal n suggests local linear models perform best, indicating nonlinearity and/or chaotic behaviour.

11.4.2 Variance Analysis of Delay Vectors

Closely related to DVS plots is the nonlinearity detection technique introduced in Khalaf and Nakayama (1998). The general idea is not to fit models, linear or otherwise, using the nearest neighbours of a delay vector, but rather to examine the variability of the set of targets corresponding to groups of close (in the Euclidean distance sense) delay vectors. For each observation x_k, $k \geqslant m+1$ construct the group, Ω_k, of nearest neighbour delay vectors given by

$$\Omega_k = \{\boldsymbol{x}(k') : k' \neq k \;\&\; d_{kk'} \leqslant \alpha A_x\}, \tag{11.6}$$

where $\boldsymbol{x}(k') = \{x_{k'-1}, x_{k'-2}, \dots, x_{k'-m}\}$, $d_{kk'} = \|\boldsymbol{x}(k') - \boldsymbol{x}(k)\|$ is the Euclidean distance, $0 < \alpha \leqslant 1$,

$$A_x = \frac{1}{N-m} \sum_{k=m+1}^{N} |x_k|$$

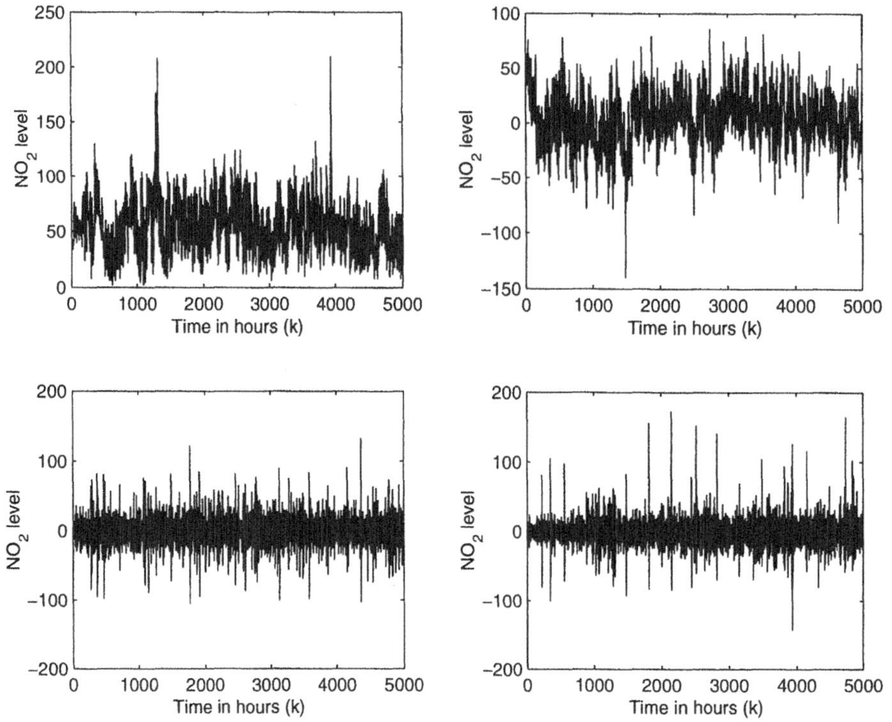

Figure 11.2 Time series plots for NO_2. Clockwise, starting from top left: raw, simulated, simulated deseasonalised, deseasonalised

and N is the length of the time series. If the series is linear, then the similar patterns $\boldsymbol{x}(k')$ belonging to a group Ω_k will map onto similar $x_{k'}$s. For nonlinear series, the patterns $\boldsymbol{x}(k')$ will not map onto similar $x_{k'}$s. This is measured by the variance σ^2 of each group Ω_k

$$\sigma_k^2 = \frac{1}{|\Omega_k|}\sum_k (x_{k'} - \mu_k)^2, \qquad \boldsymbol{x}(k') \in \Omega_k.$$

The measure of nonlinearity is taken to be the mean of σ_k^2 over all the Ω_k, denoted $\overline{\sigma_N^2}$, normalised by dividing through by σ_x^2, the variance of the entire time series

$$\overline{\sigma^2} = \frac{\overline{\sigma_N^2}}{\sigma_x^2}.$$

The larger the value of $\overline{\sigma^2}$ the greater the suggestion of nonlinearity (Khalaf and Nakayama 1998). A comparison with surrogate data is especially important with this method to get evidence of nonlinearity.

11.4.3 Dynamical Properties of NO_2 Air Pollutant Time Series

The four time series generated from the NO_2 dataset are given in Figure 11.2, with the deseasonalised series on the bottom and the simulated series on the right. The

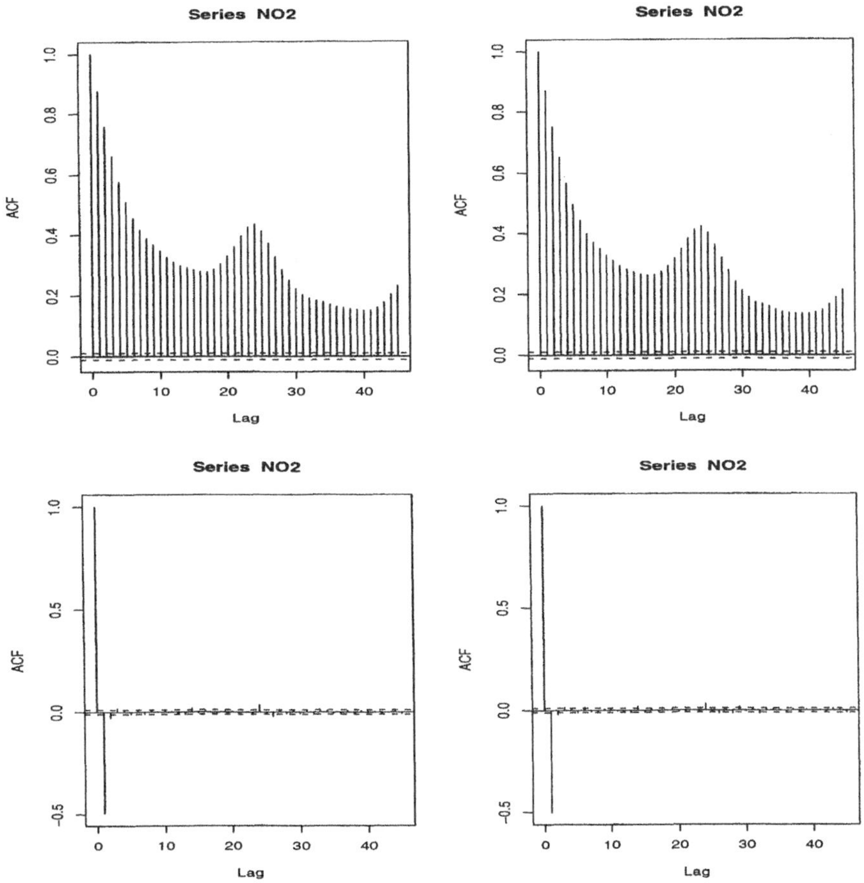

Figure 11.3 ACF plots for NO_2. Clockwise, starting from top left: raw, simulated, simulated deseasonalised, deseasonalised

sine wave structure can clearly be seen in the raw (unaltered) time series (top left), evidence confirming the relationship between NO_2 and temperature. Also note that once an air pollutant series has been simulated or deseasonalised, the condition that no readings can be below zero no longer holds. The respective ACF plots for the NO_2 series are given in Figure 11.3. The raw and simulated ACFs (top) are virtually identical – as should be the case, since the simulated time series is based on a linear AR(45) fit to the raw data, the correlations for the first 45 lags should be the same. Since generating the deseasonalised data involves application of the backshift operator, the autocorrelations are much reduced, although a 'mini-peak' can still be seen at a lag of 24 hours.

Nonlinearity detection in NO_2 signal

Figure 11.4 shows the two-dimensional attractor reconstruction for the NO_2 time series after it has been passed through a linear filter to remove some of the noise

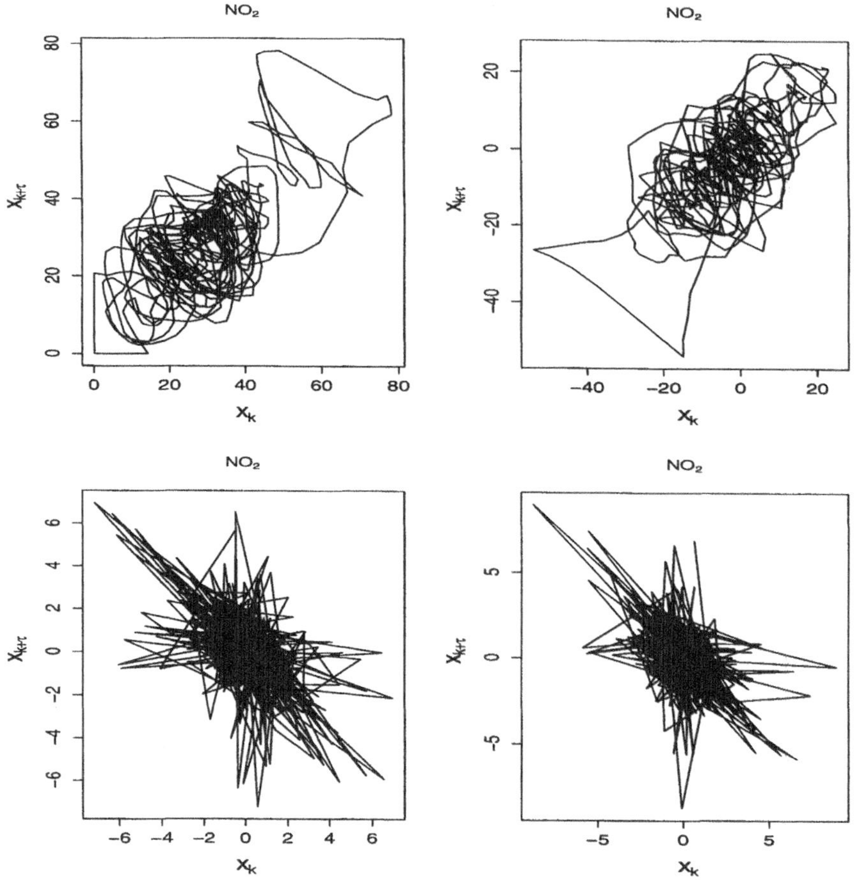

Figure 11.4 Attractor reconstruction plots for NO_2. Clockwise, starting from top left: raw, simulated, simulated deseasonalised and deseasonalised

present. This graph shows little regularity and there is little to distinguish between the raw and the simulated plots. If an attractor does exist, then it is in a higher-dimensional space or is swamped by the random noise. The DVS plots for NO_2 are given in Figure 11.5, the DVS analysis of a related air pollutant can be found in Foxall *et al.* (2001). The optimal n (that is, the value of n corresponding to the minimum of $E(n)$), is clearly less than the maximum of n for the raw data for each of the embedding dimensions (m) examined. However, the difference is not great and the minimum occurs quite close to the maximum n, so this only provides weak evidence for nonlinearity. The DVS plot for the simulated series obtains the optimal error measure at the maximum n, as is expected. The deseasonalised DVS plots follow the same pattern, except that the evidence for nonlinearity is weaker, and the best embedding dimension now is $m = 6$ rather than $m = 2$. Figure 11.6 shows the results from analysing the variance of the delay vectors for the NO_2 series. The top two plots show lesser variances for the raw series, strongly suggesting nonlinearity. However, for

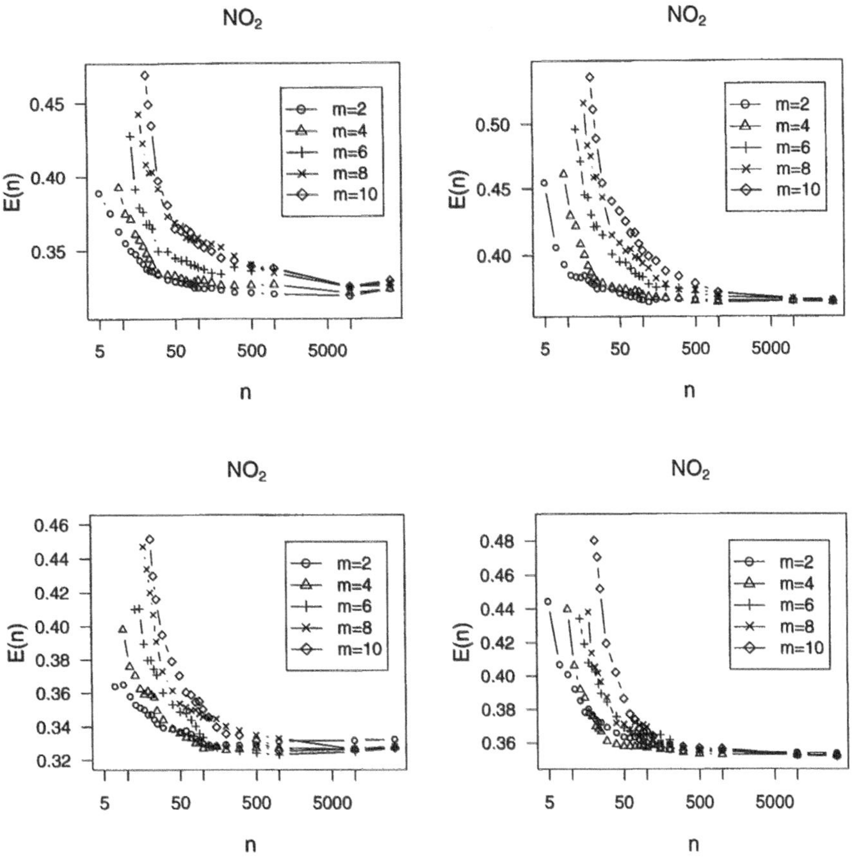

Figure 11.5 DVS plots for NO_2. Clockwise, starting from top left: raw, simulated, simulated deseasonalised and deseasonalised

Table 11.1 Performance of gradient descent algorithms in prediction of the NO_2 time series

	NGD	NNGD	Recurrent perceptron	NLMS
Predicted gain (dB)	5.78	5.81	6.04	4.75

the deseasonalised series (bottom) the variances are roughly equal, and indeed greater for higher embedding dimensions, suggesting that evidence for nonlinearity originated from the seasonality of the data.

To support the analysis, experiments on prediction of this signal were performed. The air pollution data represent hourly measurements of the concentration of nitrogen dioxide (NO_2), over the period 1994–1997, provided by the Leeds meteo station.

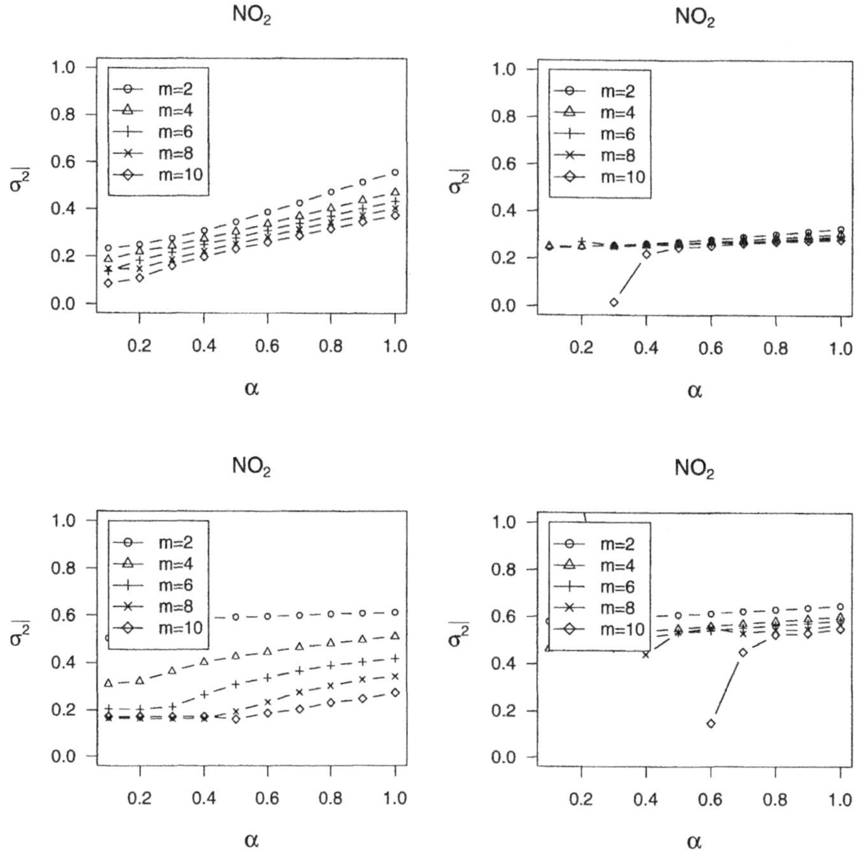

Figure 11.6 Delay vector variance plots for NO_2. Clockwise, starting from top left: raw, simulated, simulated deseasonalised and deseasonalised

In the experiments the logistic function was chosen as the nonlinear activation function of a dynamical neuron (Figure 2.6). The quantitative performance measure was the standard prediction gain, a logarithmic ratio between the expected signal and error variances $R_p = 10\log(\hat{\sigma}_s^2/\hat{\sigma}_e^2)$. The slope of the nonlinear activation function of the neuron β was set to be $\beta = 4$. The learning rate parameter η in the NGD algorithm was set to be $\eta = 0.3$ and the constant C in the NNGD algorithm was set to be $C = 0.1$. The order of the feedforward filter N was set to be $N = 10$. For simplicity, a NARMA(3,1) recurrent perceptron was used as a recurrent network. The summary of the performed experiments is given in Table 11.1. From Table 11.1, the nonlinear algorithms perform better than the linear one, confirming the analysis which detected nonlinearity in the signal. To further support the analysis given in the DVS plots, Figure 11.7(a) shows prediction gains versus number of taps for linear and nonlinear feedforward filters trained by the NGD, NNGD and NLMS algorithms, whereas Figure 11.7(b) shows prediction performance of a recurrent perceptron (Foxall *et al.* 2001). Both the nonlinear filters trained by the NGD and NNGD algorithms outperformed the linear filter trained by the NLMS algorithm. For the tap length up to $N = 10$, the NNGD was outperforming the NGD; the worse performance of the NNGD over the NGD for $N > 10$ can be explained by the insufficient approximation of the remainder of the Taylor series expansion within the derivation of the algorithm for large N. The recurrent structure achieved better performance for a smaller number of tap inputs than the standard feedforward structures.

11.5 Experiments on Heart Rate Variability

Information about heart rate variability (HRV) is extracted from the electrocardiogram (ECG). There are different approaches to the assessment of HRV from the measured data, but most of them rely upon the so-called R–R intervals, i.e. distance in time between two successive R waves in the HRV signal. Here, we use the R–R intervals that originate from ECG obtained from two patients. The first patient (A) was male, aged over 60, with a normal sinus rhythm, while patient (B) was also male, aged over 60, who suffered a miocardial infarction. In order to examine predictability of HRV signals, we use various gradient-descent-based neural adaptive filters.

11.5.1 Experimental Results

Figure 11.8(a) shows the HRV for patient A, while Figure 11.8(b) shows HRV for patient B. Prediction was performed using a logistic activation function Φ of a dynamical neuron with $N = 10$. The quantitative performance measure was the standard prediction gain $R_{\mathrm{p}} = 10\log(\hat{\sigma}_{\mathrm{s}}^2/\hat{\sigma}_{\mathrm{e}}^2)$. The slope of the nonlinear activation function of the neuron β was set to be $\beta = 4$. Due to the saturation type logistic nonlinearity, input data were prescaled to fit within the range of the neuron activation function. Both the standard NGD and the data-reuse modifications of the NGD algorithm were used. The number of data-reuse iterations L was set to be $L = 10$. The performance comparison between the NGD algorithm and a data-reusing NGD algorithm is shown in Figure 11.9. The plots show the prediction gain versus the tap length and the prediction horizon (number of steps ahead in prediction). In all the cases from Figure 11.9, the data-reusing algorithms outperformed the standard algorithms for short-term prediction. The standard algorithms showed better prediction results for long-term prediction. As expected, the performance deteriorates with the order of prediction ahead. In the next experiment we compare the performance of a recurrent perceptron trained with the fixed learning rate $\eta = 0.3$ and a recurrent perceptron trained by the NRTRL algorithm on prediction of the HRV signal. In the experiment the MA and the AR part of the recurrent perceptron vary from 1 to 15, while prediction horizon varies from 1 to 10. The results of the experiment are shown in Figures 11.10 and 11.11. From Figure 11.10, for a relatively large input line and feedback tap delay lines, there is a saturation in performance. This confirms that the recurrent structure was able to capture the dynamics of the HRV signal. The prediction performance deteriorates with the prediction step, and due to the recurrent nature of the filter, the performance is not good for a NARMA recurrent perceptron with

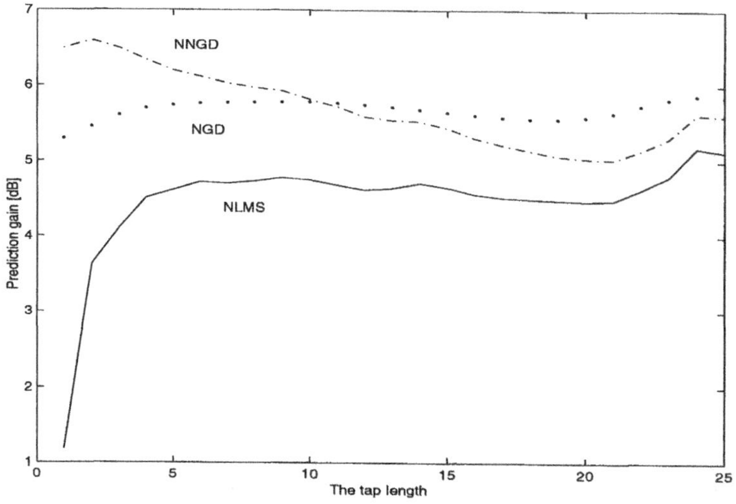

(a) Performance of the NGD, NNGD and NLMS algorithms in the prediction of NO_2 time series

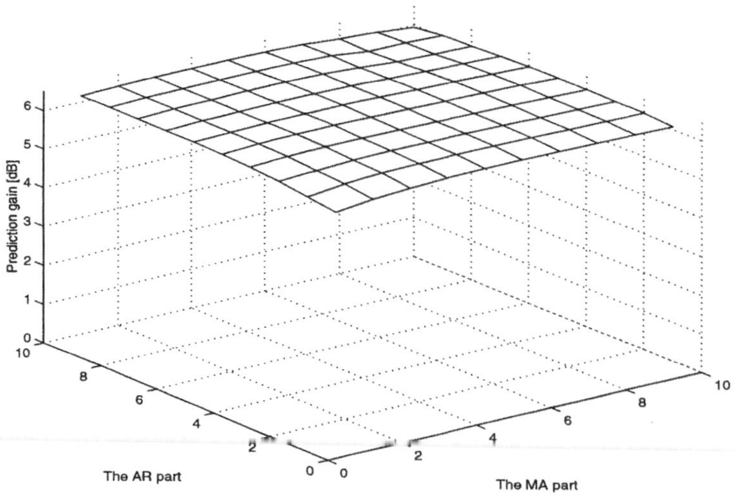

(b) Performance of the recurrent perceptron in the prediction of NO_2 time series

Figure 11.7 Performance comparison of various structures for prediction of NO_2 series

a small order of the AR and MA part. Figure 11.11 shows the results of an experiment similar to the previous one, with the exception that the employed algorithm was the NRTRL algorithm. The NARMA(p, q) recurrent perceptron trained with this

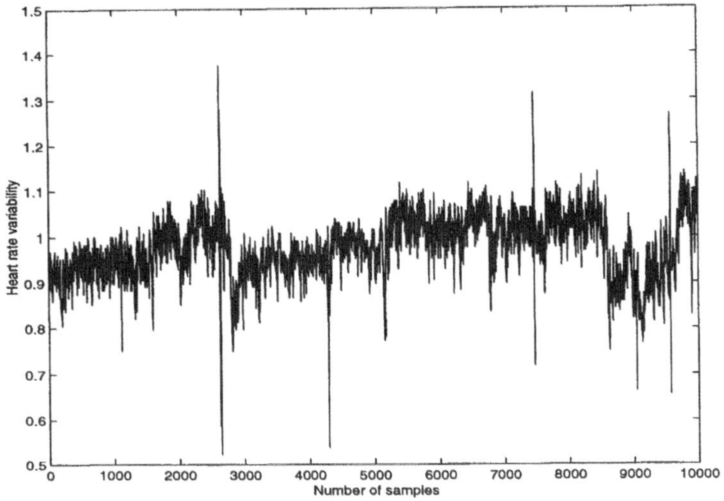

(a) HRV signal for patient A

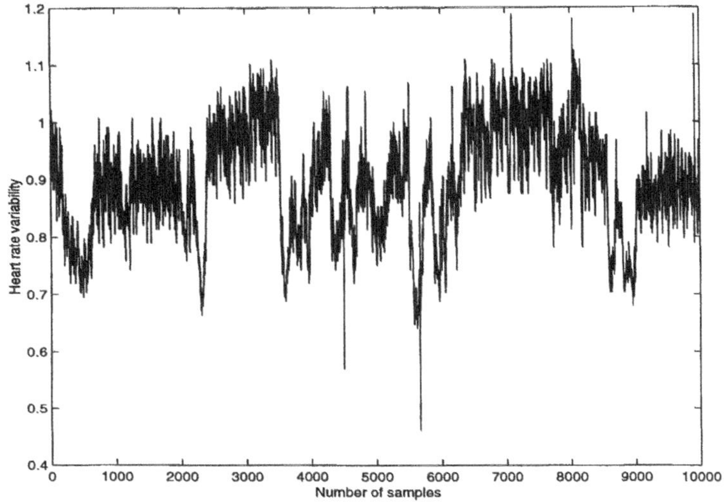

(b) HRV signal for patient B

Figure 11.8 Heart rate variability signals for patients A and B

algorithm persistently outperformed the standard recurrent perceptron trained by the RTRL.

Figure 11.12 shows performance of the recurrent perceptron with fixed η in prediction of HRV time series (patient B), for different prediction horizons. Similar arguments as for patient A are applicable.

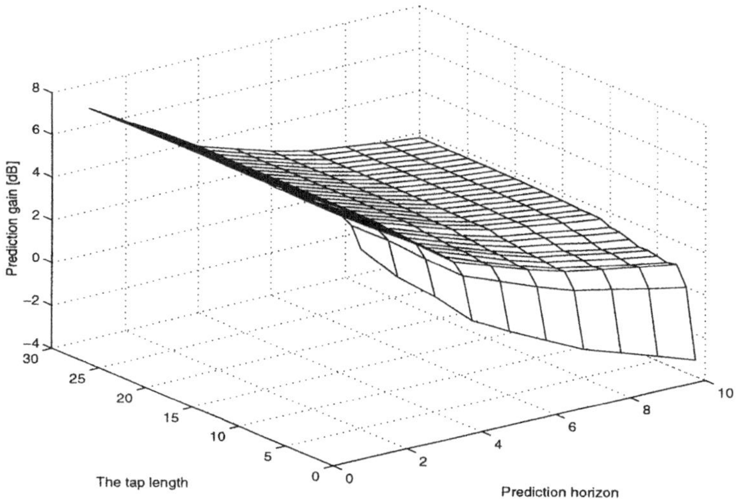

(a) Performance of the NGD algorithm in prediction of HRV time series, patient A

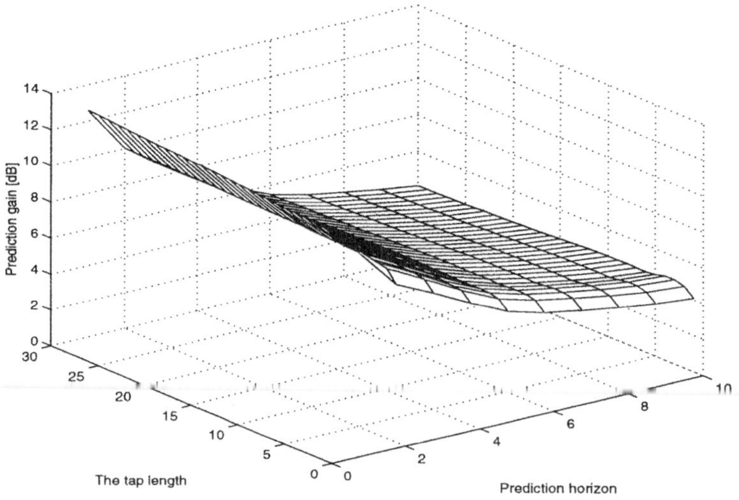

(b) Performance of the NGD algorithm in prediction of HRV time series, patient B

Figure 11.9 Performance comparison between standard and data-reusing algorithms for prediction of HRV signals

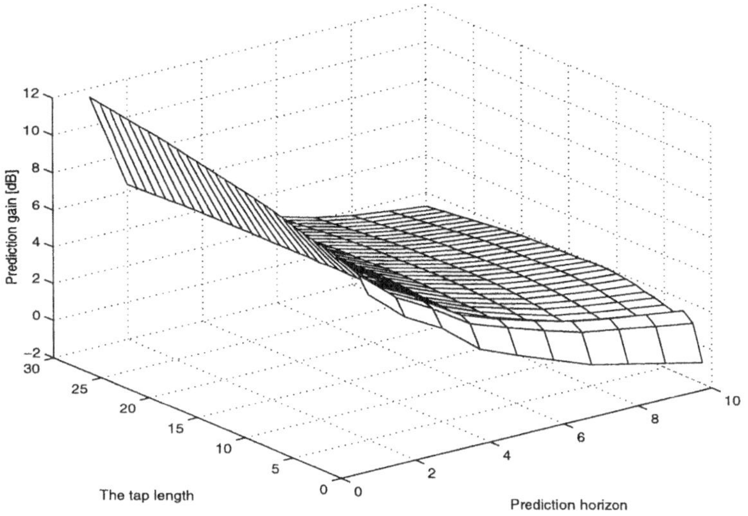

(c) Performance of the data-reusing NGD algorithm in prediction of HRV time series, patient A, $L = 10$

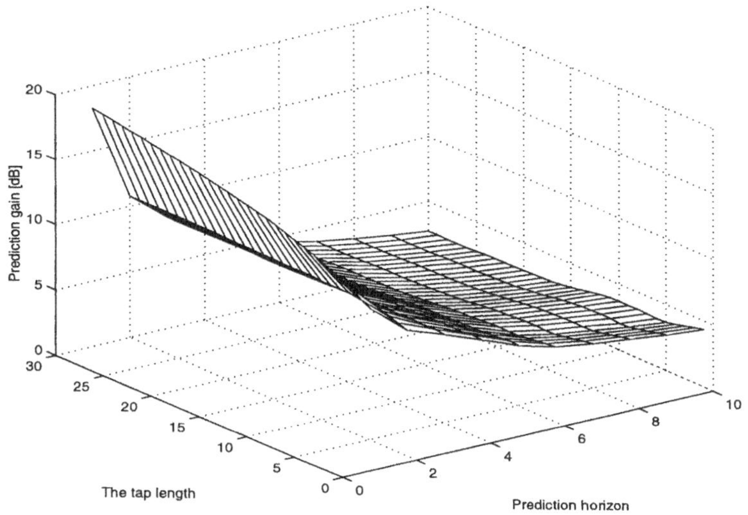

(d) Performance of the data-reusing NGD algorithm in prediction of HRV time series, patient B, $L = 10$

Figure 11.9 *Cont.*

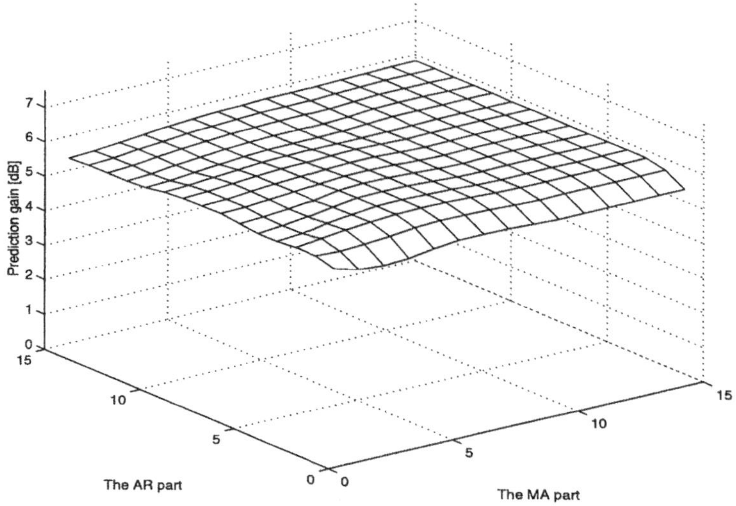

(a) Performance of the recurrent perceptron with fixed learning rate in prediction of HRV time series, patient A, prediction horizon is 1

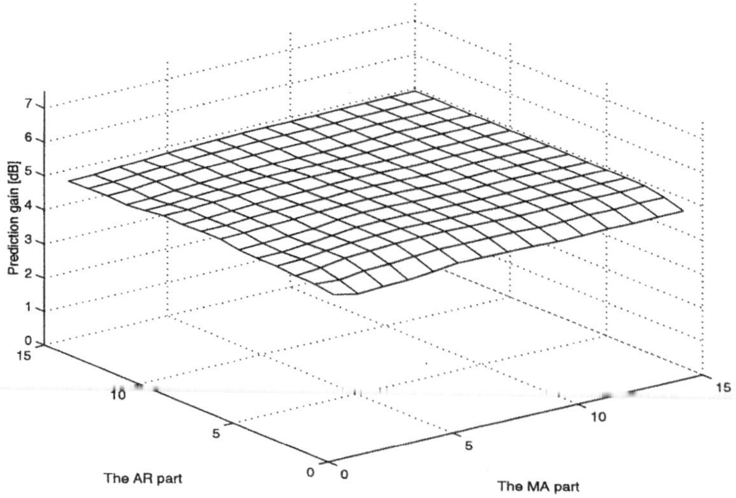

(b) Performance of the recurrent perceptron with fixed learning rate in prediction of HRV time series, patient A, prediction horizon is 2

Figure 11.10 Performance of a NARMA recurrent perceptron on prediction of HRV signals for different prediction horizons

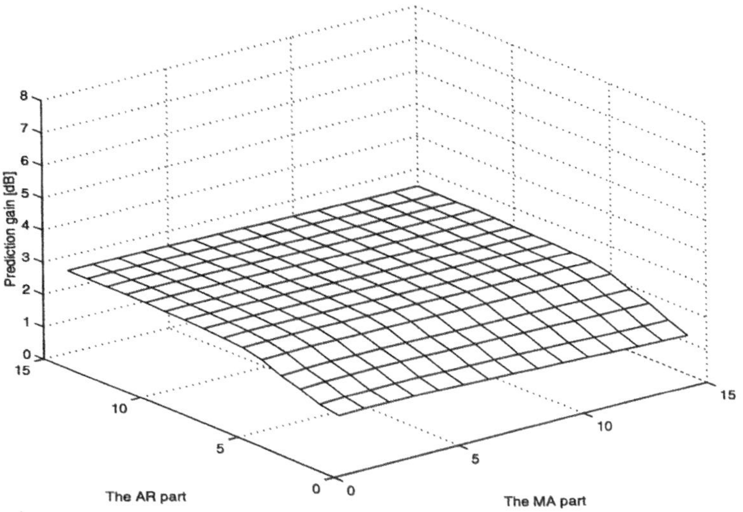

(c) Performance of the recurrent perceptron with fixed learning rate in prediction of HRV time series, patient A, prediction horizon is 5

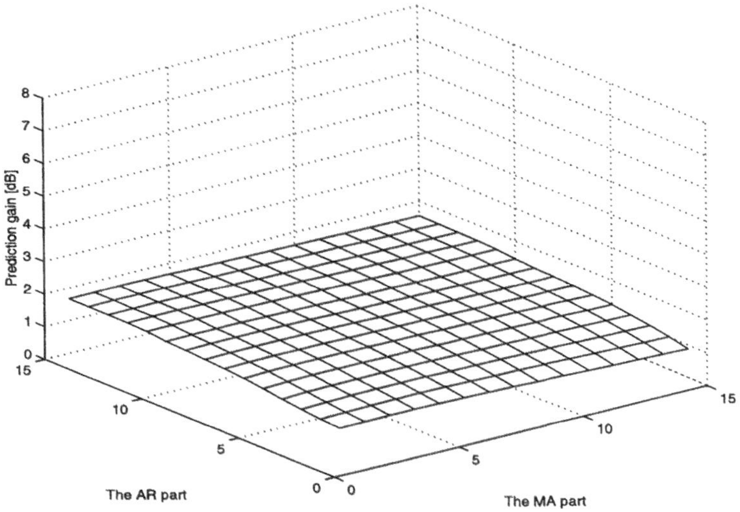

(d) Performance of the recurrent perceptron with fixed learning rate in prediction of HRV time series, patient A, prediction horizon is 10

Figure 11.10 *Cont.*

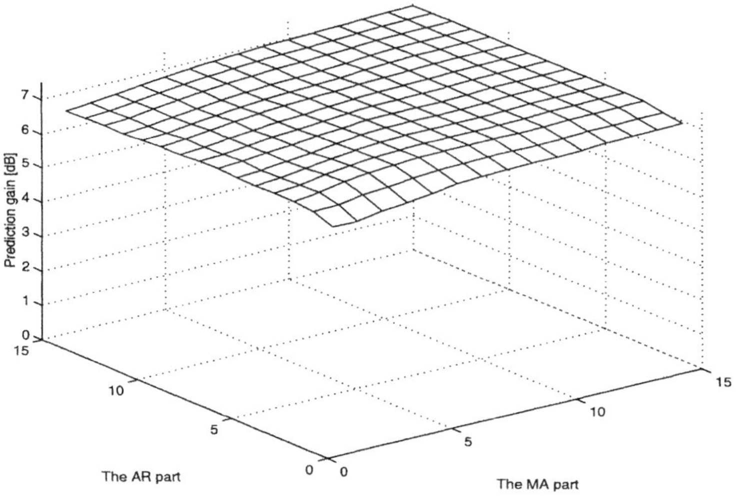

(a) Performance of the recurrent perceptron trained with the NRTRL algorithm in prediction of HRV time series, patient A, prediction horizon is 1

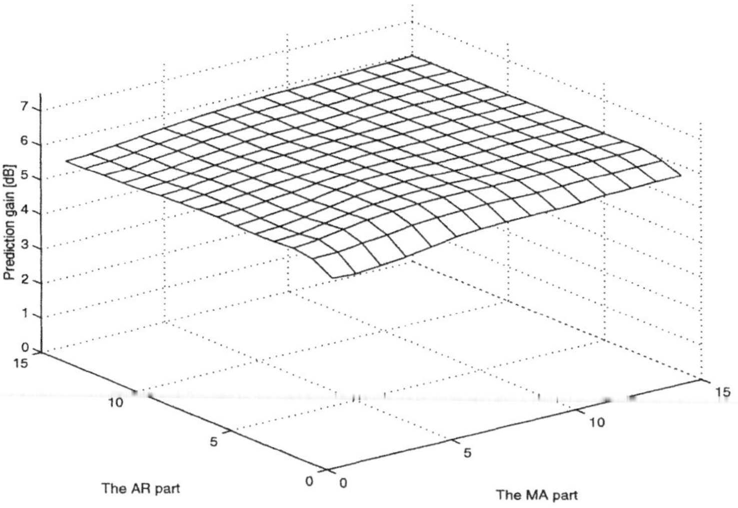

(b) Performance of the recurrent perceptron trained with the NRTRL algorithm in prediction of HRV time series, patient A, prediction horizon is 2

Figure 11.11 Performance of the NRTRL algorithms on prediction of HRV, for different prediction horizons

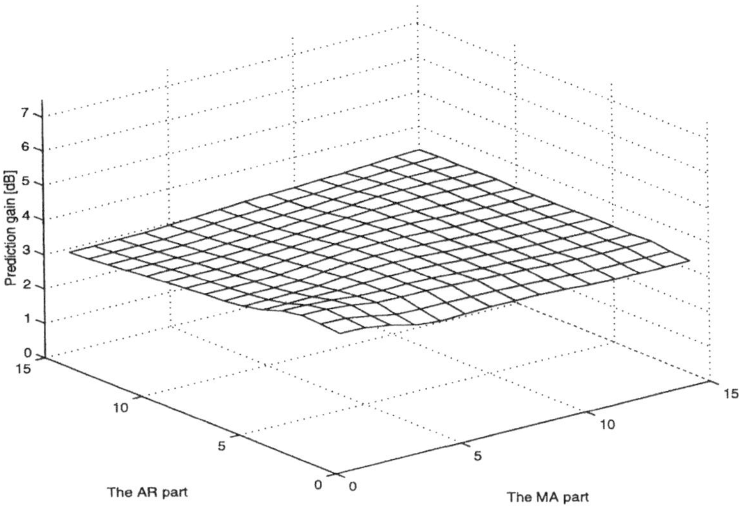

(c) Performance of the recurrent perceptron trained with the NRTRL algorithm in prediction of HRV time series, patient A, prediction horizon is 5

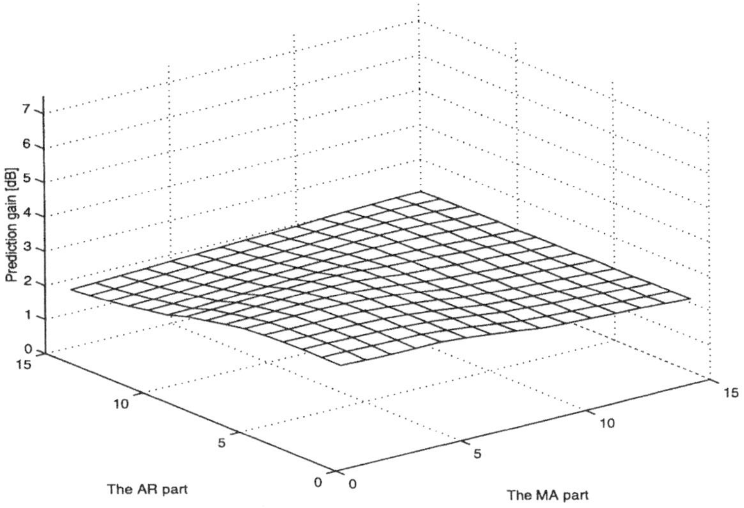

(d) Performance of the recurrent perceptron trained with the NRTRL algorithm in prediction of HRV time series, patient A, prediction horizon is 10

Figure 11.11 *Cont.*

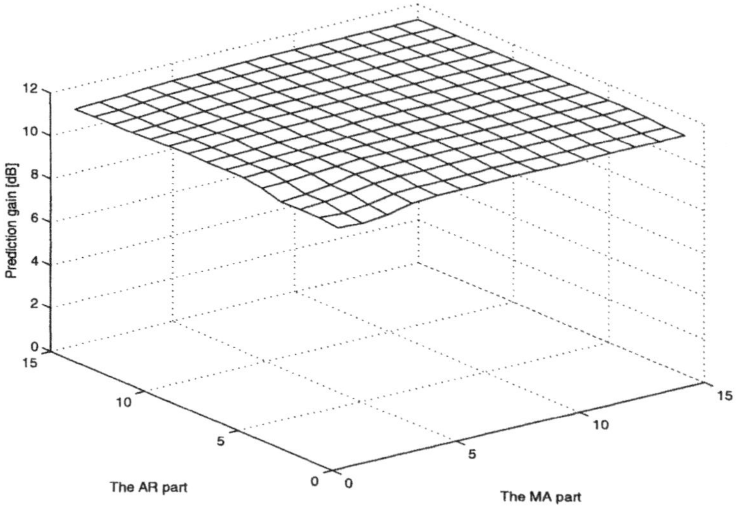

(a) Performance of the recurrent perceptron with fixed learning rate in prediction of HRV time series, patient B, prediction horizon is 1

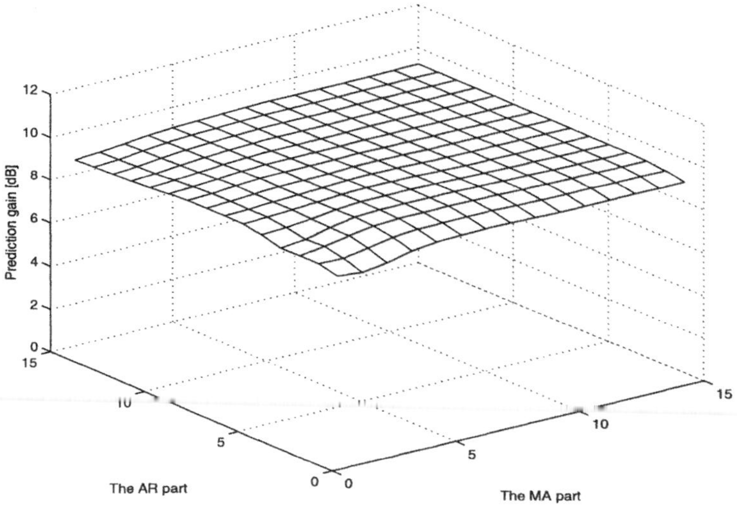

(b) Performance of the recurrent perceptron with fixed learning rate in prediction of HRV time series, patient B, prediction horizon is 2

Figure 11.12 Performance of a recurrent perceptron for prediction of HRV signals for different prediction horizons

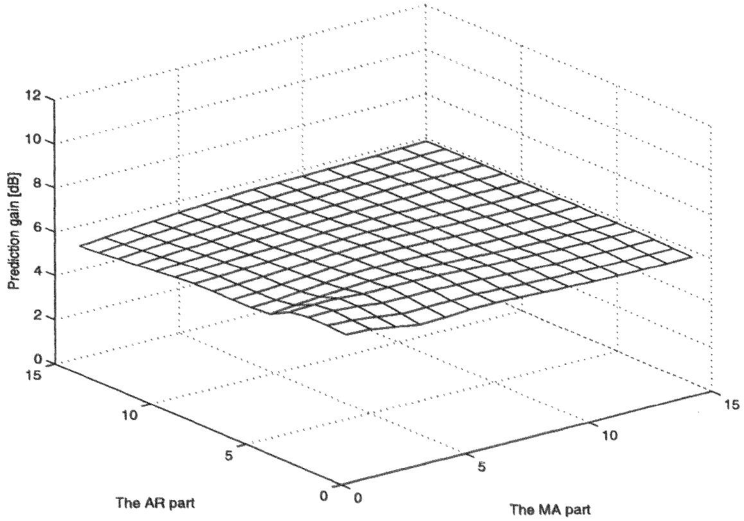

(c) Performance of the recurrent perceptron with fixed learning rate in prediction of HRV time series, patient B, prediction horizon is 5

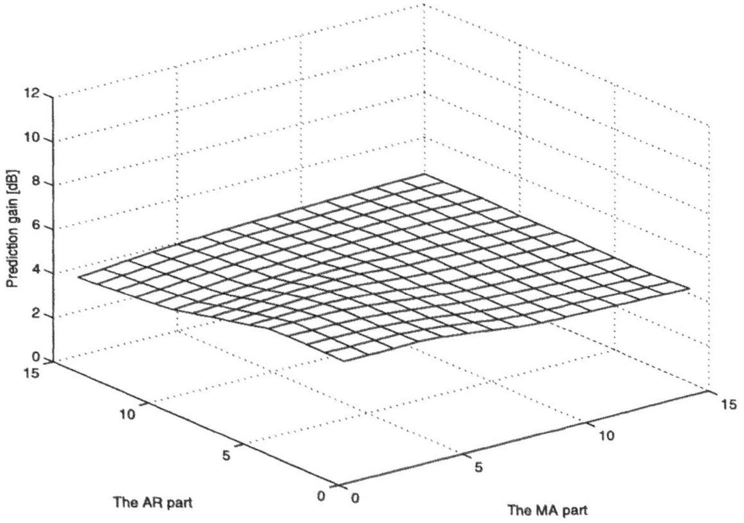

(d) Performance of the recurrent perceptron with fixed learning rate in prediction of HRV time series, patient B, prediction horizon is 10

Figure 11.12 *Cont.*

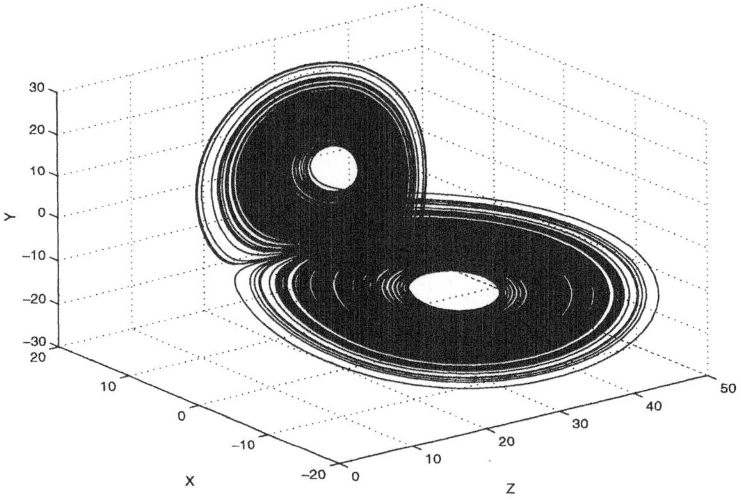

(a) The Lorenz attractor

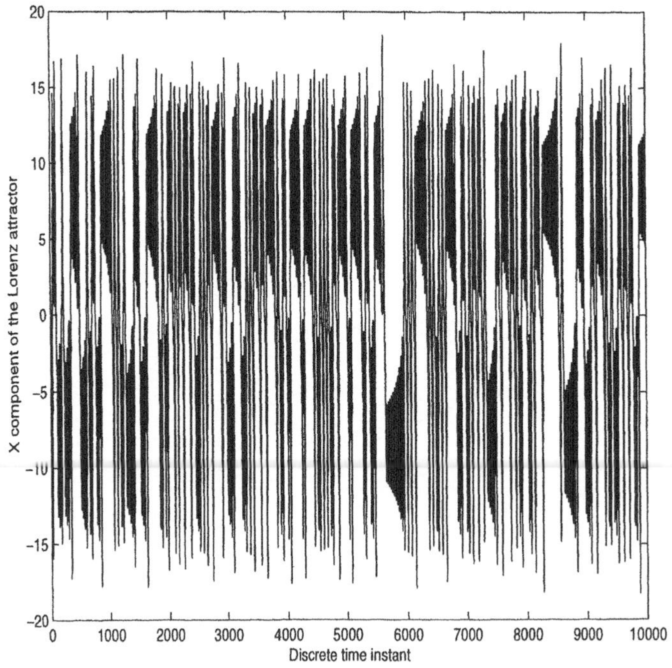

(b) The x-component of Lorenz attractor

Figure 11.13 The Lorenz attractor and its x-component

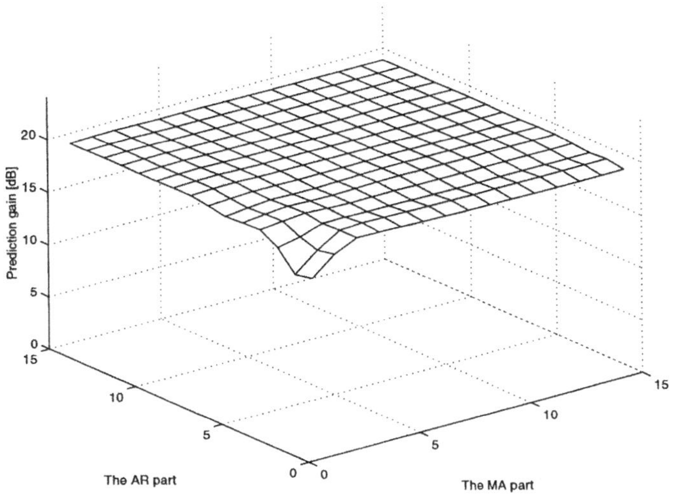

(a) Performance of the recurrent perceptron with fixed learning rate in prediction of the x-component of Lorenz system, prediction horizon is 1

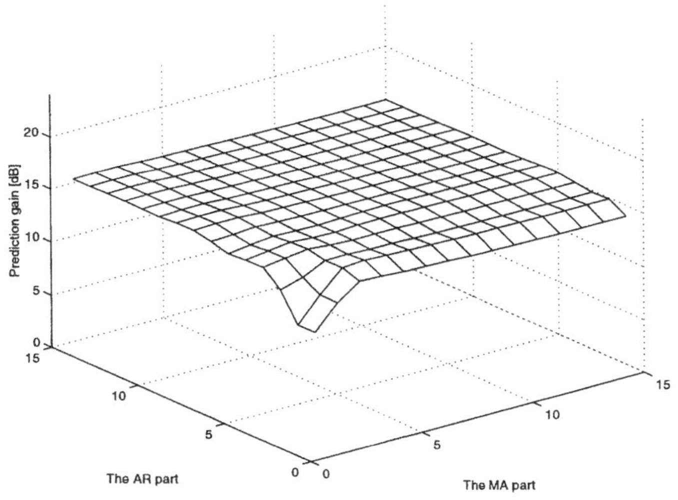

(b) Performance of the recurrent perceptron with fixed learning rate in prediction of the x-component of Lorenz system, prediction horizon is 2

Figure 11.14 Performance of algorithms for prediction of the x-component of the Lorenz system

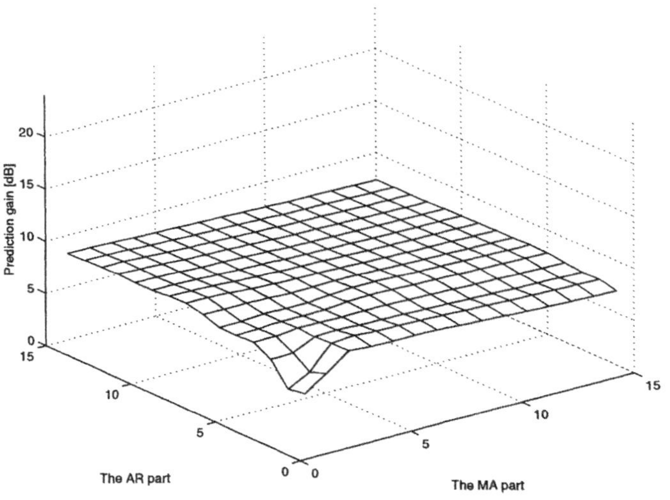

(c) Performance of the recurrent perceptron with fixed learning rate in prediction of the x-component of Lorenz system, prediction horizon is 5

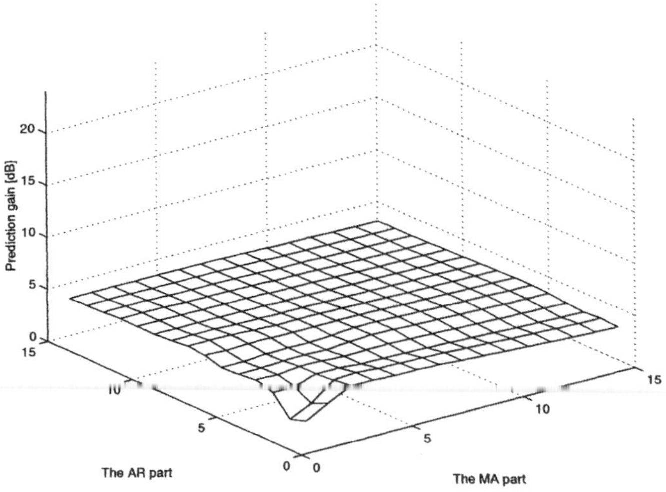

(d) Performance of the recurrent perceptron with fixed learning rate in prediction of the x-component of Lorenz system, prediction horizon is 10

Figure 11.14 *Cont.*

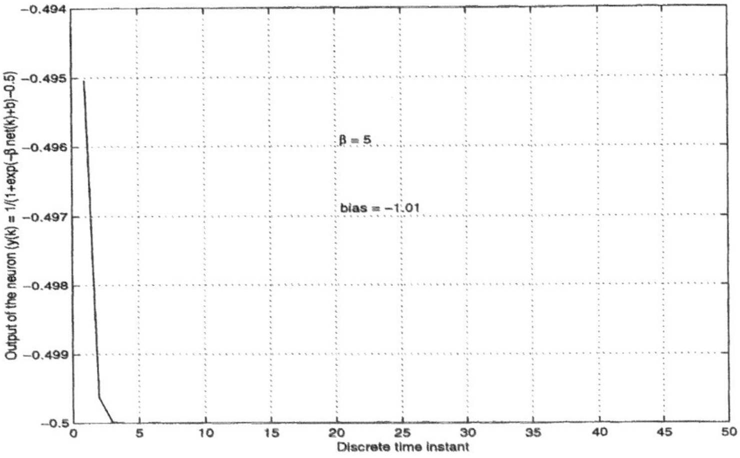

(a) Output of the recurrent perceptron with the logistic activation function, $\boldsymbol{w} = [1, 1.5, 1.5]$

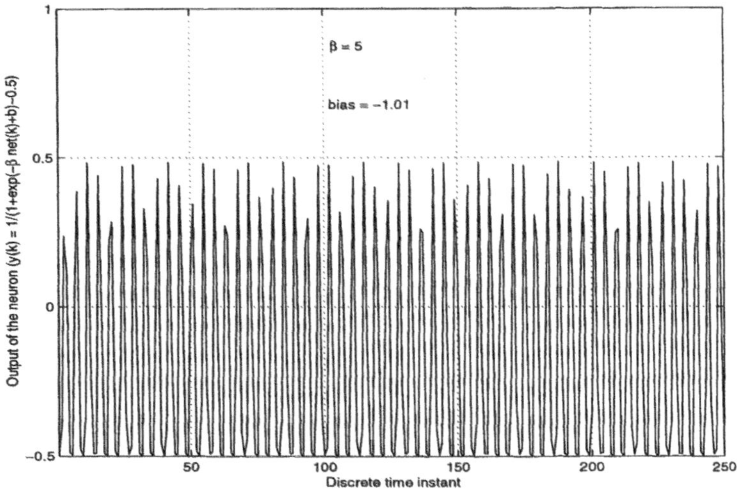

(b) Output of the recurrent perceptron with the logistic activation function, $\boldsymbol{w} = [-1, -1.5, -1.5]$

Figure 11.15 Output of a perceptron for different activation functions

11.6 Prediction of the Lorenz Chaotic Series

The Lorenz attractor is a system of three nonlinear coupled differential equations, given by (11.2). In order to perform experiments on the Lorenz system, the set of

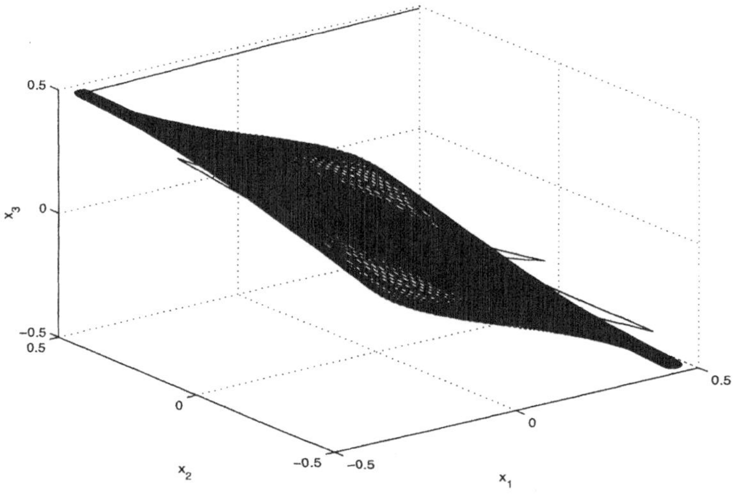

(a) Attractor of the recurrent perceptron with the logistic activation function, bias = 0.01

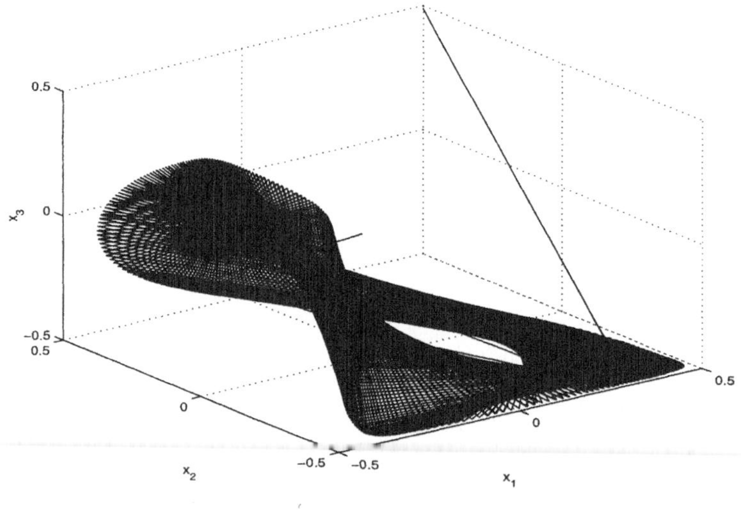

(b) Attractor of the recurrent perceptron with the logistic activation function, bias = -1.01

Figure 11.16 Attractor diagrams for a recurrent perceptron and various biases

nonlinear Equations (11.2) was integrated using the Runge–Kutta method of numerical integration, starting from the initial state $[5, 20, -5]^{\mathrm{T}}$. Thus, we obtained three discrete time series, sampled with sampling frequency $f_s = 40$ Hz. Figure 11.13 shows the Lorenz attractor and its x-component. Different GD-based neural adaptive filters were employed in prediction of the x-component of the Lorenz system. In applied neural adaptive filters the parameters of the filters were the same as for the previous experiments. Due to the saturation type of logistic nonlinearity, input data were prescaled to fit within the range of a neuron activation function. In the experiment the MA and the AR part of the recurrent perceptron vary from 1 to 15, while the prediction horizon varies from 1 to 10. Figure 11.14 shows the performance of the recurrent perceptron with fixed η in prediction of the x-component of the Lorenz system, for different prediction horizons. Again, the recurrent structure was able to capture the dynamics of the modelled signal. As expected, due to the nature of the signal, the nonlinear structures were able to achieve good short term prediction performance, whereas the long term prediction performance was bad.

11.7 Bifurcations in Recurrent Neural Networks

Presence of feedback in a neural network structure may result in a very complex output behaviour. The output of the RNN is bounded, due to the saturation type of nonlinear activation function of an output neuron. Further, behaviour of an RNN can be described by the set of nonlinear difference equations. Evidence of chaotic motion within the dynamics of an RNN has been given in Dror and Tsodyks (2000) and Chen and Aihara (1997). Even though the recurrent perceptron has a simple structure, for certain values of the network parameters, the dynamics of a recurrent perceptron can exhibit a very complex behaviour. Here, we look at the value of the slope β of a neuron activation function and the value of the bias term b. Figure 11.15(a) shows the output of the recurrent perceptron with the logistic activation function shifted by -0.5, i.e. $y = (1/(1 + \mathrm{e}^{-\beta x})) - 0.5$, with $\beta = 5$ and $q = 3$,[1] weight vector $\boldsymbol{w} = [1, 1.5, 1.5]^{\mathrm{T}}$ and bias $= -1.01$. Figure 11.15(b) shows the output of the recurrent perceptron with the logistic activation function shifted by -0.5, i.e. $y = (1/(1 + \mathrm{e}^{-\beta x})) - 0.5$, with $\beta = 5$, $q = 3$, weight vector $\boldsymbol{w} = [-1, -1.5, -1.5]^{\mathrm{T}}$ and bias $= -1.01$. In both cases, the dynamics of the recurrent system were autonomous, hence performing a kind of fixed point iteration and producing the so-called bifurcation maps. From the figure, the recurrent structures produce complex dynamical behaviour. To further depict the ability of recurrent neural networks to produce complex dynamics, we plot the attractors for the above simulated bifurcation diagrams. Figure 11.16(a) shows the attractor of the recurrent perceptron with the logistic activation function shifted by -0.5, i.e. $y = (1/(1 + \exp(-\beta x))) - 0.5$, with $\beta = 5$, $q = 3$, weight vector $\boldsymbol{w} = [-1, -1.5, -1.5]^{\mathrm{T}}$ and bias $= -0.01$. Figure 11.16(b) shows the attractor of the recurrent perceptron with the logistic activation function shifted by -0.5, i.e. $y = (1/(1+\exp(-\beta x))) - 0.5$, with $\beta = 5$, $q = 3$, weight vector $\boldsymbol{w} = [-1, -1.5, -1.5]^{\mathrm{T}}$ and bias $= -1.01$. The x_1-, x_2- and x_3-axes from Figure 11.16 represent, respectively, the outputs $y(k)$, $y(k-1)$ and $y(k-2)$ from a recurrent perceptron. The attractors show regularity and change the shape according to the change in the input parameters. Further reading on chaos in neural networks can be found in Haykin and Principe (1998).

[1] This q is within the NARMA(p, q) model of a recurrent perceptron, i.e. the order of the feedback.

11.8 Summary

To demonstrate some practical issues and difficulties with modelling and prediction of real-time signals, an analysis of nonlinearity, predictability and performance of nonlinear adaptive algorithms has been undertaken. These issues, especially the detection of nonlinearity, are often neglected by practitioners in the area. The experimental results confirm the superiority of nonlinear architectures and algorithms over the linear ones for modelling a nonlinear signal. Finally, the ability of recurrent structures to exhibit complex dynamical behaviour, such as bifurcations and attractors is briefly demonstrated. In the experiments, recurrent perceptrons are used due to their simplicity and due to the fact that to make a fair comparison with the feedforward and linear structures, a simple recurrent structure ought to be used.

12

Exploiting Inherent Relationships Between Parameters in Recurrent Neural Networks

12.1 Perspective

Optimisation of complex neural network parameters is a rather involved task. It becomes particularly difficult for large-scale networks, such as modular networks, and for networks with complex interconnections, such as feedback networks. Therefore, if an inherent relationship between some of the free parameters of a neural network can be found, which holds at every time instant for a dynamical network, it would help to reduce the number of degrees of freedom in the optimisation task of learning in a particular network.

We derive such relationships between the gain β in the nonlinear activation function of a neuron Φ and the learning rate η of the underlying learning algorithm for both the gradient descent and extended Kalman filter trained recurrent neural networks.

The analysis is then extended in the same spirit for modular neural networks. Both the networks with parallel modules and networks with nested (serial) modules are analysed. A detailed analysis is provided for the latter, since the former can be considered a linear combination of modules that consist of feedforward or recurrent neural networks.

For all these cases, the static and dynamic equivalence between an arbitrary neural network described by β, η and $\boldsymbol{W}(k)$ and a referent network described by $\beta^{\mathrm{R}} = 1$, η^{R} and $\boldsymbol{W}^{\mathrm{R}}(k)$ are derived. A deterministic relationship between these parameters is provided, which allows one degree of freedom less in the nonlinear optimisation task of learning in this framework. This is particularly significant for large-scale networks of any type.

12.2 Introduction

When using neural networks, many of their parameters are chosen empirically. Apart from the choice of topology, architecture and interconnection, the parameters that influence training time and performance of a neural network are the learning rate η, gain of the activation function β and set of initial weights $\boldsymbol{W}_0$. The optimal values for these parameters are not known *a priori* and generally they depend on external quantities, such as the training data. Other parameters that are also important in this context are

- steepness of the sigmoidal activation function, defined by $\gamma\beta$; and
- dimensionality of the input signal to the network and dimensionality and character of the feedback for recurrent networks.

It has been shown (Thimm and Fiesler 1997a,b) that the distribution of the initial weights has almost no influence on the training time or the generalisation performance of a trained neural network. Hence, we concentrate on the relationship between the parameters of a learning algorithm (η) and those of a nonlinear activation function (β).

To improve performance of a gradient descent trained network, Jacobs (1988) proposed that the acceleration of convergence of learning in neural networks be achieved through the learning rate adaptation. His arguments were that

1. every adjustable learning parameter of the cost function should have its own learning rate parameter; and
2. every learning rate parameter should vary from one iteration to the next.

These arguments are intuitively sound. However, if there is a dependence between some of the parameters in the network, this approach would lead to suboptimal learning and oscillations, since coupled parameters would be trained using different learning rates and different speed of learning, which would deteriorate the performance of the network. To circumvent this problem, some heuristics on the values of the parameters have been derived (Haykin 1994). To shed further light onto this problem and offer feasible solutions, we therefore concentrate on finding relationships between coupled parameters in recurrent neural networks. The derived relationships are also valid for feedforward networks, since recurrent networks degenerate into feedforward networks when the feedback is removed.

Let us consider again a common choice for the activation function,

$$\Phi(\gamma, \beta, x) = \frac{\gamma}{1 + \mathrm{e}^{-\beta x}}. \tag{12.1}$$

This is a $\Phi : \mathbb{R} \to (0, \gamma)$ function. The parameter β is called the *gain* and the product $\gamma\beta$ the *steepness* (slope) of the activation function.[1] The reciprocal of gain is also referred to as the *temperature*. The gain γ of a node in a neural network is a constant that amplifies or attenuates the net input to the node. In Kruschke and Movellan (1991), it has been shown that the use of gradient descent to adjust the gain of the node increases learning speed.

Let us consider again the general gradient-descent-based weight adaptation algorithm, given by

$$\boldsymbol{W}(k) = \boldsymbol{W}(k-1) - \eta\nabla_{\boldsymbol{W}} E(k), \tag{12.2}$$

[1] The gain and steepness are identical for activation functions with $\gamma = 1$. Hence, for such networks, we often use the term slope for β.

where $E(k) = \frac{1}{2}e^2(k)$ is a cost function, $\boldsymbol{W}(k)$ is the weight vector/matrix at the time-instant k and η is a learning rate. The gradient $\nabla_{\boldsymbol{W}} E(k)$ in (12.2) comprises the first derivative of the nonlinear activation function (12.1), which is a function of β (Narendra and Parthasarathy 1990). For instance, for a simple nonlinear FIR filter shown in Figure 12.1, the weight update is given by

$$\boldsymbol{w}(k+1) = \boldsymbol{w}(k) + \eta\Phi'(\boldsymbol{x}^{\mathrm{T}}(k)\boldsymbol{w}(k))e(k)\boldsymbol{x}(k). \tag{12.3}$$

For a function $\Phi(\beta, x) = \Phi(\beta x)$, which is the case for the logistic, tanh and arctan nonlinear functions,[2] Equation (12.3) becomes

$$\boldsymbol{w}(k+1) = \boldsymbol{w}(k) + \eta\beta\Phi'(\boldsymbol{x}^{\mathrm{T}}(k)\boldsymbol{w}(k))e(k)\boldsymbol{x}(k). \tag{12.4}$$

From (12.4), if β increases, so too will the step on the error performance surface for a fixed η. It seems, therefore, advisable to keep β constant, say at unity, and to control the features of the learning process by adjusting the learning rate η, thereby having one degree of freedom less, when all of the parameters in the network are adjustable. Such reduction may be very significant for nonlinear optimisation algorithms employed for parameter adaptation in a particular recurrent neural network.

A fairly general gradient algorithm that continuously adjusts parameters η, β and γ can be expressed by

$$\left.\begin{aligned} y(k) &= \Phi(\boldsymbol{X}(k), \boldsymbol{W}(k)), \\ e(k) &= s(k) - y(k), \\ \boldsymbol{W}(k+1) &= \boldsymbol{W}(k) - \frac{\eta(k)}{2}\frac{\partial e^2(k)}{\partial \boldsymbol{W}(k)}, \\ \eta(k+1) &= \eta(k) - \frac{\rho}{2}\frac{\partial e^2(k)}{\partial \eta(k)}, \\ \beta(k+1) &= \beta(k) - \frac{\theta}{2}\frac{\partial e^2(k)}{\partial \beta(k)}, \\ \gamma(k+1) &= \gamma(k) - \frac{\zeta}{2}\frac{\partial e^2(k)}{\partial \gamma(k)}, \end{aligned}\right\} \tag{12.5}$$

where ρ is a small positive constant that controls the adaptive behaviour of the step size sequence $\eta(k)$, whereas small positive constants θ and ζ control the adaptation

---

[2] For the logistic function

$$\sigma(\beta, x) = \frac{1}{1 + \mathrm{e}^{-\beta x}} = \sigma(\beta x),$$

its first derivative becomes

$$\frac{\mathrm{d}\sigma(\beta, x)}{\mathrm{d}x} = -\frac{\beta \mathrm{e}^{-\beta x}}{(1 + \mathrm{e}^{-\beta x})^2},$$

whereas for the tanh function

$$\tanh(\beta, x) = \frac{\mathrm{e}^{\beta x} - \mathrm{e}^{-\beta x}}{\mathrm{e}^{\beta x} + \mathrm{e}^{-\beta x}} = \tanh(\beta x),$$

we have

$$\frac{\mathrm{d}\tanh(\beta x)}{\mathrm{d}x} = \beta\frac{\mathrm{d}\tanh(\beta x)}{\mathrm{d}\beta x}.$$

The same principle is valid for the Gaussian and inverse tangent activation functions.

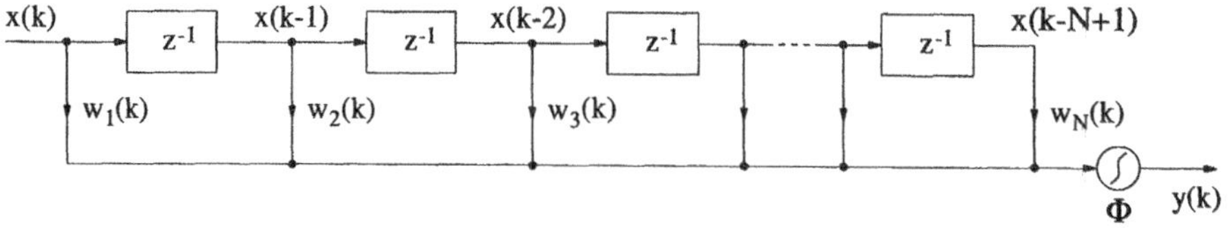

Figure 12.1 A simple nonlinear adaptive filter

of the gain of the activation function β and gain of the node γ, respectively. We will concentrate only on adaptation of β and η.

The selection of learning rate η is critical for the gradient descent algorithms (Mathews and Xie 1993). An η that is small as compared to the reciprocal of the input signal power will ensure small misadjustment in the steady state, but the algorithm will converge slowly. A relatively large η, on the other hand, will provide faster convergence at the cost of worse misadjustment and steady-state characteristics. Therefore, an ideal choice would be an adjustable η which would be relatively large in the beginning of adaptation and become gradually smaller when approaching the global minimum of the error performance surface (optimal values of weights).

We illustrate the above ideas on the example of a simple nonlinear FIR filter, shown in Figure 12.1, for which the output is given by

$$y(k) = \Phi(\boldsymbol{x}^{\mathrm{T}}(k)\boldsymbol{w}(k)). \tag{12.6}$$

We can continually adapt the step size using a gradient descent algorithm so as to reduce the squared estimation error at each time instant. Extending the approach from Mathews and Xie (1993) to the nonlinear case, we obtain

$$\left.\begin{aligned} e(k) &= s(k) - \Phi(\boldsymbol{x}^{\mathrm{T}}(k)\boldsymbol{w}(k)), \\ \boldsymbol{w}(k) &= \boldsymbol{w}(k-1) + \eta(k-1)e(k-1)\Phi'(k-1)\boldsymbol{x}(k-1), \\ \eta(k) &= \eta(k-1) - \frac{\rho}{2}\frac{\partial}{\partial\eta(k-1)}e^2(k) \\ &= \eta(k-1) - \frac{\rho}{2}\frac{\partial^{\mathrm{T}}e^2(k)}{\partial\boldsymbol{w}(k)}\frac{\partial\boldsymbol{w}(k)}{\partial\eta(k-1)} \\ &= \eta(k-1) + \rho e(k)e(k-1)\Phi'(k)\Phi'(k-1)\boldsymbol{x}^{\mathrm{T}}(k-1)\boldsymbol{x}(k), \end{aligned}\right\} \tag{12.7}$$

where $\Phi'(k) = \Phi'(\boldsymbol{x}^{\mathrm{T}}(k)\boldsymbol{w}(k))$, $\Phi'(k-1) = \Phi'(\boldsymbol{x}^{\mathrm{T}}(k-1)\boldsymbol{w}(k-1))$ and ρ is a small positive constant that controls the adaptive behaviour of the step size sequence $\eta(k)$.

If we adapt the step size for each weight individually, we have

$$\eta_i(k) = \eta_i(k-1) + \rho e(k)e(k-1)\Phi'(k)\Phi'(k-1)x_i(k)x_i(k-1), \qquad i = 1, \dots, N, \tag{12.8}$$

and

$$w_i(k+1) = w_i(k) + \eta_i(k)e(k)x_i(k), \qquad i = 1, \dots, N. \tag{12.9}$$

These expressions become much more complicated for large and recurrent networks.

As an alternative to the continual learning rate adaptation, we might consider continual adaptation of the gain of the activation function $\Phi(\beta x)$. The gradient descent algorithm that would update the adaptive gain can be expressed as

$$\left.\begin{aligned} e(k) &= s(k) - \Phi(\boldsymbol{w}^{\mathrm{T}}(k)\boldsymbol{x}(k)),\\ \boldsymbol{w}(k) &= \boldsymbol{w}(k-1) + \eta(k-1)e(k-1)\Phi'(k-1)\boldsymbol{x}(k-1),\\ \beta(k) &= \beta(k-1) - \frac{\theta}{2}\frac{\partial}{\partial\beta(k-1)}e^2(k)\\ &= \beta(k-1) - \frac{\theta}{2}\frac{\partial^{\mathrm{T}} e^2(k)}{\partial \boldsymbol{w}(k)}\frac{\partial \boldsymbol{w}(k)}{\partial\beta(k-1)}\\ &= \beta(k-1) + \theta\eta(k-1)e(k)e(k-1)\Phi'(k)\Phi''_{\beta}(k-1)\boldsymbol{x}^{\mathrm{T}}(k-1)\boldsymbol{x}(k). \end{aligned}\right\} \tag{12.10}$$

For the adaptation of $\beta(k)$ there is a need to calculate the second derivative of the activation function, which is rather computationally involved. Such an adaptive gain algorithm was, for instance, analysed in Birkett and Goubran (1997). The proposed function was

$$\sigma(x, a) = \begin{cases} x, & |x| \leqslant a,\\ \operatorname{sgn}(x)\left[(1-a)\tanh\left(\dfrac{|x|-a}{1-a}\right) + a\right], & |x| > a, \end{cases} \tag{12.11}$$

where x is the input signal and a defines the adaptive linear region of the sigmoid. This activation function is shown in Figure 12.2. Parameter a is updated according to the stochastic gradient rule. The benefit of this algorithm is that the slope and region of linearity of the activation function can be adjusted. Although this and similar approaches are an alternative to the learning rate adaptation, researchers have not taken into account that parameters β and η might be coupled. If a relationship between them can be derived, then we choose adaptation of the parameter that is less computationally expensive to adapt and less sensitive to adaptation errors. As shown above, adaptation of the gain β is far more computationally expensive that adaptation of η. Hence, there is a need to mathematically express the dependence between the two and reduce the computational load for training neural networks.

Thimm *et al.* (1996) provided the relationship between the gain β of the logistic activation function,

$$\Phi(\beta, x) = \frac{1}{1 + \mathrm{e}^{-\beta x}}, \tag{12.12}$$

and the learning rate η for a class of general feedforward neural networks trained by backpropagation. They prove that changing the gain of the activation function is equivalent to simultaneously changing the learning rate and the weights. This simplifies the backpropagation learning rule by eliminating one of its parameters (Thimm *et al.* 1996). This concept has been successfully applied to compensate for the non-standard gain of optical sigmoids for optical neural networks. Relationships between η and β for recurrent and modular networks were derived by Mandic and Chambers (1999a,e).

Basic modular architectures are the parallel and serial architecture. Parallel architectures provide linear combinations of neural network modules, and learning algorithms for them are based upon minimising the linear combination of the output

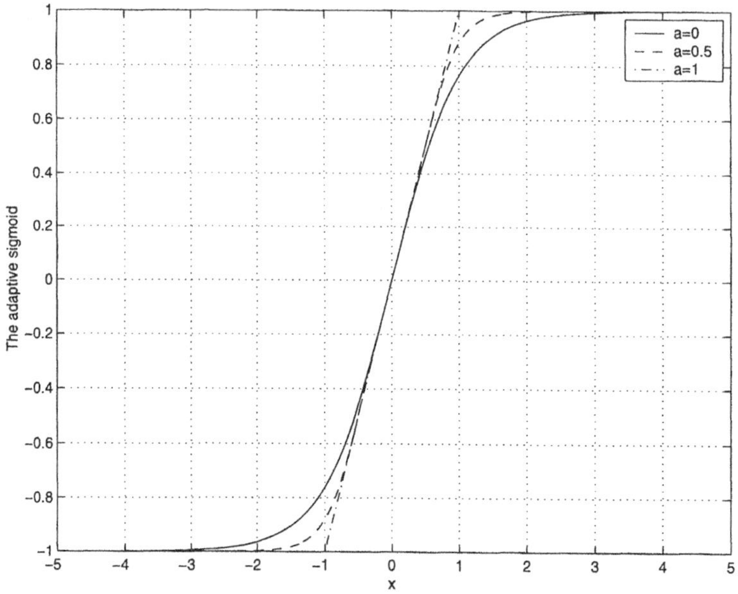

Figure 12.2 An adaptive sigmoid

errors of particular modules. Hence, the algorithms for training such networks are extensions of standard algorithms designed for single modules. Serial (nested) modular architectures are more complicated, an example of which is a pipelined recurrent neural network (PRNN). This is an emerging architecture used in nonlinear time series prediction (Haykin and Li 1995; Mandic and Chambers 1999f). It consists of a number of nested small-scale recurrent neural networks as its modules, which means that a learning algorithm for such a complex network has to perform a nonlinear optimisation task on a number of parameters. We look at relationships between the learning rate and the gain of the activation function for this architecture and for various learning algorithms.

12.3 Overview

A relationship between the learning rate η in the learning algorithm and the gain β in the nonlinear activation function, for a class of recurrent neural networks (RNNs) trained by the real-time recurrent learning (RTRL) algorithm is provided. It is shown that an arbitrary RNN can be obtained via the referent RNN, with some deterministic rules imposed on its weights and the learning rate. Such relationships reduce the number of degrees of freedom when solving the nonlinear optimisation task of finding the optimal RNN parameters. This analysis is further extended for modular neural architectures.

We define the conditions of static and dynamic equivalence between a referent network, with $\beta = 1$, and an arbitrary network with an arbitrary β. Since the dynamic equivalence is dependent on the chosen learning algorithm, the relationships are provided for a variety of both the gradient descent (GD) and the extended recursive least-squares (ERLS) classes of learning algorithms and a general nonlinear activation function of a neuron.

By continuity, the derived results are also valid for feedforward networks and their linear and nonlinear combinations.

12.4 Static and Dynamic Equivalence of Two Topologically Identical RNNs

As the aim is to eliminate either the gain β or the learning rate η from the paradigm of optimisation of the RNN parameters, it is necessary to derive the relationship between a network with arbitrarily chosen parameters β and η and the referent network, so as the outputs of the networks are identical for every time instant. An obvious choice for the referent network is the network with the gain of the activation function $\beta = 1$. Let us therefore denote all the entries in the referent network, which are different from those of an arbitrary network, with the superscript 'R' joined to a particular variable, i.e. $\beta^{\mathrm{R}} = 1$.

For two networks to be equivalent, it is necessary that their outputs are identical and that this is valid for both the trained network and while on the run, i.e. while tracking some dynamical process. We therefore differentiate between the equivalence in the static and dynamic sense. We define the static and dynamic equivalence between two networks below.

Definition 12.4.1. By static equivalence, we consider the equivalence of the outputs of an arbitrary network and the referent network with fixed weights, for a given input vector $\boldsymbol{u}(k)$, at a fixed time instant k.

Definition 12.4.2. By dynamic equivalence, we consider the equivalence of the outputs between an arbitrary network and the referent network for a given input vector $\boldsymbol{u}(k)$, with respect to the learning algorithm, while the networks are running.

The static equivalence is considered for already trained networks, whereas both static and dynamic equivalence are considered for networks being adapted on the run. We can think of the static equivalence as an analogue to the forward pass in computation of the outputs of a neural network, whereas the dynamic equivalence can be thought of in terms of backward pass, i.e. weight update process. We next derive the conditions for either case.

12.4.1 Static Equivalence of Two Isomorphic RNNs

In order to establish the static equivalence between an arbitrary and referent RNN, the outputs of their neurons must be the same, i.e.

$$y_n(k) = y_n^{\mathrm{R}}(k) \quad \Leftrightarrow \quad \Phi(\boldsymbol{u}_n^{\mathrm{T}}(k)\boldsymbol{w}_n(k)) = \Phi^{\mathrm{R}}(\boldsymbol{u}_n^{\mathrm{T}}(k)\boldsymbol{w}_n^{\mathrm{R}}(k)), \tag{12.13}$$

where the index 'n' runs over all neurons in an RNN, and $\boldsymbol{w}_n(k)$ and $\boldsymbol{u}_n(k)$ are, respectively, the set of weights and the set of inputs which belong to the neuron n. For a general nonlinear activation function, we have

$$\Phi(\beta, \boldsymbol{w}_n, \boldsymbol{u}_n) = \Phi(1, \boldsymbol{w}_n^{\mathrm{R}}, \boldsymbol{u}_n) \quad \Leftrightarrow \quad \beta \boldsymbol{w}_n = \boldsymbol{w}_n^{\mathrm{R}}. \tag{12.14}$$

To illustrate this, consider, for instance, the logistic nonlinearity, given by

$$\frac{1}{1+\mathrm{e}^{-\beta \boldsymbol{u}_n^{\mathrm{T}} \boldsymbol{w}_n}} = \frac{1}{1+\mathrm{e}^{-\boldsymbol{u}_n^{\mathrm{T}} \boldsymbol{w}_n^{\mathrm{R}}}} \quad \Leftrightarrow \quad \beta \boldsymbol{w}_n = \boldsymbol{w}_n^{\mathrm{R}}, \tag{12.15}$$

where the time index (k) is neglected, since all the vectors above are constant during the calculation of the output values. As the equality (12.14) should be valid for every neuron in the RNN, it is therefore valid for the complete weight matrix $\boldsymbol{W}$ of the RNN.

The essence of the above analysis is given in the following lemma, which is independent of the underlying learning algorithm for the RNN, which makes it valid for two isomorphic[3] RNNs of any topology and architecture.

Lemma 12.4.3 (see Mandic and Chambers 1999e). *For a recurrent neural network, with weight matrix $\boldsymbol{W}$ and gain of the activation function β, to be equivalent in the static sense to the referent network, characterised by $\boldsymbol{W}^{\mathrm{R}}$ and $\beta^{\mathrm{R}} = 1$, with the same topology and architecture (isomorphic), the following condition must be satisfied:*

$$\beta \boldsymbol{W}(k) = \boldsymbol{W}^{\mathrm{R}}(k). \tag{12.16}$$

12.4.2 Dynamic Equivalence of Two Isomorphic RNNs

The equivalence of two RNNs, includes both the static equivalence and dynamic equivalence. As in the learning process (12.2), the learning rate η is multiplied by the gradient of the cost function, we shall investigate the role of β in the gradient of the cost function for the RNN. We are interested in a general class of nonlinear activation functions where

$$\frac{\partial \Phi(\beta, x)}{\partial x} = \frac{\partial \Phi(\beta x)}{\partial(\beta x)} \frac{\partial(\beta x)}{\partial x} = \beta \frac{\partial \Phi(\beta x)}{\partial(\beta x)} = \beta \frac{\partial \Phi(1, x)}{\partial x}. \tag{12.17}$$

In our case, it becomes

$$\Phi'(\beta, \boldsymbol{w}, \boldsymbol{u}) = \beta \Phi'(1, \boldsymbol{w}^{\mathrm{R}}, \boldsymbol{u}). \tag{12.18}$$

Indeed, for a simple logistic function (12.12), we have

$$\Phi'(x) = \frac{\beta \mathrm{e}^{-\beta x}}{(1+\mathrm{e}^{-\beta x})^2} = \beta \Phi'(x^{\mathrm{R}}),$$

where $x^{\mathrm{R}} = \beta x$ denotes the argument of the referent logistic function (with $\beta^{\mathrm{R}} = 1$), so that the network considered is equivalent in the static sense to the referent network. The results (12.17) and (12.18) mean that wherever Φ' occurs in the dynamical equation of the RTRL-based learning process, the first derivative (or gradient when applied to all the elements of the weight matrix $\boldsymbol{W}$) of the referent function equivalent in the static sense to the one considered, becomes multiplied by the gain β.

The following theorem provides both the static and dynamic interchangeability of the gain in the activation function β and the learning rate η for the RNNs trained by the RTRL algorithm.

---

[3] Isomorphic networks have identical topology, architecture and interconnections.

Theorem 12.4.4 (see Mandic and Chambers 1999e). *For a recurrent neural network, with weight matrix $\boldsymbol{W}$, gain of the activation function β and learning rate in the RTRL algorithm η, to be equivalent in the dynamic sense to the referent network, characterised by $\boldsymbol{W}^{\mathrm{R}}$, $\beta^{\mathrm{R}} = 1$ and η^{R}, with the same topology and architecture (isomorphic), the following conditions must hold.*

(i) *The networks are equivalent in the static sense, i.e.*

$$\boldsymbol{W}^{\mathrm{R}}(k) = \beta \boldsymbol{W}(k). \tag{12.19}$$

(ii) *The learning rate η of the network considered and the learning rate η^{R} of the referent network are related by*

$$\eta^{\mathrm{R}} = \beta^2 \eta. \tag{12.20}$$

Proof. From the equivalence in the static sense, the weight update equation for the referent network can be written as

$$\boldsymbol{W}^{\mathrm{R}}(k) = \boldsymbol{W}^{\mathrm{R}}(k-1) + \beta \Delta \boldsymbol{W}(k), \tag{12.21}$$

which gives

$$\Delta \boldsymbol{W}^{\mathrm{R}}(k) = \beta \Delta \boldsymbol{W}(k) = \beta\left(\eta e(k) \frac{\partial y_1(k)}{\partial \boldsymbol{W}(k)}\right) = \eta \beta e(k) \boldsymbol{\Pi}_1(k), \tag{12.22}$$

where $\boldsymbol{\Pi}_1(k)$ is the matrix with elements $\pi^1_{n,l}(k)$.

Now, in order to derive the conditions of dynamical equivalence between an arbitrary and the referent RNN, the relationship between the appropriate matrices $\boldsymbol{\Pi}_1(k)$ and $\boldsymbol{\Pi}^{\mathrm{R}}_1(k)$ must be established. That implies that, for all the neurons in the RNN, the matrix $\boldsymbol{\Pi}(k)$, which comprises all the terms $\partial y_j / \partial w_{n,l}$, $\forall w_{n,l} \in \boldsymbol{W}$, $j = 1, 2, \ldots, N$, must be interrelated to the appropriate matrix $\boldsymbol{\Pi}^{\mathrm{R}}(k)$, which represents the referent network.

We shall prove this relationship by induction. For convenience, let us denote $\mathrm{net}(k) = \boldsymbol{u}^{\mathrm{T}}(k)\boldsymbol{w}(k)$ and $\mathrm{net}^{\mathrm{R}}(k) = \boldsymbol{u}^{\mathrm{T}}(k)\boldsymbol{w}^{\mathrm{R}}(k)$.

Given:

$$\begin{aligned}
\boldsymbol{W}^{\mathrm{R}}(k) &= \beta \boldsymbol{W}(k) && \text{(static equivalence)},\\
\Phi'(\mathrm{net}^{\mathrm{R}}(k)) &= \frac{1}{\beta}\Phi'(\mathrm{net}(k)) && \text{(activation function derivative)},\\
y^{\mathrm{R}}_j(k) &= \Phi(\mathrm{net}^{\mathrm{R}}(k)) && \\
&= \Phi(\mathrm{net}(k)) && \\
&= y_j(k), \qquad j = 1, \ldots, N && \text{(activation)}.
\end{aligned}$$

Induction base: the recursion (D.11) starts as

$$\begin{aligned}(\pi^j_{n,l}(k=1))^{\mathrm{R}} &= \Phi'(\mathrm{net}^{\mathrm{R}}(k))\left[\sum_{m=1}^{N} w^{\mathrm{R}}_{j,m+p+1}(k=0)\pi^m_{n,l}(k=0) + \delta_{nj}u_l(k=0)\right] \\ &= \frac{1}{\beta}\Phi'(\mathrm{net}(k))\delta_{nj}u_l(k=0) \\ &= \frac{1}{\beta}\pi^j_{n,l}(k=1),\end{aligned}$$

which gives $\boldsymbol{\Pi}^{\mathrm{R}}(k=1) = (1/\beta)\boldsymbol{\Pi}(k=1)$.

Induction step:

$$(\pi^j_{n,l}(k))^{\mathrm{R}} = \frac{1}{\beta}\pi^j_{n,l}(k) \quad \text{and} \quad \boldsymbol{\Pi}^{\mathrm{R}}(k) = \frac{1}{\beta}\boldsymbol{\Pi}(k) \quad \text{(assumption)}.$$

Now, for the $(k+1)$st step we have

$$\begin{aligned}(\pi^j_{n,l}(k+1))^{\mathrm{R}} &= \Phi'(\mathrm{net}^{\mathrm{R}})\left[\sum_{m=1}^{N} w^{\mathrm{R}}_{j,m+p+1}(k)\pi^m_{n,l}(k) + \delta_{nj}u_l(k)\right] \\ &= \frac{1}{\beta}\Phi'(\mathrm{net})\left[\sum_{m=1}^{N} \beta w_{j,m+p+1}(k)\frac{1}{\beta}\pi^m_{n,l}(k) + \delta_{nj}u_l(k)\right] \\ &= \frac{1}{\beta}\pi^j_{n,l}(k+1),\end{aligned}$$

which means that

$$\boldsymbol{\Pi}^{\mathrm{R}}(k+1) = \frac{1}{\beta}\boldsymbol{\Pi}(k+1).$$

Based upon the established relationship, the learning process for the referent RNN can be expressed as

$$\Delta \boldsymbol{W}^{\mathrm{R}}(k) = \beta\Delta\boldsymbol{W}(k) = \beta\eta e(k)\boldsymbol{\Pi}_1(k) = \beta^2\eta e(k)\boldsymbol{\Pi}^{\mathrm{R}}_1(k) = \eta^{\mathrm{R}}e(k)\boldsymbol{\Pi}^{\mathrm{R}}_1(k). \tag{12.23}$$

Hence, the referent network with the learning rate $\eta^{\mathrm{R}} = \beta^2\eta$ and gain $\beta^{\mathrm{R}} = 1$ is equivalent in the dynamic sense, with respect to the RTRL algorithm, to an arbitrary RNN with gain β and learning rate η. □

12.5 Extension to a General RTRL Trained RNN

It is now straightforward to show that the conditions for static and dynamic equivalence derived so far are valid for a general recurrent neural network trained by a gradient algorithm. For instance, for a general RTRL trained RNN, the cost function comprises squared error terms over all the output neurons, i.e.

$$E(k) = \frac{1}{2}\sum_{j\in C} e^2_j(k), \tag{12.24}$$

where $\mathcal{C}$ denotes the set of those neurons whose outputs are included in the cost function. The crucial equations for the static and dynamic equivalence remain intact. It should, however, be noted that the weight update equation in the RTRL algorithm (12.23) now becomes

$$\begin{aligned}\Delta \boldsymbol{W}^{\mathrm{R}}(k) &= \beta \Delta \boldsymbol{W}(k) = \beta\eta \sum_{i\in\mathcal{C}} e_i(k)\boldsymbol{\Pi}(k) \\ &= \beta^2\eta \sum_{i\in\mathcal{C}} e_i(k)\boldsymbol{\Pi}^{\mathrm{R}}(k) = \eta^{\mathrm{R}} \sum_{i\in\mathcal{C}} e_i(k)\boldsymbol{\Pi}^{\mathrm{R}}(k), \end{aligned} \tag{12.25}$$

which is structurally the same as the one used in the proof of Theorem 12.4.4. Hence the relationships between β and η derived for a single-output neuron RNN are valid for a general RNN with any number of output neurons.

12.6 Extension to Other Commonly Used Activation Functions

Other frequently used nonlinear activation functions of a neuron, such as the hyperbolic tangent $\Phi(x) = \tanh(\beta x)$ and inverse tangent $\Phi(x) = \arctan(\beta x)$, have the same functional dependence between β, x and their derivatives as the logistic function. Since for the dynamic equivalence of two isomorphic networks, it is crucial to consider the first derivative of the underlying activation function, let us recall that for all of the nonlinear activation functions mentioned, the first derivative of a function with an arbitrary gain can be connected to the derivative with respect to variable x as

$$\beta \frac{\mathrm{d}\Phi(\beta, x)}{\mathrm{d}(\beta x)} = \frac{\mathrm{d}\Phi(\beta, x)}{\mathrm{d}x}, \tag{12.26}$$

which means that the results derived for the logistic activation function are also valid for a large class of nonlinear activation functions with the first derivative that satisfies (12.26). This can be formally expressed as follows (Mandic and Krcmar 2000).

Corollary 12.6.1. *Theorem 12.4.4 holds for every differentiable nonlinear activation function with the first derivative that satisfies*

$$\beta \frac{\mathrm{d}\Phi(\beta, x)}{\mathrm{d}(\beta x)} = \frac{\mathrm{d}\Phi(\beta, x)}{\mathrm{d}x}. \tag{12.27}$$

12.7 Extension to Other Commonly Used Learning Algorithms for Recurrent Neural Networks

After having derived the relationship between the learning rate η and the gain in the activation function β for the RTRL trained recurrent neural network of an arbitrary size, and knowing that the dynamical equivalence of an arbitrary and the referent network is highly dependent on the learning algorithm chosen, let us consider two other frequently used learning algorithms for a general RNN, namely the backpropagation through time (BPTT) algorithm and the recurrent backpropagation (RBP) algorithm.

In both cases, the derivation of the conditions of the static equivalence between an arbitrary and the referent network follow the same spirit as for the RTRL trained RNN, and will be omitted. Moreover, the analysis given in Section 12.6, holds for these algorithms, and we will therefore consider the whole class of functions for which (12.26) holds, without going into detail for a particular function belonging to the class.

12.7.1 *Relationships Between β and η for the Backpropagation Through Time Algorithm*

The backpropagation through time algorithm for training a recurrent neural network stems from the common backpropagation algorithm. It may be derived by unfolding the temporal operation of the network into a multilayer feedforward network, the topology of which grows by one such block at every time step (Haykin 1994; Werbos 1990). As our intention is to show the conditions of dynamical equivalence between an arbitrary and referent RNN, with respect to the parameters β and η, we will concentrate only on the points that are specific to BPTT with regards to the RTRL algorithm. Therefore, the correction to the element $w_{i,j}(k)$ of the weight matrix $\boldsymbol{W}$ for the BPTT trained RNN can be expressed as

$$\Delta w_{i,j}(k) = -\eta \sum_{k=k_0+1}^{k_1} \delta_j(k) x_i(k-1), \tag{12.28}$$

where the BPTT is performed back to time (k_0+1) and the gradients δ are calculated as (Haykin 1994; Werbos 1990)

$$\delta_j(k) = \begin{cases} \Phi'(v_j(k))e_j(k), & k = k_1, \\ \Phi'(v_j(k))\left[e_j(k) + \sum_{i\in\mathcal{A}} w_{i,j}\delta_i(k-1)\right], & k_0 < k < k_1. \end{cases} \tag{12.29}$$

The symbol $\mathcal{A}$ denotes the set of output neurons and $e_j(k)$, $j \in \mathcal{A}$, are the corresponding errors. The first line in (12.29) is structurally the same as in the standard backpropagation algorithm, whereas in the second line in (12.29), there is a sum of product terms $w_{i,j}\delta_i(k-1)$, which is different from the backpropagation algorithm, since it comprises the unfolded feedback terms. Examining these product terms, which are structurally the same as appropriate terms in (D.11), it follows that (Mandic and Chambers 1999a)

$$w_{i,j}\delta_i = w_{i,j}^{\mathrm{R}}\delta_i^{\mathrm{R}}. \tag{12.30}$$

Now, from the results obtained for the RTRL trained RNN and relationships for feedforward networks trained by backpropagation (Thimm *et al.* 1996), the conditions of static and dynamic equivalence of an arbitrary and the referent BPTT trained RNN are expressed in the following corollary.

Corollary 12.7.1. *For a recurrent neural network, with weight matrix $\boldsymbol{W}$, gain in the activation function β and learning rate in the BPTT algorithm η, to be equivalent to the referent network, characterised by $\boldsymbol{W}^{\mathrm{R}}$, $\beta^{\mathrm{R}} = 1$ and η^{R}, with the same topology and architecture (isomorphic), the following conditions must hold.*

(i) *The networks are equivalent in the static sense, i.e.*

$$\boldsymbol{W}^{\mathrm{R}}(k) = \beta \boldsymbol{W}(k). \tag{12.31}$$

(ii) *The learning rate* η *of the network considered and the learning rate* η^{R} *of the equivalent referent network are related by*

$$\eta^{\mathrm{R}} = \beta^2 \eta. \tag{12.32}$$

As seen from Corollary 12.7.1, the conditions for the static and dynamic equivalence of an arbitrary and the referent RNN trained by the BPTT algorithm are the same as for the RTRL trained RNN.

12.7.2 Results for the Recurrent Backpropagation Algorithm

Without going deeply into the core of the RBP algorithm (Haykin 1994; Pineda 1989), let us only consider the key relation for the gradient calculation, which can be expressed as

$$\delta_n(k) = \Phi'\biggl(\sum_i \delta_i(k) w_{i,n}(k) + e_n(k)\biggr), \qquad n = 1, \dots, N, \tag{12.33}$$

which is structurally the same as the appropriate term for the RTRL algorithm. Therefore, we can express the conditions of the static and dynamic equivalence of an arbitrary and the referent RBP trained RNN, as follows.

Corollary 12.7.2. *The conditions of static and dynamic equivalence of an arbitrary and the referent RNN trained by the recursive backpropagation algorithm are the same as for the RTRL and BPTT trained networks.*

By continuity, the analysis provided so far is valid for an RNN of any size, number of hidden layers and character of feedback.

12.7.3 Results for Algorithms with a Momentum Term

A momentum term can be introduced into the update equation of gradient learning algorithms as (Haykin 1994)

$$\Delta \boldsymbol{W}(k) = \alpha \Delta \boldsymbol{W}(k-1) - \eta \nabla_{\boldsymbol{W}(k-1)} E(k), \tag{12.34}$$

where α is a momentum constant. This introduces an infinite impulse response element into the learning rule, which in some cases helps to increase the rate of convergence of a gradient algorithm (An *et al.* 1994). To preserve stability of (12.34), $0 < \alpha < 1$.

As (12.34) is linear in $\boldsymbol{W}$, the analysis for the plain gradient algorithm for neural networks, applies also for the momentum algorithm, as shown in Mandic and Krcmar (2000).

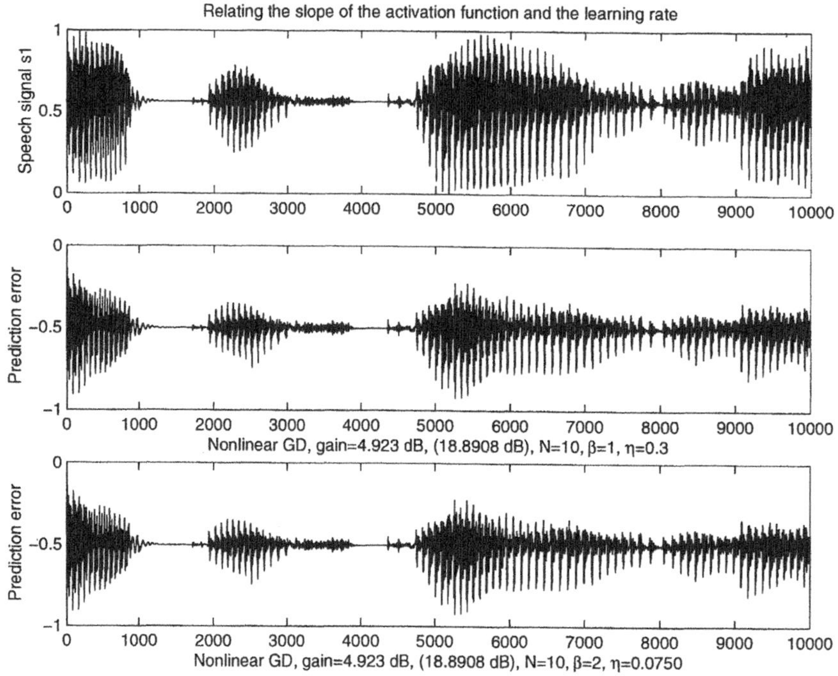

Figure 12.3 Verification of the β–η–$\boldsymbol{W}$ relationships based upon the prediction of the speech signal s1

Corollary 12.7.3. *The conditions of static and dynamic equivalence of an arbitrary and the referent recurrent neural network trained by a gradient descent algorithm with a momentum factor are the same as for the previously analysed algorithms, providing*

$$\alpha^{\mathrm{R}} = \alpha, \tag{12.35}$$

where α^{R} *is the momentum constant of a referent network, whereas* α *is the momentum constant of an arbitrary network.*

12.8 Simulation Results

To support the analysis, simulation results that illustrate the derived relationships are provided. A tap delay line of length $N = 10$ was used to feed a neuron that performed prediction, trained by a direct gradient descent algorithm.[4] The input signals were two speech signals denoted by s1 and s2. Firstly, the data were fed into a referent network with $\beta = 1$, the network was adapted and the outputs were recorded. Afterwards the learning rate of an arbitrary network was adjusted according to the β–η relationship and the whole procedure was repeated. The results of simulations for signal s1 are shown in Figure 12.3, whereas Figure 12.4 shows the results of simulations for signal

[4] The aim was not to build a good predictor, but to demonstrate the validity of the derived results, hence the quality of the performance was not important for this experiment.

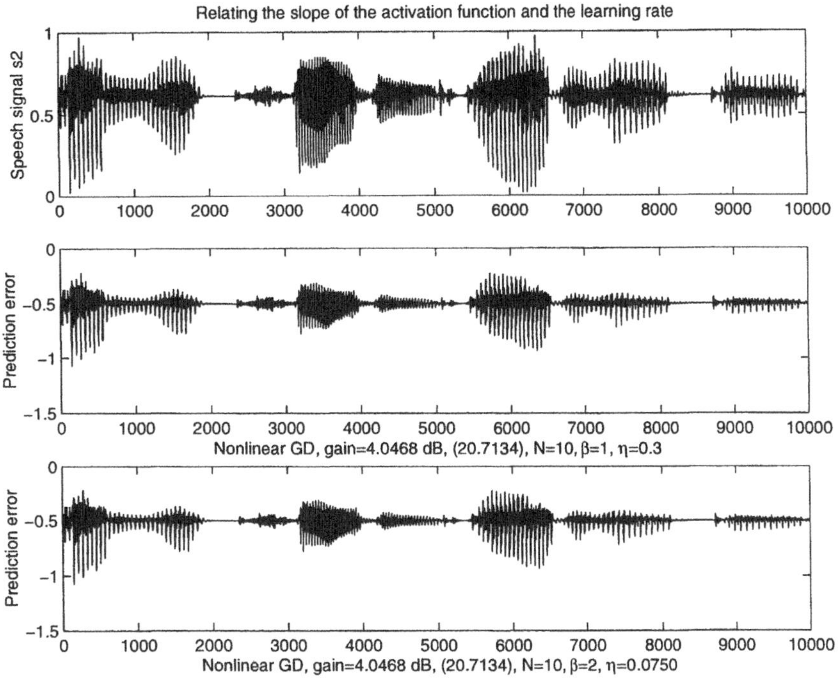

Figure 12.4 Verification of the β–η–$\boldsymbol{W}$ relationships based upon the prediction of the speech signal s2

s2. It was first ensured that the conditions of static equivalence $\boldsymbol{W}^{\mathrm{R}}(k) = \beta\boldsymbol{W}(k)$ were preserved. In both Figures 12.3 and 12.4, the top graph shows the input signal to the adaptive neuron. The middle graph shows the output prediction error for the referent network with $\beta = 1$, whereas the bottom graph shows the output prediction error for an arbitrary network. For the gain $\beta = 2$ for the arbitrary network used to generate the results in Figure 12.3, after the conditions of static and dynamic equivalence were satisfied, the learning rate became $\eta = \eta^{\mathrm{R}}/\beta^2 = 0.3/4 = 0.075$, and likewise for the situation shown in Figure 12.4. The output prediction error and the ratio between the signal and error variance (and the signal and error power in brackets) in decibels were identical for the referent and arbitrary network, as expected.

12.9 Summary of Relationships Between β and η for General Recurrent Neural Networks

The relationship between the gain β in a general activation function of a neuron and the step size η in the RTRL-based training of a general RNN is provided. Both static and dynamic equivalence of an arbitrary RNN and the referent network, with respect to β and η have been provided. As the conditions of dynamical equivalence are dependent on the choice of the underlying learning algorithm, the same analysis has been undertaken for two frequently used learning algorithms for training RNNs, namely the BPTT and the RBP algorithms. It has been shown that a general RNN can be replaced with the referent isomorphic RNN, with gain $\beta^{\mathrm{R}} = 1$ and modified learning rate η^{R}, hence providing one degree of freedom less in a nonlinear optimisation paradigm of training RNNs. Although the analysis has been undertaken on the example of the logistic nonlinearity, it is shown that it holds for a large class of $C^1(-\infty, +\infty)$ functions with the first derivative which satisfies

$$\beta \frac{\mathrm{d}\Phi(\beta, x)}{\mathrm{d}(\beta x)} = \frac{\Phi(\beta, x)}{\mathrm{d}x}.$$

12.10 Relationship Between η and β for Modular Neural Networks: Perspective

We next provide the relationship between the learning rate and the gain of a nonlinear activation function of a neuron within the framework of modular neural networks. This leads to reduction in the computational complexity of learning algorithms which continuously adapt the weights of a modular network because there is a smaller number of independent parameters to optimise. Although the analysis is provided for networks with recurrent modules, and particular algorithms are analysed for nested modular recurrent networks, the results obtained are valid for any modular networks trained by the algorithms under consideration.

Based upon results for ordinary recurrent neural networks, we extend our analysis to modular neural networks and provide the static and dynamic equivalence between an arbitrary modular network described by β, η and $\boldsymbol{W}(k)$ and a referent modular network described by $\beta^{\mathrm{R}} = 1$, η^{R} and $\boldsymbol{W}^{\mathrm{R}}(k)$. We show that there is a deterministic relationship between them, which allows one degree of freedom less in the nonlinear optimisation task of learning in such a framework. The relationships are provided for both the gradient descent (GD) and the extended recursive least-squares (ERLS) learning algorithms (Baltersee and Chambers 1998; Mandic *et al.* 1998), and a general nonlinear activation function of a neuron.

12.11 Static Equivalence Between an Arbitrary and a Referent Modular Neural Network

By static equivalence between two modular neural networks, we consider the equivalence between the outputs of the neurons in an arbitrary modular network and the outputs of the corresponding neurons in a referent modular network (Mandic and Chambers 1999a), for a fixed time-instant k, i.e.

$$\begin{aligned} y_{i,n}(k) = y^{\mathrm{R}}_{i,n}(k) \quad &\Leftrightarrow \quad \Phi(\beta, \boldsymbol{w}_{i,n}(k), \boldsymbol{u}_{i,n}(k)) \\ &= \Phi(1, \boldsymbol{w}^{\mathrm{R}}_{i,n}(k), \boldsymbol{u}_{i,n}(k)) \quad \Leftrightarrow \quad \beta \boldsymbol{w}_{i,n}(k) = \boldsymbol{w}^{\mathrm{R}}_{i,n}(k), \end{aligned} \tag{12.36}$$

where $\boldsymbol{w}_{i,n}(k)$ and $\boldsymbol{u}_{i,n}(k)$ are, respectively, the set of weights and the set of inputs which belong to the neuron n, $n = 1, \dots, N$, in module i, $i = 1, \dots, M$. This is valid irrespectively whether the modules are parallel or nested. For the ith neuron in the nth module and the logistic nonlinearity we have

$$\frac{1}{1 + \mathrm{e}^{-\beta \boldsymbol{u}^{\mathrm{T}}_{i,n} \boldsymbol{w}_{i,n}}} = \frac{1}{1 + \mathrm{e}^{-\boldsymbol{u}^{\mathrm{T}}_{i,n} \boldsymbol{w}^{\mathrm{R}}_{i,n}}} \quad \Leftrightarrow \quad \beta \boldsymbol{w}_{i,n} = \boldsymbol{w}^{\mathrm{R}}_{i,n}, \tag{12.37}$$

where the time index k is neglected and illustrates the extension of the previous analysis result for a general modular neural network.

The following lemma, which is independent of the underlying learning algorithm, comprises the above analysis.

Lemma 12.11.1 (see Mandic and Chambers 1999a). *An arbitrary modular network with weight matrix $\boldsymbol{W}(k)$, and a gain in the activation function β, is equivalent in the static sense to the isomorphic referent modular network, characterised by $\boldsymbol{W}^{\mathrm{R}}(k)$ and $\beta^{\mathrm{R}} = 1$, if*

$$\beta \boldsymbol{W}(k) = \boldsymbol{W}^{\mathrm{R}}(k). \tag{12.38}$$

12.12 Dynamic Equivalence Between an Arbitrary and a Referent Modular Network

While by static equivalence between two networks we consider the calculation of the outputs of the networks, for a given weight matrix and input vector, by dynamic equivalence, we consider the equality between two networks in the sense of adaptation of the weights. Hence, it is dependent on the underlying learning algorithm. Since we have to concentrate on a particular algorithm for a particular neural network, it is convenient to use the already described pipelined recurrent neural network. The overall cost function of the PRNN is given by[5] (Haykin and Li 1995)

$$E(k) = \frac{1}{2}\sum_{i=1}^{M} \lambda^{i-1} e_i^2(k), \tag{12.39}$$

where $e_i(k)$ is the instantaneous output error from module i and a weighting factor $\lambda \in (0, 1]$, is introduced which determines the weighting of the individual modules (Haykin and Li 1995). Obviously, this function introduces some sense of forgetting for nested modules and importance for parallel modules. The following analysis is hence independent of the chosen configuration.

For a general class of nonlinear activation functions, the weight updating includes the first derivative of a nonlinear activation function of a neuron, where

$$\Phi'(\beta, \boldsymbol{w}, \boldsymbol{u}) = \beta \Phi'(1, \boldsymbol{w}^{\mathrm{R}}, \boldsymbol{u}). \tag{12.40}$$

For the logistic function, for instance, we have

$$\Phi'(x) = \frac{\beta \mathrm{e}^{-\beta x}}{(1 + \mathrm{e}^{-\beta x})^2} = \beta \Phi'(x^{\mathrm{R}}), \tag{12.41}$$

where $x^{\mathrm{R}} = \beta x$ denotes the argument of the referent logistic function (with $\beta^{\mathrm{R}} = 1$), so that the network considered is equivalent in the static sense to the referent network.

---

[5] The same form of the cost function is used for parallel modular networks. If the modules are of equal importance, then $\lambda = 1/M$.

12.12.1 Dynamic Equivalence for a GD Learning Algorithm

Since recurrent modules degenerate into feedforward ones when the feedback is removed, for generality, we provide the analysis for recurrent modules. All the results so derived are also valid for the feedforward modules. From the equivalence in the static sense, the weight update equation for the referent network, can be written as

$$W^{\mathrm{R}}(k) = W^{\mathrm{R}}(k-1) + \beta \Delta W(k), \tag{12.42}$$

which gives

$$\Delta W^{\mathrm{R}}(k) = \beta \Delta W(k) = \beta\biggl(\eta \sum_{i=1}^{M} e_i(k) \frac{\partial y_{i,1}(k)}{\partial W(k)}\biggr) = \eta\beta \sum_{i=1}^{M} e_i(k) \Pi_{i,1}(k), \tag{12.43}$$

where $\Pi_{i,1}(k)$ is the matrix with elements $\pi_{n,l}^{i,1}(k)$ (D.23)–(D.30).

The matrix $\Pi(k)$, which comprises all the terms

$$\frac{\partial y_{i,j}}{\partial w_{n,l}}, \qquad \forall w_{n,l} \in W, \quad i = 1, \dots, M, \quad j = 1, \dots, N,$$

must be related to the appropriate matrix $\Pi^{\mathrm{R}}(k)$, which represents the gradients of the referent network. Thus, we have

$$\begin{aligned}
(\pi_{n,l}^{i,j}(k+1))^{\mathrm{R}} &= \Phi'(v_{i,j}^{\mathrm{R}}(k)) \Biggl[\sum_{m=1}^{N} w_{j,m+p+1}^{\mathrm{R}}(k) \pi_{n,l}^{i,m}(k) + \delta_{nj} u_{i,l}(k)\Biggr] \\
&= \frac{1}{\beta} \Phi'(v_{i,j}(k)) \Biggl[\sum_{m=1}^{N} \beta w_{j,m+p+1}(k) \frac{1}{\beta} \pi_{n,l}^{i,m}(k) + \delta_{nj} u_l(k)\Biggr] \\
&= \frac{1}{\beta} \pi_{n,l}^{i,j}(k+1),
\end{aligned} \tag{12.44}$$

which gives

$$\Pi^{\mathrm{R}}(k+1) = \frac{1}{\beta} \Pi(k+1). \tag{12.45}$$

The weight adaptation process for the referent PRNN can be now expressed as

$$\begin{aligned}
\Delta W^{\mathrm{R}}(k) &= \beta \Delta W(k) = \beta\eta \sum_{i=1}^{M} e_i(k) \Pi_{i,1}(k) \\
&= \beta^2 \eta \sum_{i=1}^{M} e_i(k) \Pi_{i,1}^{\mathrm{R}}(k) = \eta^{\mathrm{R}} \sum_{i=1}^{M} e_i(k) \Pi_{i,1}^{\mathrm{R}}(k),
\end{aligned} \tag{12.46}$$

which gives the required dynamic relationship and is encompassed in the following lemma.

Lemma 12.12.1 (see Mandic and Chambers 1999a). *An arbitrary modular neural network represented by β, η and $W(k)$ is equivalent in the dynamic sense* *in terms of gradient-descent-based learning to a referent isomorphic modular neural network represented by* $\beta^{\mathrm{R}} = 1$, η^{R} *and* $\boldsymbol{W}^{\mathrm{R}}(k) = \beta \boldsymbol{W}(k)$, *if*

$$\eta^{\mathrm{R}} = \beta^2 \eta. \tag{12.47}$$

12.12.2 Dynamic Equivalence Between Modular Recurrent Neural Networks for the ERLS Learning Algorithm

The extended recursive least-squares algorithm, given in Appendix D.3, as introduced in Baltersee and Chambers (1998) and Mandic *et al.* (1998) is based upon representing the dynamics of the PRNN in the state space form (Puskoris and Feldkamp 1994; Ruck *et al.* 1992)

$$\left.\begin{aligned} \boldsymbol{w}(k) &= \boldsymbol{w}(k-1) + \boldsymbol{q}(k) \quad \text{system equation,} \\ \boldsymbol{x}(k) &= \boldsymbol{h}(\boldsymbol{w}(k)) + \boldsymbol{v}(k) \quad \text{measurement equation,} \end{aligned}\right\} \tag{12.48}$$

where $\boldsymbol{w}(k)$ is the $N(N+p+1) \times 1$ vector obtained by rearranging the weight matrix $\boldsymbol{W}(k)$, $\boldsymbol{x}(k)$ is an $M \times 1$ observation vector, $\boldsymbol{q}(k) \sim \mathcal{N}(\boldsymbol{0}, \boldsymbol{Q})$ is a vector of white Gaussian noise (WGN), as well as $\boldsymbol{v}(k) \sim \mathcal{N}(\boldsymbol{0}, \boldsymbol{C})$ (Williams 1992). The first equation in (12.48) is the system equation, represented by a random walk, and satisfies the properties of the static equivalence, given in Lemma 12.11.1. The measurement equation in (12.48) is linearised using a first-order Taylor expansion, i.e.

$$\boldsymbol{h}(\boldsymbol{w}(k)) \approx \boldsymbol{h}(\hat{\boldsymbol{w}}(k \mid k-1)) + \nabla \boldsymbol{h}^{\mathrm{T}}(\hat{\boldsymbol{w}}(k \mid k-1))[\boldsymbol{w}(k) - \hat{\boldsymbol{w}}(k \mid k-1)], \tag{12.49}$$

where the gradient of $\boldsymbol{h}(\,\cdot\,)$ can be expressed as

$$\nabla \boldsymbol{h}^{\mathrm{T}} = \frac{\partial \boldsymbol{h}(\hat{\boldsymbol{w}}(k \mid k-1))}{\partial \hat{\boldsymbol{w}}(k \mid k-1)} = \boldsymbol{H}(k). \tag{12.50}$$

Furthermore, the vector $\boldsymbol{h}(k)$, which is the result of the nonlinear mapping $\boldsymbol{h}(k) = \boldsymbol{h}(\boldsymbol{w}(k))$ is actually the $M \times 1$ vector of the outputs of the PRNN modules (Baltersee and Chambers 1998; Mandic *et al.* 1998):

$$\boldsymbol{h}^{\mathrm{T}}(k) = [y_{1,1}(k), y_{2,1}(k), \dots, y_{M,1}(k)] \tag{12.51}$$

That means that the equivalence needed for the observation equation boils down to the dynamic equivalence derived for the GD learning (12.44)–(12.46).

Lemma 12.12.2 (see Mandic and Chambers 1999a). *An arbitrary modular recurrent neural network represented by* β, η *and* $\boldsymbol{W}(k)$ *is equivalent in the dynamic sense in terms of the extended recursive least-squares learning algorithm to a referent isomorphic modular recurrent neural network represented by* $\beta^{\mathrm{R}} = 1$, η^{R} *and* $\boldsymbol{W}^{\mathrm{R}}(k)$, *if*

(i) *they are equivalent in the static sense,* $\boldsymbol{W}^{\mathrm{R}}(k) = \beta \boldsymbol{W}(k)$, *and*

(ii) *the learning rates are related as* $\eta^{\mathrm{R}} = \beta^2 \eta$.

Notice that condition (i) is correspondent to the system equation of (12.48), whereas the condition (ii) corresponds to the measurement equation of (12.48).

12.12.3 Equivalence Between an Arbitrary and the Referent PRNN

Some other learning algorithms for training modular neural networks rest upon either general backpropagation, such as the BPTT algorithm (Werbos 1990), or combine general backpropagation and the direct gradient descent algorithms (RTRL), such as the RBP algorithm (Pineda 1987). Naturally, from Mandic and Chambers (1999e) and the above analysis, the relationships derived are valid for both the BPTT and RBP algorithms. From the static equivalence given in Lemma 12.11.1 and dynamic equivalence given in Lemmas 12.12.1 and 12.12.2, the following theorem encompasses a general equivalence between an arbitrary modular neural network and the referent, modular neural network.

Theorem 12.12.3 (see Mandic and Chambers 1999a). *An arbitrary modular neural network represented by β, η and $\boldsymbol{W}(k)$ is equivalent to a referent isomorphic modular neural network represented by $\beta^{\mathrm{R}} = 1$, η^{R} and $\boldsymbol{W}^{\mathrm{R}}(k)$, if*

(i) *they are equivalent in the static sense, i.e. $\boldsymbol{W}^{\mathrm{R}}(k) = \beta \boldsymbol{W}(k)$, and*

(ii) *they are equivalent in the dynamic sense, i.e. $\eta^{\mathrm{R}} = \beta^2 \eta$.*

As pointed out for common recurrent neural networks, the above analysis is also valid for the hyperbolic tangent $\gamma \tanh(\beta x) : \mathbb{R} \to (-\gamma, \gamma)$.

12.13 Note on the β–η–W Relationships and Contractivity

Sigmoid activation functions used in neural networks have been shown to be either contractive or expansive. It is important to preserve this property for functionally equivalent networks that have different parameters, since the *a posteriori* and normalised algorithms rely upon contractivity of the nonlinear activation function of a neuron. In a real neural network, for the logistic activation function $\Phi(\xi) = 1/(1 + \mathrm{e}^{-\beta\xi})$, for instance, ξ is replaced by the activation potential $\mathrm{net}(k) = \boldsymbol{x}^{\mathrm{T}}(k)\boldsymbol{w}(k)$. As the conditions of static and dynamic equivalence derived in this chapter effectively change the weights and learning rates, there is a need to address the contractivity/expansivity preservation between an arbitrary and the referent network.

A close inspection of the static equivalence equations shows that their form is

$$\boldsymbol{w}^{\mathrm{R}}(k) = \beta \boldsymbol{w}_{\mathrm{arbitrary}}(k). \tag{12.52}$$

Recall that for the referent network $\beta = 1$, the activation potentials of the referent and arbitrary network are given, respectively, by

$$\mathrm{net}^{\mathrm{R}}(k) = \boldsymbol{x}^{\mathrm{T}} \boldsymbol{w}^{\mathrm{R}}(k) \tag{12.53}$$

and

$$\mathrm{net}(k) = \boldsymbol{x}^{\mathrm{T}} \frac{\boldsymbol{w}^{\mathrm{R}}(k)}{\beta}. \tag{12.54}$$

However, the outputs of either network are identical, since for the arbitrary network

$$\Phi(\beta, \mathrm{net}(k)) = \Phi(\beta\, \mathrm{net}(k)) = \Phi\biggl(\beta \boldsymbol{x}^{\mathrm{T}} \frac{\boldsymbol{w}^{\mathrm{R}}(k)}{\beta}\biggr) = \Phi(1, \mathrm{net}^{\mathrm{R}}(k)). \tag{12.55}$$

Hence, the contractivity/expansivity is preserved upon application of the equivalence relationships between an arbitrary and a referent neural network.

12.14 Summary

The relationship between the learning rate η and the gain in the general activation function β for a nonlinear optimisation algorithm which adapts the weights of the modular neural network has been derived. This relationship is derived both in the static sense (equality of the outputs of the neurons) and the dynamic sense (equality in learning processes), for both the gradient descent (GD) and the extended recursive least-squares (ERLS) algorithms. Such a result enables the use of one degree of freedom less when adjusting variable parameters of a general neural network, and hence reduces computational complexity of learning. The results provided are shown to be easily extended for the backpropagation through time and the recurrent backpropagation algorithms, when applied in this framework.

Appendix A

The $\mathcal{O}$ Notation and Vector and Matrix Differentiation

A.1 The $\mathcal{O}$ Notation

Definition A.1.1. Let f and g be $f, g : \mathbb{R}^+ \to \mathbb{R}^+$ functions. If there exist positive numbers n_0 and c such that $f(n) \leqslant cg(n)$ for all $n \geqslant n_0$, the $\mathcal{O}$ notation can be introduced as

$$f(n) = \mathcal{O}(g(n)), \tag{A.1}$$

i.e. function f is asymptotically dominated by function g.

An algorithm is said to run in *polynomial time* if there exists $k \in \mathbb{Z}$ such that (Blondel 2000)

$$T(s) = \mathcal{O}(s^k), \tag{A.2}$$

where $T(s)$ is the execution time and s is the length of the input.

A.2 Vector and Matrix Differentiation

Let us denote vectors by lowercase bold letters and matrices by capital bold letters. Some frequently used vector and matrix differentiation rules used in this book are

- $\dfrac{\mathrm{d}(\boldsymbol{x}^{\mathrm{T}}\boldsymbol{A})}{\mathrm{d}\boldsymbol{x}} = \boldsymbol{A}$,
- $\dfrac{\mathrm{d}(\boldsymbol{x}^{\mathrm{T}}\boldsymbol{A}\boldsymbol{y})}{\mathrm{d}\boldsymbol{A}} = \boldsymbol{x}\boldsymbol{y}^{\mathrm{T}}$,
- $\dfrac{\mathrm{d}(\boldsymbol{A}\boldsymbol{x}+\boldsymbol{b})^{\mathrm{T}}\boldsymbol{C}(\boldsymbol{D}\boldsymbol{x}+\boldsymbol{e})}{\mathrm{d}\boldsymbol{x}} = \boldsymbol{A}^{\mathrm{T}}\boldsymbol{C}(\boldsymbol{D}\boldsymbol{x}+\boldsymbol{e}) + \boldsymbol{D}^{\mathrm{T}}\boldsymbol{C}^{\mathrm{T}}(\boldsymbol{A}\boldsymbol{x}+\boldsymbol{b})$,

  (i) $\dfrac{\mathrm{d}(\boldsymbol{x}^{\mathrm{T}}\boldsymbol{A}\boldsymbol{x})}{\mathrm{d}\boldsymbol{x}} = (\boldsymbol{A}+\boldsymbol{A}^{\mathrm{T}})\boldsymbol{x}$,

  (ii) $\dfrac{\mathrm{d}(\boldsymbol{x}^{\mathrm{T}}\boldsymbol{x})}{\mathrm{d}\boldsymbol{x}} = 2\boldsymbol{x}$,

- $\dfrac{\mathrm{d}^2(\boldsymbol{y}^{\mathrm{T}}\boldsymbol{x})}{\mathrm{d}\boldsymbol{x}^2} = 0,$
- $\dfrac{\mathrm{d}^2(\boldsymbol{A}\boldsymbol{x}+\boldsymbol{b})\boldsymbol{C}(\boldsymbol{D}\boldsymbol{x}+\boldsymbol{e})}{\mathrm{d}\boldsymbol{x}^2} = \boldsymbol{A}^{\mathrm{T}}\boldsymbol{C}\boldsymbol{D} + \boldsymbol{D}^{\mathrm{T}}\boldsymbol{C}^{\mathrm{T}}\boldsymbol{A}.$

Appendix B

Concepts from the Approximation Theory

Definition B.1.1 (see Mhaskar and Micchelli 1992). The function $\sigma : \mathbb{R} \to \mathbb{R}$ (not necessarily continuous) is called a *Kolmogorov* function, if for any integer $s \geqslant 1$, any compact set $K \subset \mathbb{R}^s$, any continuous function $f : K \to \mathbb{R}$ and any $\varepsilon > 0$, there exists an integer N, numbers $c_k, t_k \in \mathbb{R}$ and $\lambda_k \in \mathbb{R}^s$, $1 \leqslant k \leqslant N$, possibly depending upon s, K, f, ε, such that

$$\sup_{x \in K} \left| f(x) - \sum_{k=1}^{N} c_k \sigma(\lambda_k x - t_k) \right| < \varepsilon. \tag{B.1}$$

Definition B.1.2. A sigmoidal function with properties,

$$\lim_{x \to -\infty} \frac{\sigma(x)}{x^k} = 0, \qquad \lim_{x \to +\infty} \frac{\sigma(x)}{x^k} = 1, \tag{B.2}$$

$$|\sigma(x)| \leqslant K(1 + |x|)^k, \qquad K > 0, \tag{B.3}$$

is called the kth degree sigmoidal function.

In other words, these functions are bounded on $\mathbb{R}$ by a polynomial of degree $d \leqslant k$. It follows that function σ is a candidate for a Kolmogorov function if it is not a polynomial.

Theorem B.1.3 (see Kolmogorov 1957). *There exist fixed increasing continuous functions $\psi_{pq}(x)$, on $I = [0, 1]$ so that each continuous function f on I^n can be written in the form*

$$f(x_1, \ldots, x_n) = \sum_{q=1}^{2n+1} \Phi_q\left(\sum_{p=1}^{n} \psi_{pq}(x_p) \right), \tag{B.4}$$

where Φ_q are properly chosen continuous functions of one variable.

This result asserts that every multivariate continuous function can be represented by the superposition of a small number of univariate continuous functions.

Theorem B.1.4 (Kolmogorov–Sprecher Theorem). *For each integer $n \geqslant 2$, there exists a real monotonic increasing function $\psi(x), \psi([0,1]) = [0,1]$, dependent on n and having the following property.*

For each preassigned number $\delta > 0$, there is a rational number ε, $0 < \varepsilon < \delta$, such that every real continuous function of n variables, $\phi(\boldsymbol{x})$, defined on I^n, can be exactly represented by

$$f(\boldsymbol{x}) = \sum_{j=1}^{2n+1} \chi\left[\sum_{i=1}^{n} \lambda^i \psi(x_i + \varepsilon(j-1)) + j - 1\right], \tag{B.5}$$

where χ is a real and continuous function dependent upon f and λ is a constant independent of f.

Since no constructive method for the determination of χ is known, a direct application of the Kolmogorov–Sprecher Theorem is rather difficult.

Theorem B.1.5 (Weierstrass Theorem). *If f is a continuous real-valued function on $[a,b] \in \mathbb{R}$, then for any $\varepsilon > 0$, there exists a polynomial P on $[a,b] \in \mathbb{R}$ such that*

$$|f(x) - P(x)| < \varepsilon, \qquad \forall x \in [a,b]. \tag{B.6}$$

In other words, any continuous function on a closed and bounded interval can be uniformly approximated on that interval by polynomials to any degree of accuracy.

Definition B.1.6. Let P be a probability measure on $\mathbb{R}^m$. For measurable functions $f_1, f_2 : \mathbb{R}^m \to \mathbb{R}$ we say that f_1 approximates f_2 with accuracy $\varepsilon > 0$ and confidence $\delta > 0$ in probability if

$$P(x \in \mathbb{R}^m \mid |f_1(x) - f_2(x)| > \varepsilon) < \delta. \tag{B.7}$$

The function f_1 interpolates f_2 on p examples $x_1, \ldots, x_p$ if $f_1(x_i) = f_2(x_i)$, $i = 1, \ldots, p$.

Definition B.1.7. A function f that satisfies the condition

$$|f(x) - f(y)| \leqslant L|x - y|, \qquad x, y \in \mathbb{R}, \quad L = \text{const.}, \tag{B.8}$$

is called a Lipschitz function.

Definition B.1.8. A closure of a subset D of a topological space S, usually denoted by $\bar{D}$, is the set of points in S with the property that every neighbourhood of such a point has a nonempty intersection with D.

If S is a set, then by the closure $\bar{S}$ of S we mean the set of all points in S together with the set of all limit points of S. A set S is closed if it is identical to its closure $\bar{S}$.

Definition B.1.9. A subset D of a topological space S is called dense if $\bar{D} = S$.

Topologies considered are naturally defined by a metric. The most commonly used metrics are $\mathcal{L}_p$, and among them the supremum metric. If D is dense in S, then each element of S can be approximated arbitrarily well by elements of D. Examples are the set of rational numbers, which is dense in $\mathbb{R}$, and the set of polynomials that is dense in the space of continuous functions.

Definition B.1.10. A compact set is one in which every infinite subset contains at least one limit point.

Every closed, bounded, finite-dimensional set in a metric linear space is compact.

Definition B.1.11. A cubic spline is a spline constructed of piecewise third-order polynomials defined by a set of control points.

A spline is an interpolating polynomial which uses information from neighbouring points to obtain smoothness.

Appendix C

Complex Sigmoid Activation Functions, Holomorphic Mappings and Modular Groups

C.1 Complex Sigmoid Activation Functions

Summarising some important notions from complex analysis, let w and z be complex numbers. Some of the elementary transformations defined on these complex numbers are given in the following table.

Mapping	Formula	Domain
Inversion	$f(z) = 1/z$	
Magnification	$f(z) = az$	$a \in \mathbb{R} \neq 0$,
Magnification + Rotation	$f(z) = az$	$a \in \mathbb{C} \neq 0$,
Möbius Transformation	$f(z) = \dfrac{az+b}{cz+d}$	$a, b, c, d \in \mathbb{C}$,
Rotation	$f(z) = \mathrm{e}^{\mathrm{j}\theta}$	$\theta \in \mathbb{R}$,
Translation	$f(z) = z + a$	$a \in \mathbb{C}$.

A linear transformation can be considered as a composition of a rotation, magnification and translation.

Definition C.1.1. A differentiable function of a complex variable is *holomorphic* if its derivative is continuous. If the first derivative of a holomorphic function is nonzero at some point z_0, then that function is *conformal* at z_0.

Definition C.1.2. A complex function which is analytic at all finite points of the complex plane is called entire.

Definition C.1.3. A meromorphic function has only a finite number of isolated poles and zeros. It can be thought of as a rational function of two entire functions.

Definition C.1.4 (Liouville Theorem). A bounded entire function in the complex plane is constant.

Cauchy–Riemann Equations: for a total derivative $\mathrm{d}f/\mathrm{d}z$ of a complex function $f(x, y) = u(x, y) + \mathrm{j}v(x, y)$, where $z = x + \mathrm{j}y$ and $\mathrm{d}z = \mathrm{d}x + \mathrm{j}\,\mathrm{d}y$, to exist, the following conditions, also known as Cauchy–Riemann equations, have to be satisfied:

$$\left.\begin{aligned} \frac{\partial u}{\partial x} &= \frac{\partial v}{\partial y}, \\ \frac{\partial v}{\partial x} &= -\frac{\partial u}{\partial y}. \end{aligned}\right\} \tag{C.1}$$

Some properties of complex activation functions

For a function approximation problem, we have

$$f(x) = \sum_{i=1}^{n} c_i \sigma(x - a_i) = \sum_{i=1}^{n} \frac{c_i}{1 + \mathrm{e}^{-x}\mathrm{e}^{a_i}} = \mathrm{e}^x \sum_{i=1}^{n} c_i \frac{1}{\mathrm{e}^x + \mathrm{e}^{a_i}}, \tag{C.2}$$

which becomes

$$f(x) = z \sum_{i=1}^{n} \frac{c_i}{z + \alpha_i} = r(z) \tag{C.3}$$

with the change of variables $z = \mathrm{e}^x$ and $\alpha_i = \mathrm{e}^{a_i}$. The rational function $r(z)$ is given by

$$r(z) = \sum_{i=1}^{n} \frac{c_i}{z + \alpha_i} = \frac{P(z)}{Q(z)}, \qquad z \in \mathbb{C}. \tag{C.4}$$

Recall that any analytic function $\sigma : \mathbb{R} \to \mathbb{R}$ such as the standard sigmoid has a convergent power series expansion $\sigma(x) = \sum_{i=0}^{\infty} \sigma_i (x - a)^i$ about a point $a \in \mathbb{R}$. When we substitute x by a complex number $z = x + \mathrm{j}y$, we obtain a series $\sum_{i=0}^{\infty} \sigma_i (z - a)^i$, which converges in a disc $|z - a| < R$, where R is the radius of convergence of power series. Coefficients a_i correspond to the poles $\alpha_i = \mathrm{e}^{a_i}$, whereas scaling factors c_i represent the residues of $r(z)$ at e^{a_i} (Williamson and Helmke 1995).

C.1.1 Modular Groups

Since a Möbius transformation remains unchanged if all the coefficients (a, b, c, d) are multiplied by the same nonzero constant, we will assume $ad - bc = 1$.

If we associate a matrix

$$A = \begin{bmatrix} a & b \\ c & d \end{bmatrix} \tag{C.5}$$

with a Möbius transformation, then its determinant $\Delta = ad - bc = 1$. If A and B are matrices associated with Möbius transformations f and g, then a composition $f \circ g$ is described by the matrix product AB. In order to show that the set of all Möbius transformations forms a group under composition, we introduce the identity transformation

$$f(z) = z = \frac{1z + 0}{0z + 1},$$

which is described by the identity matrix I, which is the neutral element of the group. The matrix inverse A^{-1} is associated with the inverse of f given by

$$f^{-1}(z) = \frac{dz - b}{-cz + a}, \tag{C.6}$$

which is the last condition for the set of all Möbius transformations to form the group under composition. Now, the set of all Möbius transformations of the form,

$$f(z) = \frac{az + b}{cz + d},$$

where a, b, c, d are integers with $ad - bc = 1$ forms a modular group and is denoted by Γ. This allows us to express the Möbius transformation as

$$Az = \frac{az + b}{cz + d}. \tag{C.7}$$

The next theorem shows that the group Γ is generated by two transformations, $Tz = z + 1$ and $Sz = -1/z$ (Apostol 1997).

Theorem C.1.5 (see Apostol 1997). *The modular group Γ is generated by the two matrices*

$$T = \begin{bmatrix} 1 & 1 \\ 0 & 1 \end{bmatrix} \quad \text{and} \quad S = \begin{bmatrix} 0 & -1 \\ 1 & 0 \end{bmatrix} \tag{C.8}$$

and every $A \in \Gamma$ can be expressed in the form,

$$A = T^{n_1} S T^{n_2} S \cdots S T^{n_k}, \tag{C.9}$$

where n_i are integers.

So, for instance, a modular representation for the matrix

$$A = \begin{bmatrix} 4 & 9 \\ 11 & 25 \end{bmatrix}$$

is $A = ST^{-3}ST^{-4}ST^{2}$. However, this solution is generally not unique.

Appendix D

Learning Algorithms for RNNs

D.1 The RTRL Algorithm

The structure of a single RNN is shown in Figure D.1. The neurons (nodes) are depicted by circles and incorporate the operation Φ(sum of inputs). For the nth neuron, its weights form a $(p + F + 1) \times 1$ dimensional weight vector $\boldsymbol{w}_n^{\mathrm{T}} = [w_{n,1}, \dots, w_{n,p+F+1}]$, where p is the number of external inputs and F is the number of feedback connections, one remaining element of the weight vector $\boldsymbol{w}$ being the bias input weight. The feedback connections represent the delayed output signals of the RNN. In the case of the network shown in Figure D.1, we have $N = F$. Such a network is called a fully connected recurrent neural network (FCRNN) (Williams and Zipser 1989a). The following equations fully describe the FCRNN,

$$y_n(k) = \Phi(v_n(k)), \qquad n = 1, 2, \dots, N, \tag{D.1}$$

$$v_n(k) = \sum_{l=1}^{p+N+1} w_{n,l}(k) u_l(k), \tag{D.2}$$

$$\boldsymbol{u}_n^{\mathrm{T}}(k) = [s(k-1), \dots, s(k-p), 1, y_1(k-1), y_2(k-1), \dots, y_N(k-1)], \tag{D.3}$$

where the $(p + N + 1) \times 1$ dimensional vector $\boldsymbol{u}$ comprises both the external and feedback inputs to a neuron, with vector $\boldsymbol{u}$ having 'unity' for the constant bias input.

For the nonlinear time series prediction paradigm, there is only one output neuron of the RNN. RTRL-based training of the RNN is based upon minimising the instantaneous squared error at the output of the first neuron of the RNN (Haykin 1994; Williams and Zipser 1989a), which can be expressed as

$$\min(\tfrac{1}{2}e^2(k)) = \min(\tfrac{1}{2}[s(k) - y_1(k)]^2), \tag{D.4}$$

where $e(k)$ denotes the error at the output y_1 of the RNN and $s(k)$ is the teaching signal. Hence, the correction for the lth weight of neuron k at the time instant k can

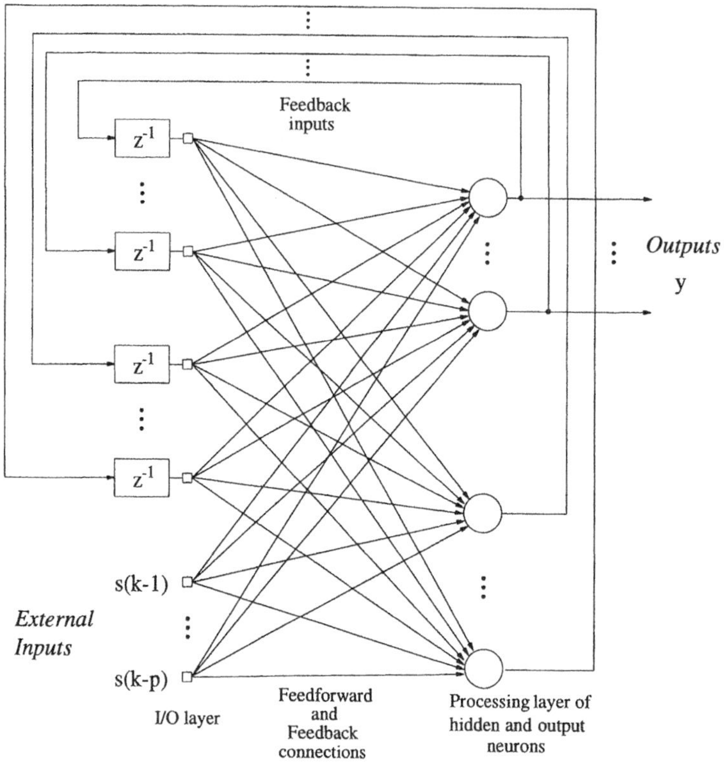

Figure D.1 Single recurrent neural network

be derived as follows:

$$\Delta w_{n,l}(k) = -\frac{\eta}{2}\frac{\partial}{\partial w_{n,l}(k)}e^2(k)$$
$$= -\eta e(k)\frac{\partial e(k)}{\partial w_{n,l}(k)}. \tag{D.5}$$

Since the external signal vector $\boldsymbol{s}$ does not depend on the elements of $\boldsymbol{W}$, the error gradient becomes

$$\frac{\partial e(k)}{\partial w_{n,l}(k)} = -\frac{\partial y_1(k)}{\partial w_{n,l}(k)}. \tag{D.6}$$

Using the chain rule, this can be rewritten as[1]

$$\frac{\partial y_1(k)}{\partial w_{n,l}(k)} = \Phi'(v_1(k))\frac{\partial v_1(k)}{\partial w_{n,l}(k)}$$
$$= \Phi'(v_1(k))\left(\sum_{\alpha=1}^{N}\frac{\partial y_\alpha(k-1)}{\partial w_{n,l}(k)}w_{1,\alpha+p+1}(k) + \delta_{n1}u_l(k)\right), \tag{D.7}$$

[1] A detailed derivation of the RTRL algorithm can be found in Williams and Zipser (1989a), Haykin (1994) and Williams and Zipser (1989b).

where

$$\delta_{nl} = \begin{cases} 1, & n = l, \\ 0, & n \neq l. \end{cases} \tag{D.8}$$

Under the assumption, also used in the RTRL algorithm (Narendra and Parthasarathy 1990; Robinson and Fallside 1987; Williams and Zipser 1989a), that when the learning rate η is sufficiently small, we have

$$\frac{\partial y_\alpha(k-1)}{\partial w_{n,l}(k)} \approx \frac{\partial y_\alpha(k-1)}{\partial w_{n,l}(k-1)}. \tag{D.9}$$

A triply indexed set of variables $\{\pi^j_{n,l}(k)\}$ can be introduced to characterise the RTRL algorithm for the RNN, as

$$\pi^j_{n,l} = \frac{\partial y_j(k)}{\partial w_{n,l}} \qquad 1 \leqslant j,\ n \leqslant N,\ 1 \leqslant l \leqslant p+1+N, \tag{D.10}$$

which is used to compute recursively the values of $\pi^j_{n,l}$ for every time step k and all appropriate j, n and l as follows,

$$\pi^j_{n,l}(k+1) = \Phi'(v_j)\left[\sum_{m=1}^{N} w_{j,m+p+1}(k)\pi^m_{n,l}(k) + \delta_{nj}u_l(k)\right], \tag{D.11}$$

with the values for j, n and l as in (D.11) and the initial conditions

$$\pi^j_{n,l}(0) = 0. \tag{D.12}$$

To simplify the presentation, we introduce three new matrices, the $N \times (N+p+1)$ matrix $\boldsymbol{\Pi}_j(k)$, the $N \times (N+p+1)$ matrix $\boldsymbol{U}_j(k)$ and the $N \times N$ diagonal matrix $\boldsymbol{F}(k)$, as (Haykin 1999b)

$$\boldsymbol{\Pi}_j(k) = \frac{\partial \boldsymbol{y}(k)}{\partial \boldsymbol{w}_j(k)}, \qquad \boldsymbol{y} = [y_1(k), \dots, y_N(k)], \quad j = 1, 2, \dots, N, \tag{D.13}$$

$$\boldsymbol{U}_j(k) = \begin{bmatrix} \boldsymbol{0} \\ \vdots \\ \boldsymbol{u}(k) \\ \vdots \\ \boldsymbol{0} \end{bmatrix} \leftarrow j\text{th row}, \quad j = 1, 2, \dots, N, \tag{D.14}$$

$$\boldsymbol{F}(k) = \operatorname{diag}[\Phi'(\boldsymbol{u}(k)^{\mathrm{T}}\boldsymbol{w}_1(k)), \dots, \Phi'(\boldsymbol{u}(k)^{\mathrm{T}}\boldsymbol{w}_N(k))]. \tag{D.15}$$

Hence, the gradient updating equation regarding the recurrent neuron can be symbolically expressed as (Haykin 1994; Williams and Zipser 1989a)

$$\boldsymbol{\Pi}_j(k+1) = \boldsymbol{F}(k)[\boldsymbol{U}_j(k) + \boldsymbol{W}_a(k)\boldsymbol{\Pi}_j(k)], \qquad j = 1, 2, \dots, N, \tag{D.16}$$

where $\boldsymbol{W}_a$ denotes the set of those entries in $\boldsymbol{W}$ which correspond to the feedback connections.

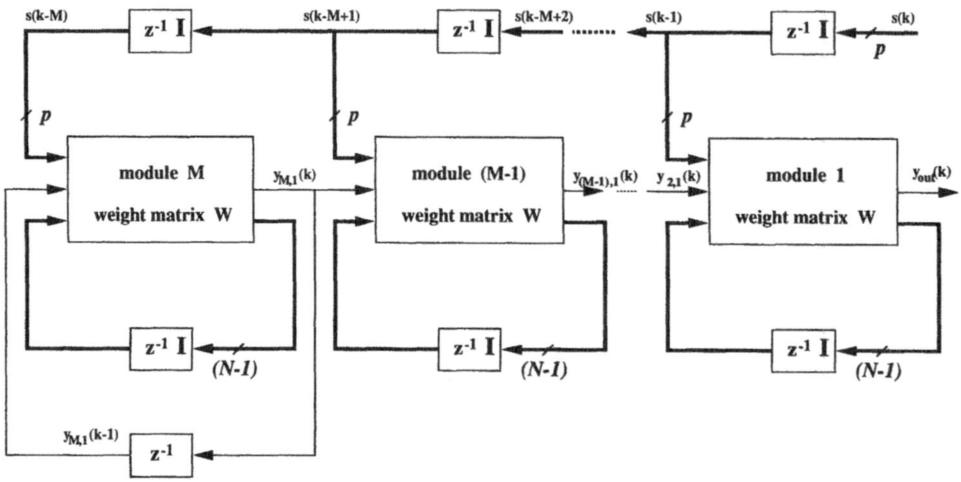

Figure D.2 Pipelined recurrent neural network

D.1.1 Teacher Forcing Modification of the RTRL Algorithm

The idea is to replace the actual output $y_1(k)$ by a teacher signal $s(k)$ in subsequent computation of the behaviour of the network. The derivation of the algorithm follows the same concept as for the RTRL algorithm. Notice that the partial derivatives of $s(k)$ with respect to $\boldsymbol{W}(k)$ are zero. Having this in mind, the equations that calculate the sensitivities $\boldsymbol{\Pi}$ have the same form as for the RTRL, except that the values of $\boldsymbol{\Pi}$ that refer to $s(k)$ are zero.

D.2 Gradient Descent Learning Algorithm for the PRNN

The PRNN is a modular neural network and consists of a certain number M of RNNs as its modules, with each module consisting of N neurons. In the PRNN configuration, the M modules, which are RNNs, are connected as shown in Figure D.2. The $(p \times 1)$-dimensional external signal vector $\boldsymbol{s}^{\mathrm{T}}(k) = [s(k-1), \dots, s(k-p)]$ is delayed by m time steps $(z^{-m}\boldsymbol{I})$ before feeding the module m. All the modules operate using the same weight matrix $\boldsymbol{W}$. The overall output signal of the PRNN is $y_{\mathrm{out}}(k) = y_{1,1}(k)$, i.e. the output of the first neuron of the first module. Thus, the overall cost function of the PRNN becomes (Baltersee and Chambers 1998; Haykin and Li 1995)

$$E(k) = \frac{1}{2}\sum_{i=1}^{M} \lambda^{i-1} e_i^2(k), \tag{D.17}$$

where $e_i(k)$ is the error from module i and a *forgetting factor* $\lambda \in (0, 1]$, is introduced which determines the weighting of the individual modules.

A full mathematical description of the PRNN is given in the following equations:

$$y_{i,n}(k) = \Phi(v_{i,n}(k)), \qquad i = 1, \dots, M, \quad n = 1, \dots, N, \tag{D.18}$$

$$v_{i,n}(k) = \sum_{l=1}^{p+N+1} w_{n,l}(k) u_{i,l}(k), \tag{D.19}$$

$$\begin{aligned} \boldsymbol{u}_i^{\mathrm{T}}(k) = [&s(k-i), \dots, s(k-i-p+1), 1, y_{i+1,1}(k), \\ &y_{i,2}(k-1), \dots, y_{i,N}(k-1)] \quad \text{for } 1 \leqslant i \leqslant M-1, \end{aligned} \tag{D.20}$$

$$\begin{aligned} \boldsymbol{u}_M^{\mathrm{T}}(k) = [&s(k-M), \dots, s(k-M-p+1), 1, y_{M,1}(k-1), \\ &y_{M,2}(k-1), \dots, y_{M,N}(k-1)] \quad \text{for } i = M. \end{aligned} \tag{D.21}$$

At the time step k, for each module i, $i = 1, \dots, M$, the one-step forward prediction error $e_i(k)$ associated with a module, is

$$e_i(k) = s(k-i+1) - y_{i,1}(k). \tag{D.22}$$

The cost function for the PRNN is

$$E(k) = \frac{1}{2} \sum_{i=1}^{M} \lambda^{i-1} e_i^2(k), \tag{D.23}$$

i.e. a weighted sum of squared errors at the output of every module of the PRNN. Hence, the correction for the lth weight of neuron n at the time instant k is derived as follows:

$$\begin{aligned} \Delta w_{n,l}(k) &= -\frac{\eta}{2} \frac{\partial}{\partial w_{n,l}(k)} \left(\sum_{i=1}^{M} \lambda^{i-1} e_i^2(k) \right) \\ &= -\eta \sum_{i=1}^{M} \lambda^{i-1} e_i(k) \frac{\partial e_i(k)}{\partial w_{n,l}(k)}. \end{aligned} \tag{D.24}$$

Now,

$$\frac{\partial e_i(k)}{\partial w_{n,l}(k)} = -\frac{\partial y_{i,1}(k)}{\partial w_{n,l}(k)}, \tag{D.25}$$

which can be rewritten as

$$\frac{\partial y_{i,1}(k)}{\partial w_{n,l}(k)} = \Phi'(v_{i,1}(k)) \frac{\partial v_{i,1}(k)}{\partial w_{n,l}(k)}. \tag{D.26}$$

Now, inserting (D.18)–(D.21) into (D.26) yields (Baltersee and Chambers 1998; Haykin and Li 1995)

$$\Phi'(v_{i,1}(k)) \frac{\partial v_{i,1}(k)}{\partial w_{n,l}(k)} = \Phi'(v_{i,1}(k)) \left(\sum_{\alpha=1}^{p+N+1} \left(\frac{\partial w_{1,\alpha}(k)}{\partial w_{n,l}(k)} u_{i,\alpha}(k) + \frac{\partial u_{i,\alpha}(k)}{\partial w_{n,l}(k)} w_{1,\alpha}(k) \right) \right). \tag{D.27}$$

The first term in (D.27) is zero except for $n = 1$ and $l = \alpha$ (Williams and Zipser 1989a), and the only elements of the input vector $\boldsymbol{u}$ that depend on the elements of

W are the feedback values. Therefore, Equation (D.27) can be simplified to

$$\Phi'(v_{i,1}(k))\frac{\partial v_{i,1}(k)}{\partial w_{n,l}(k)} = \Phi'(v_{i,1}(k))\bigg(\sum_{\alpha=1}^{N}\frac{\partial y_{i,\alpha}(k-1)}{\partial w_{n,l}(k)}w_{1,\alpha+p+1}(k) + \delta_{n1}u_{i,l}(k)\bigg), \tag{D.28}$$

where

$$\delta_{nl} = \begin{cases} 1, & n = l, \\ 0, & n \neq l. \end{cases} \tag{D.29}$$

A quadruply indexed set of variables $\{\pi_{n,l}^{ij}(k)\} = \partial y_{i,j}(k)/\partial w_{n,l}$ can be introduced to characterise the RTRL algorithm for the PRNN, as

$$\pi_{n,l}^{ij} = \frac{\partial y_{i,j}(k)}{\partial w_{n,l}}, \qquad 1 \leqslant i \leqslant M, \quad 1 \leqslant j, \quad n \leqslant N, \quad 1 \leqslant l \leqslant p+1+N. \tag{D.30}$$

The variant of the RTRL algorithm suitable for the PRNN is used to compute recursively the values of $\pi_{n,l}^{ij}$ for every time step k and all appropriate i, j, n and l as follows,

$$\pi_{n,l}^{ij}(k+1) = \Phi'(v_{i,j})\bigg[\sum_{m=1}^{N} w_{j,m+p+1}(k)\pi_{n,l}^{im}(k) + \delta_{nj}u_{i,l}(k)\bigg], \tag{D.31}$$

with the values for i, j, n and l as in (D.31) and the initial conditions,

$$\pi_{n,l}^{ij}(0) = 0, \tag{D.32}$$

where $u_{i,l}$ and $v_{i,j}$ are, respectively, the lth input to the ith module of the PRNN and the internal activation function of the jth neuron in the ith module. Updating process (D.30) can be written in the same matrix form as (D.16).

D.3 The ERLS Algorithm

The extended recursive least-squares (ERLS) algorithm, which is presented here (Baltersee and Chambers 1998; Mandic *et al.* 1998), is based upon the idea of the extended Kalman filter (EKF)(Haykin 1996a; Iiguni and Sakai 1992; Kay 1993). It is presented for the general case of the PRNN and boils down to the ERLS algorithm for the RNN for a PRNN with one module. The cost function of the PRNN becomes

$$E_{\mathrm{ERLS}}(k) = \frac{1}{2}\sum_{l=1}^{k}\xi^{k-l}E(l), \tag{D.33}$$

which is to be minimised with respect to the elements of the weight matrix W. The newly introduced constant $\xi \in (0,1]$ represents a *forgetting factor* so that the resulting learning algorithm becomes suitable for the prediction task of non-stationary signals. The ERLS algorithm is used to solve the nonlinear minimisation problem of (D.33). In order to derive the ERLS algorithm, the vector state–vector observation Kalman filter equations are considered (Haykin *et al.* 1997; Kay 1993),

$$\boldsymbol{w}(k) = \boldsymbol{a}(\boldsymbol{w}(k-1)) + \boldsymbol{u}(k), \tag{D.34}$$

$$\boldsymbol{x}(k) = \boldsymbol{h}(\boldsymbol{w}(k)) + \boldsymbol{v}(k), \tag{D.35}$$

where $\boldsymbol{w}(k)$ becomes the $N(N+p+1)\times 1$ weight vector, $\boldsymbol{x}(k)$ is the $M\times 1$ observation (signal) vector, $\boldsymbol{u}(k)$ is a white Gaussian noise vector, $\boldsymbol{u} \sim \mathcal{N}(\boldsymbol{0}, \boldsymbol{Q})$ and $\boldsymbol{v}(k)$ is observation noise, WGN vector $\boldsymbol{v} \sim \mathcal{N}(\boldsymbol{0}, \boldsymbol{C})$.[2] Furthermore, we have the nonlinear mapping functions,

$$\boldsymbol{a} : \mathbb{R}^{N(N+p+1)} \to \mathbb{R}^{N(N+p+1)} \tag{D.36}$$

and

$$\boldsymbol{h} : \mathbb{R}^{N(N+p+1)} \to \mathbb{R}^{M}, \tag{D.37}$$

which, respectively, map the space spanned over the weighting vector $\boldsymbol{w}$ onto the same space and the weight vector space onto the 'output of the PRNN'-dimensional space. For prediction of speech, however, the function $\boldsymbol{a}(\,\cdot\,)$ is unknown, so that the state Equation (D.34) may be approximated by the random walk model (Haykin 1994; Puskoris and Feldkamp 1994; Ruck *et al.* 1992)

$$\boldsymbol{w}(k) = \boldsymbol{w}(k-1) + \boldsymbol{u}(k). \tag{D.38}$$

As for the EKF, the nonlinear mapping function $\boldsymbol{h}(\,\cdot\,)$ is linearised using the first-order Taylor expansion around the estimate of $\boldsymbol{w}(k)$, based on the previous data, i.e. $\hat{\boldsymbol{w}}(k \mid k-1)$, which yields

$$\boldsymbol{h}(\boldsymbol{w}(k)) \approx \boldsymbol{h}(\hat{\boldsymbol{w}}(k \mid k-1) + \nabla \boldsymbol{h}^{\mathrm{T}}[\boldsymbol{w}(k) - \hat{\boldsymbol{w}}(k \mid k-1)], \tag{D.39}$$

where the gradient of $\boldsymbol{h}(\,\cdot\,)$ can be written as

$$\nabla \boldsymbol{h}^{\mathrm{T}} = \frac{\partial \boldsymbol{h}(\hat{\boldsymbol{w}}(k \mid k-1))}{\partial \hat{\boldsymbol{w}}(k \mid k-1)} = \boldsymbol{H}(k) \tag{D.40}$$

so that the observation equation becomes

$$\boldsymbol{x}(k) = \boldsymbol{H}(k)\boldsymbol{w}(k) + \boldsymbol{v}(k) + [\boldsymbol{h}(\hat{\boldsymbol{w}}(k \mid k-1)) - \boldsymbol{H}(k)\hat{\boldsymbol{w}}(k \mid k-1)]. \tag{D.41}$$

Moreover, the correlation matrix of the process state noise vector $\boldsymbol{u}(k)$ equals a scaled version of the minimum mean square error matrix of the EKF (Haykin 1994; Kay 1993)

$$\boldsymbol{Q}(k) = E[\boldsymbol{u}(k)\boldsymbol{u}^{\mathrm{T}}(k)] = (\xi^{-1} - 1)\boldsymbol{M}(k), \tag{D.42}$$

where ξ is the forgetting factor of (D.33). Using (D.34), (D.41), (D.42) and the definition of the EKF in Kay (1993), the final equations of the ERLS algorithm for the PRNN become (Baltersee and Chambers 1998; Mandic *et al.* 1998)

$$\boldsymbol{K}(k) = \xi^{-1}\boldsymbol{M}(k-1)\boldsymbol{H}^{\mathrm{T}}[\boldsymbol{C}(k) + \xi^{-1}\boldsymbol{H}(k)\boldsymbol{M}(k-1)\boldsymbol{H}^{\mathrm{T}}(k)]^{-1}, \tag{D.43}$$

$$\hat{\boldsymbol{w}}(k) = \hat{\boldsymbol{w}}(k-1) + \boldsymbol{K}(k)[\boldsymbol{x}(k) - \boldsymbol{h}(\hat{\boldsymbol{w}}(k-1))], \tag{D.44}$$

$$\boldsymbol{M}(k) = \xi^{-1}[\boldsymbol{I} - \boldsymbol{K}(k)\boldsymbol{H}(k)]\boldsymbol{M}(k-1). \tag{D.45}$$

[2] Although the observation noise vector $\boldsymbol{v}$ is to be, generally speaking, described by its covariance matrix, $\boldsymbol{C}$, we will assume that that matrix is diagonal, i.e. that the observation noise $\boldsymbol{v}$ is satisfactorily described by its variance vector, together with its mean value.

For the PRNN, the $(M \times 1)$-dimensional vector $\boldsymbol{x}(k)$ becomes

$$\boldsymbol{x}^{\mathrm{T}}(k) = [s(k), s(k-1), \dots, s(k-M+1)], \tag{D.46}$$

which is the input signal itself. Furthermore, the $(M \times 1)$-dimensional vector, $\boldsymbol{h}(k) = \boldsymbol{h}(\boldsymbol{w}(k))$ becomes

$$\boldsymbol{h}^{\mathrm{T}}(k) = [y_{1,1}(k), y_{2,1}(k), \dots, y_{M,1}(k)]. \tag{D.47}$$

Now, since by (D.34), (D.38)

$$\hat{\boldsymbol{w}}(k \mid k-1) = \hat{\boldsymbol{w}}(k-1 \mid k-1) = \hat{\boldsymbol{w}}(k-1)$$

the gradient matrix $\boldsymbol{H} = \nabla \boldsymbol{h}$ becomes

$$\boldsymbol{H}(k) = \frac{\partial \boldsymbol{h}(\hat{\boldsymbol{w}}(k-1))}{\partial \hat{\boldsymbol{w}}(k-1)} \tag{D.48}$$

the elements of which are available from Mandic *et al.* (1998) and Baltersee and Chambers (1998)

$$\frac{\partial y_{M,j}(k)}{\partial w_{n,l}(k)} \approx \Phi'(v_{M,j}(k))\left[\sum_{\alpha=1}^{N} \frac{\partial y_{M,\alpha}(k-1)}{\partial w_{n,l}(k-1)} w_{j,\alpha+p+1}(k) + \delta_{nj} u_{M,l}(k)\right] \tag{D.49}$$

for $i = M$, and

$$\begin{aligned}\frac{\partial y_{i,j}(k)}{\partial w_{n,l}(k)} \approx \Phi'(v_{i,j}(k))\Bigg[&\frac{\partial y_{i+1,j}(k)}{\partial w_{n,l}(k)} w_{j,p+2}(k) \\ &+ \sum_{\alpha=2}^{N} \frac{\partial y_{i,\alpha}(k-1)}{\partial w_{n,l}(k-1)} w_{j,\alpha+p+1}(k) + \delta_{nj} u_{i,l}(k)\Bigg]\end{aligned} \tag{D.50}$$

for $i \neq M$, where δ_{nj} is given by (D.8), so that the derivation of the ERLS algorithm for the PRNN is now complete. The ERLS algorithm for an RNN can be obtained from the derived algorithm, for the number of modules $M = 1$.

Appendix E

Terminology Used in the Field of Neural Networks

The field of artificial neural networks has developed alongside many disciplines, such as neurobiology, mathematics, statistics, economics, computer science, engineering and physics, to mention but a few. Consequently, the terminology used in the field varies from discipline to discipline. An initiative from the IEEE Neural Networks Council to standardise the terminology has resulted in recommended terminology and several definitions (Eberhart 1990). We present four of them.

Activation Function. Algorithm for computing the activation value of a neurode as a function of its net input. Net input is typically the sum of weighted inputs to the neurode.

Feedforward Network. Network ordered into layers with no feedback paths. The lowest layer is the input layer, the highest is the output layer. The outputs of a given layer go only to higher layers and its inputs come only from lower layers.

Supervised Learning. Learning procedure in which a network is presented with a set of input pattern and target pairs. The network can compare its output to the target and adapt itself according to the learning rules.

Unsupervised Learning. Learning procedure in which the network is presented with a set of input patterns. The network adapts itself according to the statistical associations in the input patterns.

A later effort by Fiesler (1994), which appeared in *Computer Standards and Interfaces*, further tackles this issue. In particular, Fiesler considers functions important for neural networks, such as

1. neuron functions (or transfer functions) which specify the output of a neuron, given its inputs (this includes nonlinearity);

2. learning rules (or learning laws) which define how weights (and offsets) will be updated;

3. clamping functions, which determine if and when certain neurons will be insusceptible to incoming information, i.e. they retain their present activation value; and

4. ontogenic functions, which specify changes in the neural network topology.

Hence, a neural network is described by its topology, constraints, initial state and transition function. Topology includes the frame and interconnection structure. In our work we refer to frame as topology and interconnection structure as architecture.

Ljung and Sjöberg have offered a link between terms from the areas of control theory and neural networks (Ljung and Sjöberg 1992), of which we present several frequently used ones:

model structure	net,
model order	number of hidden units,
estimation	training, learning,
iteration	training cycle,
recursive gradient algorithm	backpropagation,
overfit	overtraining.

Appendix F

On the *A Posteriori* Approach in Science and Engineering

F.1 History of *A Posteriori* Techniques

In the Oxford Interactive Encyclopedia, 1997, the notions of *a priori* and *a posteriori* are defined as follows: '*a priori* is a term from epistemology meaning knowledge or concepts which can be gained independently of all experience. It is contrasted with *a posteriori* knowledge, in which experience plays an essential role'. *A posteriori* techniques have been considered for more than two millennia now. Probably the oldest written study on *a posteriori* reasoning techniques in logic was by Aristotle, sometime between 343 BC and 323 BC (Aristotle 1975). He developed *a posteriori* conclusion techniques. In the late sixteenth century, Galileo composed a manuscript between 1589 and 1591, nowadays known as MS27, while he was teaching or preparing to teach at the University of Pisa (Wallace 1992). The manuscript was a study, based upon Aristotle's famous books *Prior Analytics* and *Posterior Analytics*[1] (Aristotle 1975), upon which logic had been taught at universities ever since the fourth century BC. Galileo's work, based upon the canons introduced by Aristotle in *Posterior Analytics*, includes *Discourse on Bodies on or in Water* and *Letters on Sunspots*. He studies the role of the notions of foreknowledge and demonstration in science in general. 'A science can give a real definition of its total subject *a posteriori* only, because the real definition is not foreknown in the science, therefore it is sought, therefore it is demonstrable' (Wallace 1992). 'We know something either *a posteriori* or *a priori*, *a posteriori* through demonstration of the fact, *a priori* through demonstration of the reasoned fact. *A posteriori* is referred to as demonstration *from an effect* or *conjectural*' (Wallace 1992).

Galileo also adapted the *regressus* procedure to accommodate it with his experimental techniques. The *regressus*, after Galileo, required that cause and effect be convertible.

---

[1] Aristotle used the word *analytics* to represent today's meaning of logic.

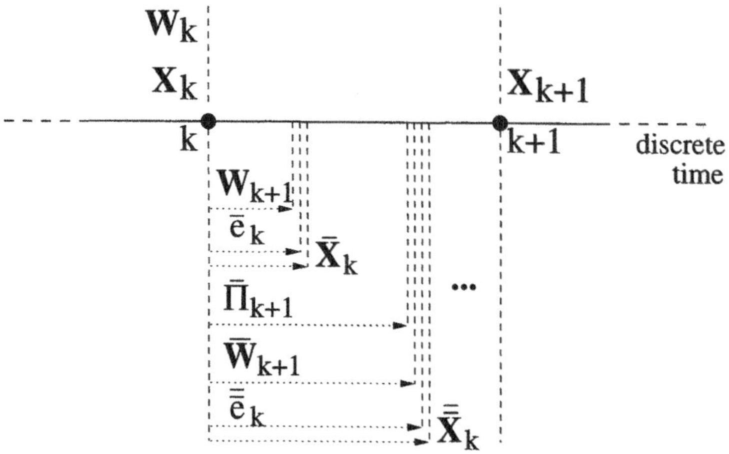

Figure F.1 Time management of the *a posteriori* approach

F.2 The Usage of *A Posteriori*

The notion *a posteriori* is also used in the work based upon Bayes' theorem, in solving linear differential equations, and in Kalman filtering. In modern technical disciplines, *a priori* quantities are estimated from past measurements of a process in hand or presupposed by experience, while *a posteriori* quantities are measured or computed from observations (Stark and Woods 1986).

The *a posteriori* quantities in learning algorithms, such as an error or output, are based upon the newly calculated set of weights $\boldsymbol{W}(k+1)$, which is available sometime between the consecutive discrete time instants k and $(k+1)$. The *a posteriori* quantities will be denoted with the same symbol as the corresponding *a priori* ones, with the addition of a bar, i.e. $\bar{e}(k)$ is the *a posteriori* error at the time instant k, calculated as $\bar{e}(k) = d(k) - \Phi(\boldsymbol{X}^{\mathrm{T}}(k)\boldsymbol{W}(k+1))$, where $\boldsymbol{X}(k)$ is some general input vector and $\boldsymbol{W}(k)$ some general set of weights. The corresponding *a priori* error $e(k)$ is calculated as $e(k) = d(k) - \Phi(\boldsymbol{X}^{\mathrm{T}}(k)\boldsymbol{W}(k))$.

For prediction applications, for instance, there may not be sufficient information to track the trajectory in the state space uniquely, or noise in the observable time series may add too much uncertainty to be able to satisfy the prediction requirements.

A way to avoid these problems is to involve some sort of regularisation through, for instance, the use of the prior knowledge of the system. *A posteriori* techniques naturally use such prior knowledge, by data-reusing, and the equations which describe such algorithms exhibit normalisation, which additionally stabilised the system.

F.2.1 A Posteriori Techniques in the RNN Framework

To further illustrate this concept in adaptive real-time algorithms, let us consider the situation shown in Figure F.1. All relevant time instants in Figure F.1 are denoted in the index of the appropriate variable, e.g. $\boldsymbol{X}_k$ instead of $\boldsymbol{X}(k)$. At the discrete time interval k, the RNN has vectors $\boldsymbol{X}(k)$ and $\boldsymbol{W}(k)$ (and inherently $\boldsymbol{\Pi}(k)$) available for

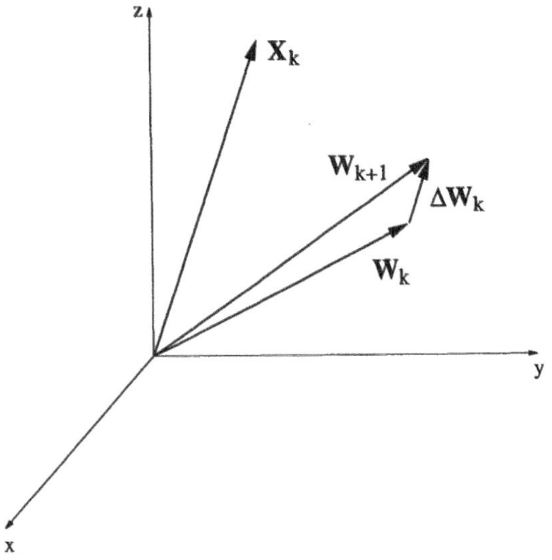

Figure F.2 Weight updating by the LMS

further calculation. For either the *a priori* or *a posteriori* approach, there is a need to calculate $\boldsymbol{W}(k+1)$, which must be done before the time instant $(k+1)$. Therefore, speaking in terms of discrete time, it appears before $(k+1)$ and thus belongs to k, but this is not strictly true, since $\boldsymbol{W}(k+1)$ is not available at k. Reiterating this idea, all the variables necessary for *a posteriori* techniques are shown in Figure F.1, together with the notion of the time of their availability.

F.2.2 The Geometric Interpretation of A Posteriori Error Learning

For the sake of simplicity we treat only the linear feedforward case with only one node. The equations which specify the algorithm can be written as

$$\left.\begin{aligned} \boldsymbol{w}_{i+1}(k) &= \boldsymbol{w}_i(k) + \eta e_i(k)\boldsymbol{x}(k), \qquad i = 1, \dots, L, \\ e_i(k) &= d(k) - \boldsymbol{x}^{\mathrm{T}}(k)\boldsymbol{w}_i(k), \end{aligned}\right\} \tag{F.1}$$

where $\boldsymbol{w}_1(k) = \boldsymbol{w}(k)$ and $\boldsymbol{w}(k+1) = \boldsymbol{w}_{L+1}(k)$. For $L = 1$ Equation (F.1) degenerates into the LMS algorithm. From (F.1), it is obvious that the direction of the vectors $\Delta\boldsymbol{w}_i(k)$ is the same as the direction of the input vector $\boldsymbol{x}(k)$ (collinear). This is further depicted in Figure F.2 (Widrow and Lehr 1990). As the LMS gives only an approximate solution of the Wiener filtering problem, the quadratic surface defined by $e^2(k)$ has a solution set which is a linear variety, instead of a single minimum. Hence, the space of solution is a hyperplane (Schnaufer and Jenkins 1993), whose dimension is one less that the space upon it rests, and vector $\boldsymbol{x}(k)$ is perpendicular to the solution hyperplane $S(k)$. Figure F.3 provides the geometric interpretation of the relation between the LMS, NLMS and *a posteriori* (data-reusing) LMS. The direction of vectors $\boldsymbol{w}$ is $\boldsymbol{w}(k) + \mathrm{span}(\boldsymbol{x}(k))$ (Schnaufer and Jenkins 1993). As the LMS algorithm is an approximative algorithm which uses instantaneous estimates

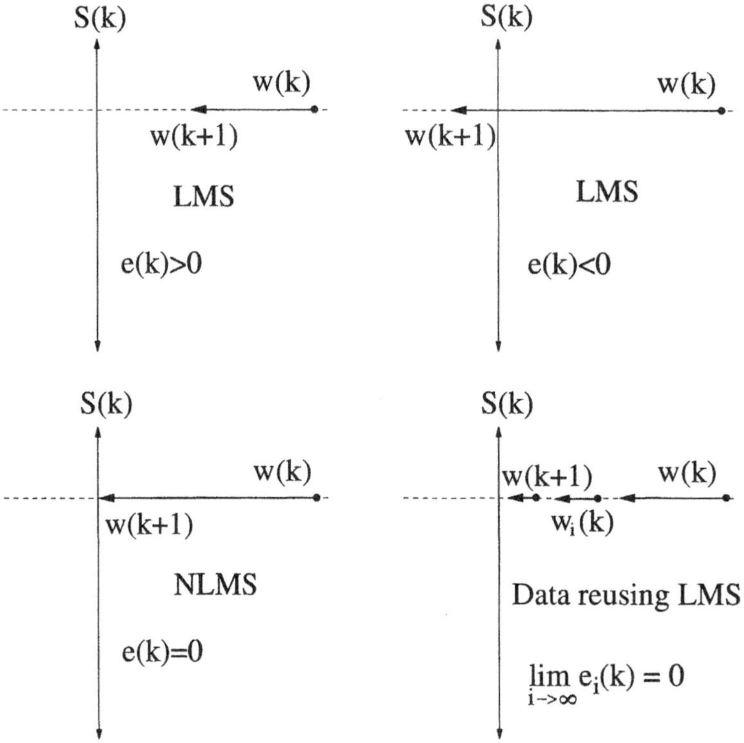

Figure F.3 Geometric interpretation of data-reusing techniques

instead of statistical expectations, the output error of a filter trained by LMS is either positive (the weight update short of an optimal weight update) or negative (the weight update exceeds the optimal weight update), which is shown in the top two diagrams of Figure F.3. The NLMS algorithm, on the other hand is defined to minimise the instantaneous output error ($e(k) = 0$), which is geometrically shown in the bottom left diagram in the figure. The *a posteriori* (data-reuse) algorithms start either from the situation described by $e(k) > 0$ or $e(k) < 0$, i.e. from either of the top two diagrams in Figure F.3. The iterative weight updates then approach the performance of normalised algorithms as the number of iterations increases, as shown in the bottom right diagram in the figure. From Figure F.3, it is evident that repeating a data-reusing technique for a sufficient number of times, the process approaches the normalised (zero-error) solution.

Appendix G

Contraction Mapping Theorems

G.1 Fixed Points and Contraction Mapping Theorems

When numerically solving $F : \mathbb{R} \to \mathbb{R}$ for its zeros, i.e. to find an x^* such that $F(x^*) = 0$, a convenient approach is to rearrange $F(x) = 0$ into $x = K(x)$ and undertake the iteration (Dennis and Schnabel 1983),

$$x_{i+1} = K(x_i), \qquad x_0 \text{ chosen}. \tag{G.1}$$

If (G.1) converges to x^*, then, by continuity,

$$x^* = \lim_{i\to\infty} K(x_i) = K(x^*). \tag{G.2}$$

Hence, x^* is a fixed point of K and thus a zero of F.

G.1.1 Contraction Mapping Theorem in $\mathbb{R}$

Theorem G.1.1. *If*

(i) $x \in [a, b] \Rightarrow K(x) \in [a, b]$,

(ii) $\exists\gamma < 1$ *such that* $|K(x) - K(y)| \leqslant \gamma|x - y|$, $\forall x, y \in [a, b]$,

then the equation $x = K(x)$ *has a unique solution* $x^* \in [a, b]$ *and the iteration*

$$x_{i+1} = K(x_i) \tag{G.3}$$

converges to x^* *for any* $x_0 \in [a, b]$.

Proof. Existence: From (i), we have $a - K(a) \leqslant 0$ and $b - K(b) \leqslant 0$, as shown in Figure G.1. The intermediate value theorem (IVT) shows that there is some solution $x^* \in [a, b]$ such that $x^* = K(x^*)$.

Uniqueness: If $\bar{x} \in [a, b]$ is also a solution of (G.3), then

$$|\bar{x} - x^*| = |K(\bar{x}) - K(x^*)| \leqslant \gamma|\bar{x} - x^*|. \tag{G.4}$$

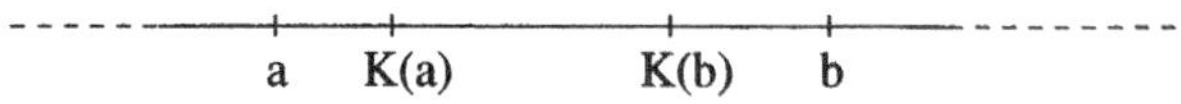

Figure G.1 The contraction mapping

Since $\gamma < 1$, this is a contradiction unless $\bar{x} \equiv x^*$.

Convergence: We have

$$|x_i - x^*| = |K(x_{i-1}) - K(x^*)| \leqslant \gamma |x_{i-1} - x^*|. \tag{G.5}$$

Thus $|x_i - x^*| \leqslant \gamma^i |x_0 - x^*|$ and $\lim_{i\to\infty} \gamma^i = 0$. Hence $\{x_i\} \xrightarrow{i} x^*$.

Condition (ii) from Theorem G.1.1 is checked by showing that $|K'(x)| \leqslant \gamma < 1$, $\forall x \in (a, b)$. Namely the mean value theorem (MVT) shows that $\exists \xi \in (a, b)$ such that

$$|K(x) - K(y)| = |K'(\xi)(x-y)| = |K'(\xi)||x-y| \leqslant \gamma |x-y|, \qquad x, y \in [a, b], \quad \xi \in (a, b). \tag{G.6}$$

G.1.2 Contraction Mapping Theorem in $\mathbb{R}^N$

Theorem G.1.2. *Let $\mathcal{M}$ be a closed subset of $\mathbb{R}^N$ such that*

(i) $K : \mathcal{M} \to \mathcal{M}$,

(ii) $\exists \gamma < 1$ *such that* $\|\boldsymbol{K}(\boldsymbol{x}) - \boldsymbol{K}(\boldsymbol{y})\| \leqslant \gamma \|\boldsymbol{x} - \boldsymbol{y}\|$, $\forall \boldsymbol{x}, \boldsymbol{y} \in \mathcal{M}$,

then equation

$$\boldsymbol{x} = \boldsymbol{K}(\boldsymbol{x}) \tag{G.7}$$

has a unique solution $\boldsymbol{x}^* \in \mathcal{M}$ and the iteration

$$\boldsymbol{x}_{i+1} = \boldsymbol{K}(\boldsymbol{x}_i) \tag{G.8}$$

converges to $\boldsymbol{x}^*$ for any starting value $\boldsymbol{x}_0 \in \mathcal{M}$.

G.2 Lipschitz Continuity and Contraction Mapping

We first give the definition of a Lipschitz continuous function.

Definition G.2.1. Let X be a complete metric space with metric d containing a closed nonempty set Ω and let $g : \Omega \to \Omega$. The function g is said to be Lipschitz continuous with Lipschitz constant $\gamma \in \mathbb{R}$ if

$$\forall x, y \in \Omega, \qquad d[g(x), g(y)] \leqslant \gamma d(x, y).$$

From Definition G.2.1 we differentiate the following cases.

(i) For $0 \leqslant \gamma < 1$, g is a contraction mapping on Ω and γ defines the rate of convergence.

(ii) For $\gamma = 1$, g is a nonexpansive mapping.

(iii) For $\gamma > 1$, g is a Lipschitz continuous mapping on Ω.

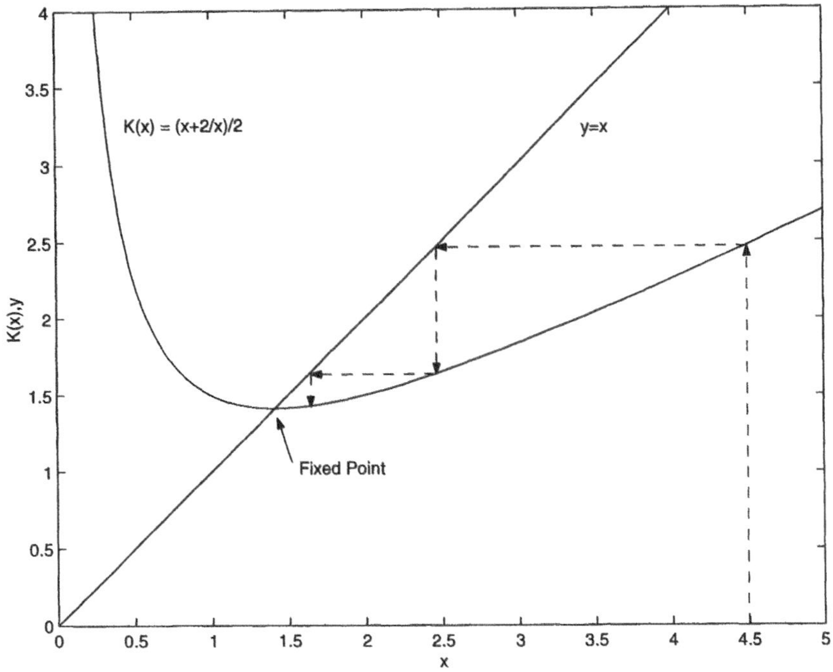

Figure G.2 Babylonian iteration

The following application of the IVT gives an important criterion for the *existence* of the fixed point.

Lemma G.2.2 (Brower's Fixed Point Theorem (Devaney 1989)). *Let $\Omega = [a, b]^N$ be a closed set of $\mathbb{R}^N$ and $f : \Omega \rightarrow \Omega$ be a continuous vector-valued function. Then f has at least one fixed point in Ω.*

The behaviour of state trajectories (orbits) in the vicinity of fixed points defines the character of fixed points. For an asymptotically stable (or *attractive*) fixed point x^* of a function F, there exists a neighbourhood $\mathbb{O}(x^*)$ of x^* such that $\lim_{k\to\infty} F(x_k) = x^*$, for all $x_k \in \mathbb{O}(x^*)$. In this case, each eigenvalue λ of the Jacobian of F at x^*, is less than unity in magnitude. Eigenvalues of the Jacobian of F which are greater than unity in magnitude give rise to an expansion, whereas eigenvalues smaller than unity provide a contraction. In the former case of F an expansion x^* is a *repulsive* point or repellor. If some eigenvalues of the Jacobian of F are greater and some smaller than unity, x^* is called a saddle point.

G.3 Historical Perspective

Evidence from clay tablets held within the Yale University Babylonian collection suggests that iterative techniques for finding square roots go back several thousand years (Kreith and Chakerian 1999). One such clay tablet depicts how to calculate the diagonals of a square. Babylonians used a positional, base 60 number system. On this

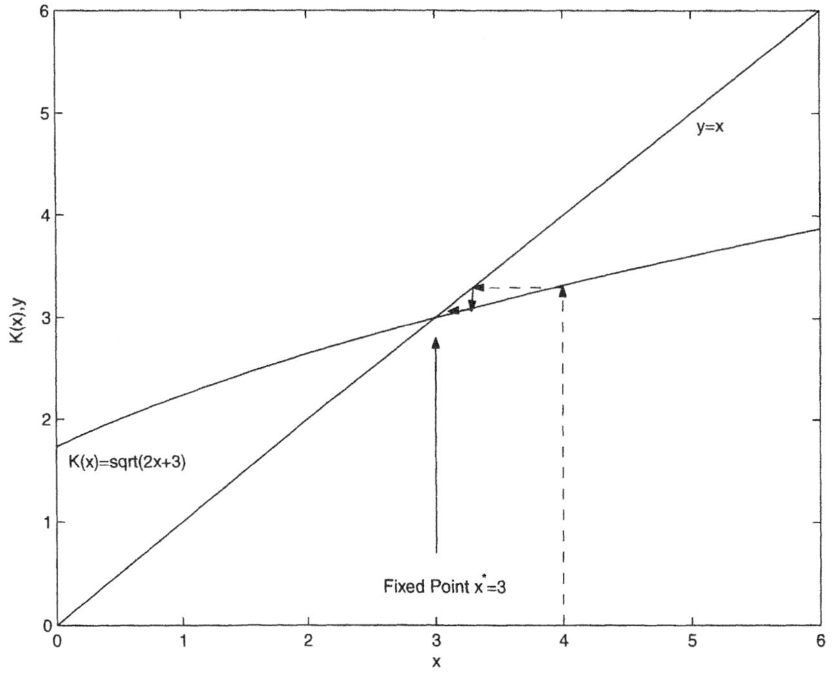

Figure G.3 FPI solution for SQRT

tablet, $30\sqrt{2}$ is calculated as $42 + \frac{25}{60} + \frac{35}{60^2} \approx 42.426\,389$ geometrically via iteration of doubled squares.

In terms of modern mathematics, the Babylonians employed an iterator

$$x = K(x) = \frac{1}{2}\left(x + \frac{2}{x}\right) \tag{G.9}$$

so that successive applications of $K(x)$, which start from $x_0 = 4.5$, give $x_1 = K(x_0) = \frac{3}{2}, x_2 = K(x_1) = \frac{17}{12}, x_3 = K(x_2) = \frac{577}{408}$ (Kreith and Chakerian 1999), as shown in Figure G.2. The point of intersection of curves $K(x)$ and $y = x$ can be found by solving $\frac{1}{2}(x + 2/x) = x$, which has the value $x = \sqrt{2}$ and is a *fixed point* of K and a solution of $K(x) = x$.

The next example shows how to set up a fixed point iteration.

Example G.3.1. Find the roots of function $F(x) = x^2 - 2x - 3$.

Solution. Roots of function $F(x) = x^2 - 2x - 3$ can be found by rearranging $x = K(x) = \sqrt{2x+3}$. The roots of function F are -1 and 3. The FPI which started from $x_0 = 4$ gives the sequence of iterates 3.3116, 3.1037, 3.0344, 3.0114, 3.0038, 3.0013, 3.0004, 3.0001, 3.000, whereby $x^* = 3$ is the fixed point and a solution of the equation. The fixed point iteration is illustrated in Figure G.3. □

Fixed point iterations for another two characteristic cases, namely an oscillatory convergence and divergence are given, respectively, in Figure G.4(a) and Figure G.4(b). For the case depicted in Figure G.4(b), the value of the first derivative

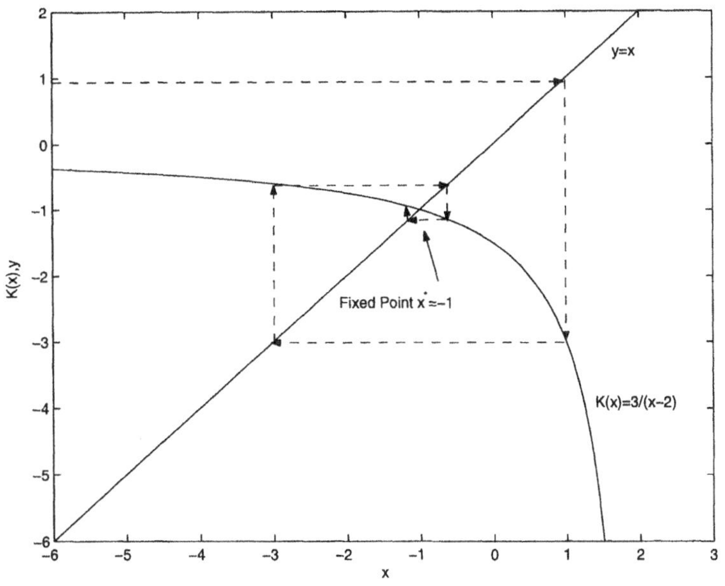

(a) Oscillatory FPI for $K(x) = 3/(x - 2)$

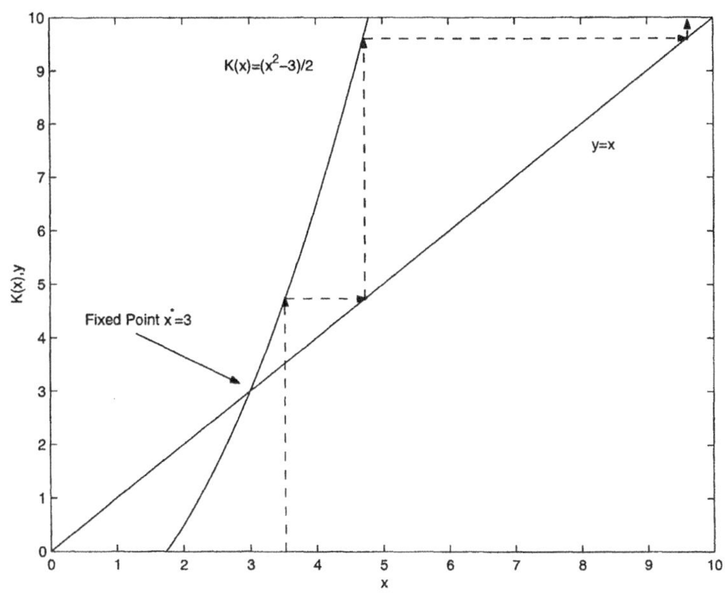

(b) A repellor for $K(x) = (x^2 - 3)/2$

Figure G.4 FPI for an oscillatory case and the case of a repellor of function $K(x) = (x^2 - 3)/2$ is greater than unity for $x > x^* = 3$. Hence, by CMT (Theorem G.1.1), function K is an expansion and a fixed point iteration for starting values $x_0 > 3$ diverges. Notice, however, that an FPI with a starting value $x_0 < 3$ would converge to fixed point $x^* = 3$ by virtue of CMT. Hence, fixed point $x^* = 3$ is an unstable fixed point of function K. For the case of function $K(x) = \sqrt{2x+3}$ depicted in Figure G.3, an FPI which starts from $x_0 < 3$ would still converge to fixed point $x^* = 3$ ($|K'(x)| < 1$), hence point $x^* = 3$ is a stable fixed point of function K.

Appendix H

Linear GAS Relaxation

H.1 Relaxation in Linear Systems

The problem of global asymptotic stability (GAS) of an Nth-order time-variant difference equation,

$$y(k) = \boldsymbol{a}^{\mathrm{T}}(k)\boldsymbol{y}(k-1) = a_1(k)y(k-1) + \cdots + a_N(k)y(k-N), \tag{H.1}$$

is important in the theory of linear systems (Barnett and Storey 1970; Golub and Van Loan 1996; Haykin 1996a; Kailath 1980; LaSalle 1986). Equation (H.1) represents an autonomous system, in fact, it is a relaxation equation (Basar and Bernhard 1995) which stems from a general linear system representation,

$$\boldsymbol{Y}(k+1) = \boldsymbol{A}(k)\boldsymbol{Y}(k) + \boldsymbol{B}(k)\boldsymbol{u}(k), \tag{H.2}$$

for the zero exogenous input vector $\boldsymbol{u}(k) = \mathbf{0}$, $\forall k > 0$ (Kailath 1980; LaSalle 1986). The matrix form of Equation (H.1) now becomes

$$\begin{bmatrix} y(k+1) \\ y(k) \\ \vdots \\ y(k-N+1) \end{bmatrix} = \begin{bmatrix} a_1(k) & a_2(k) & \cdots & a_N(k) \\ 1 & 0 & \cdots & 0 \\ \vdots & \vdots & \ddots & \vdots \\ 0 & \cdots & 1 & 0 \end{bmatrix} \begin{bmatrix} y(k) \\ y(k-1) \\ \vdots \\ y(k-N) \end{bmatrix}$$

or

$$\boldsymbol{Y}(k+1) = \boldsymbol{A}(k)\boldsymbol{Y}(k)$$

and

$$y(k+1) = [1\ 0\ \cdots\ 0]\boldsymbol{Y}(k+1). \tag{H.3}$$

The matrix $\boldsymbol{A}$ is a Frobenius matrix, which is a special form of the companion matrix. The fundamental theorem of matrices (Horn and Johnson 1985; Wilkinson 1965) states that every matrix $\boldsymbol{A}$ can be reduced by a similarity transformation to a sum of Frobenius matrices (Ciarlet 1989; Wilkinson 1965). That is why it is important to consider the stability results for the Frobenius matrix $\boldsymbol{A}$ (H.3), since a stability

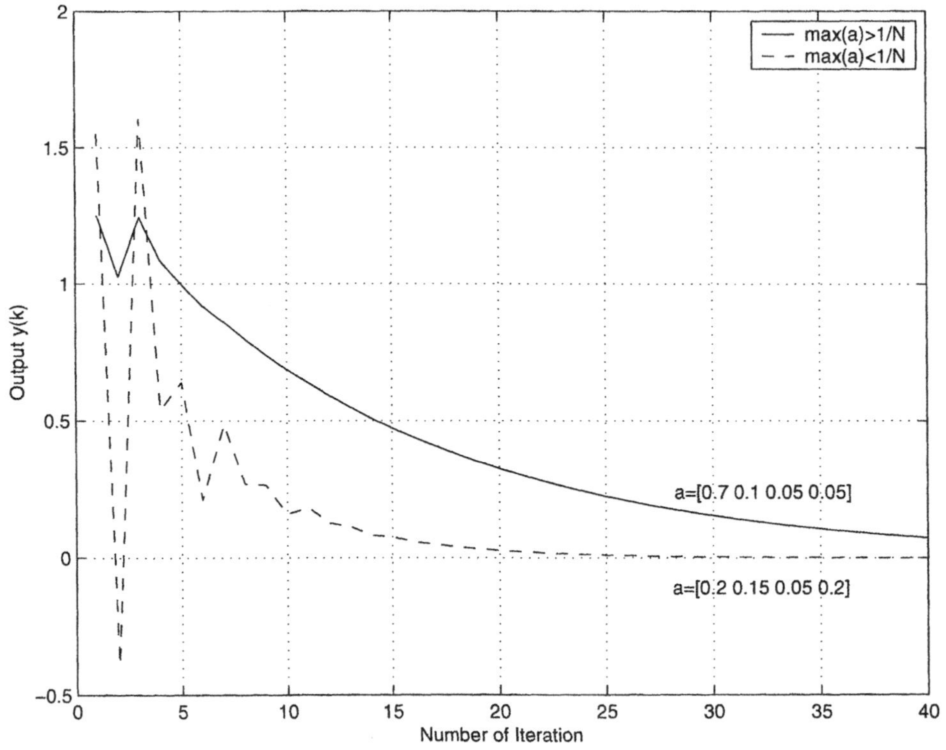

Figure H.1 The monotonic and nonmonotonic convergence

result of a general system $\boldsymbol{Cx} = \boldsymbol{y}$ can be obtained through the stability result of (H.3). Moreover, the analysis of convergence and stability of some learning algorithms for adaptive systems can be undertaken using this approach (Mandic and Chambers 2000a). The following analysis gives the stability conditions for relaxation (H.1), for systems with positive coefficients. The analysis of robust relaxation for nonlinear dynamical systems embarks upon this result and can be found in Mandic and Chambers (2000c). The bound on the size of the parameter vector for convergence in the CMT sense is provided below (Mandic and Chambers 2000d). In order to preserve contractivity of relaxation (H.1), we have

$$\begin{aligned} |y(k)| &= |a_1(k)y(k-1) + \cdots + a_N(k)y(k-N)| \\ &\leqslant a_1|y(k-1)| + \cdots + a_N|y(k-N)| \\ &< (a_1(k) + \cdots + a_N(k))|y(k-1)| \\ &\leqslant N \max_{a_i(k),\ i=1,\dots,N} a_i(k)|y(k-1)|. \end{aligned} \tag{H.4}$$

From (H.4),

$$\max_{a_i(k),\ i=1,\dots N} a_i(k) > \frac{1}{N} \quad \Leftrightarrow \quad \|\boldsymbol{a}(k)\|_\infty > \frac{1}{N}.$$

Figure H.1 shows two cases, the strict aperiodic convergence and a pseudoperiodic convergence. The solid line in Figure H.1 decays monotonically towards zero for lags $k > 4$. The dashed line in Figure H.1 decays in an oscillatory way towards zero for lags $k > 4$. □

Observation H.1.1. *The system (H.3) with the constraint* $\|\boldsymbol{a}(k)\|_1 < 1$ *converges towards zero in the fixed point iteration (FPI) sense, exhibiting linear convergence, with the convergence rate* $\|\boldsymbol{a}(k)\|$.

This is straightforward to show using results from Appendix G.

H.1.1 Stability Result for $\sum_{i=1}^{m} a_i = 1$

Let us consider the case with constant parameter vector $\boldsymbol{a} = [a_1, \dots, a_N]^{\mathrm{T}}$, where $\|\boldsymbol{a}\|_1 = 1$. In that case, matrix $\boldsymbol{A}$ from (H.3) becomes a stochastic matrix (Golub and Van Loan 1996; Lipschutz 1965; Stark and Woods 1986), since each of its rows is a probability vector, and process (H.3) can be rewritten as

$$\boldsymbol{Y}(k+1) = \boldsymbol{A}\boldsymbol{Y}(k) = \boldsymbol{A}^2\boldsymbol{Y}(k-1) = \cdots = \boldsymbol{A}^k\boldsymbol{Y}(0), \tag{H.5}$$

whereby the dynamics of (H.3) are fully described by its initial state $\boldsymbol{Y}(0)$ and the system matrix $\boldsymbol{A}$. Since the product of two stochastic matrices is a stochastic matrix, there is a unique fixed vector $\boldsymbol{t} = [t_1, \dots, t_N]^{\mathrm{T}}$ such that (Golub and Van Loan 1996; Lipschutz 1965)

$$\boldsymbol{t}\boldsymbol{A} = \boldsymbol{t}. \tag{H.6}$$

Vector $\boldsymbol{t}$ is a probability vector, i.e. $\|\boldsymbol{t}\|_1 = 1$. Therefore, FPI gives

$$\boldsymbol{A}^k = \begin{bmatrix} a_1 & a_2 & \cdots & a_N \\ 1 & 0 & \cdots & 0 \\ \vdots & \vdots & \ddots & \vdots \\ 0 & \cdots & 1 & 0 \end{bmatrix}^k \xrightarrow{k\to\infty} \begin{bmatrix} t_1 & t_2 & \cdots & t_N \\ t_1 & t_2 & \cdots & t_N \\ \vdots & \vdots & \ddots & \vdots \\ t_1 & t_2 & \cdots & t_N \end{bmatrix}. \tag{H.7}$$

Observation H.1.2 (see Mandic and Chambers 2000d). *The process (H.1) with the constant nonnegative coefficient vector* $\boldsymbol{a} = [a_1, \dots, a_N]^{\mathrm{T}}$, *converges to*

(i) $|y_\infty| = \left|\sum_{i=1}^{N} t_i y(k-i)\right| \geqslant 0$ *for* $\|\boldsymbol{a}\|_1 = 1$,

(ii) $y_\infty = 0$ *for* $\|\boldsymbol{a}\|_1 < 1$,

from any finite initial state $\boldsymbol{Y}(0)$.

H.2 Examples

We next present some results which depict the need for a strict analysis of relaxation type equations. The initial condition for all examples below was $Y = [1\ 2]^{\mathrm{T}}$.

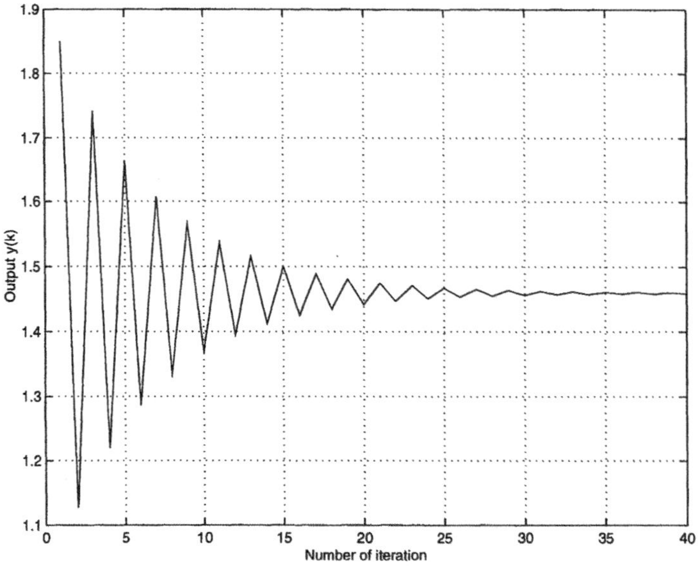

(a) The output $y(k)$

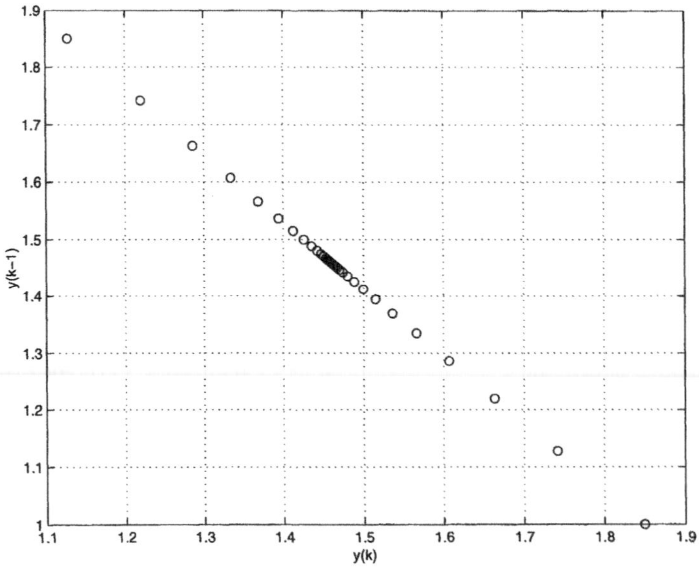

(b) Convergence of $\boldsymbol{Y}$ in the phase plane

Figure H.2 Convergence of the process $\boldsymbol{Y}(k) = \boldsymbol{A}\boldsymbol{Y}(k-1)$

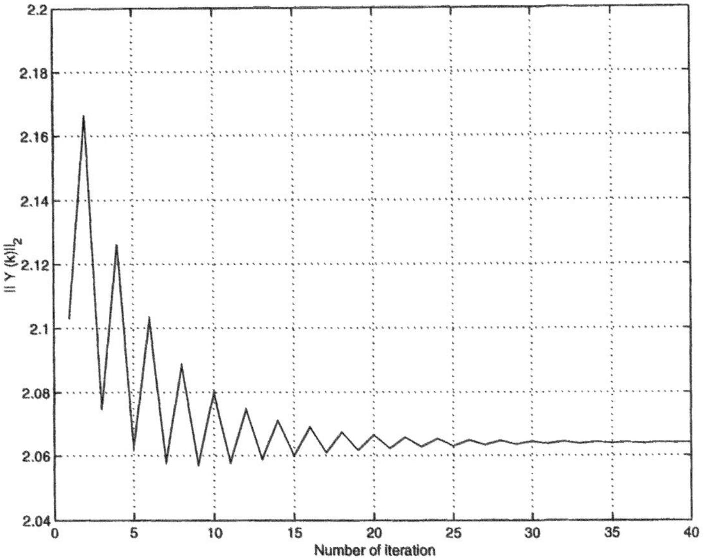

(c) Convergence of $\boldsymbol{Y}$ in the $\|\cdot\|_2$ norm

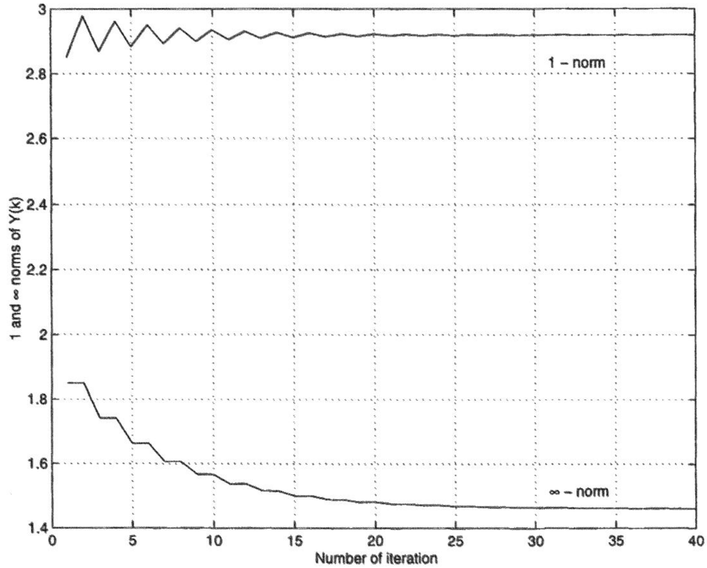

(d) Convergence of $\boldsymbol{Y}$ in the $\|\cdot\|_1$ and $\|\cdot\|_\infty$ norm

Figure H.2 *Cont.*

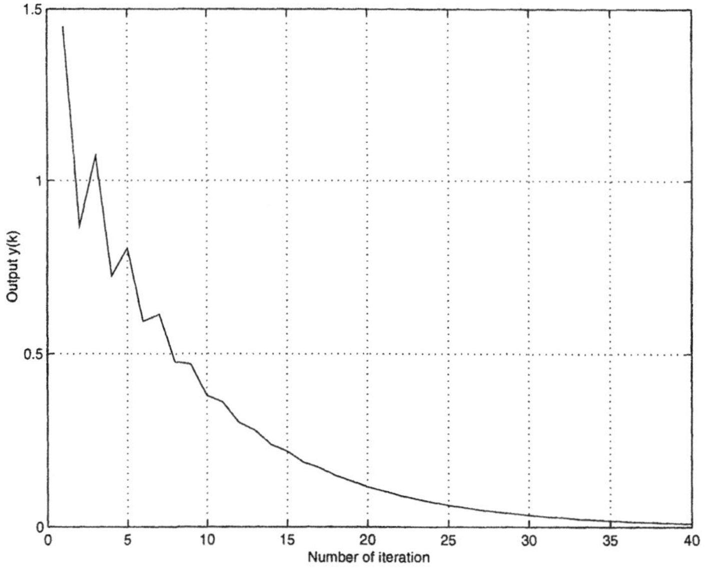

(a) The output $y(k)$

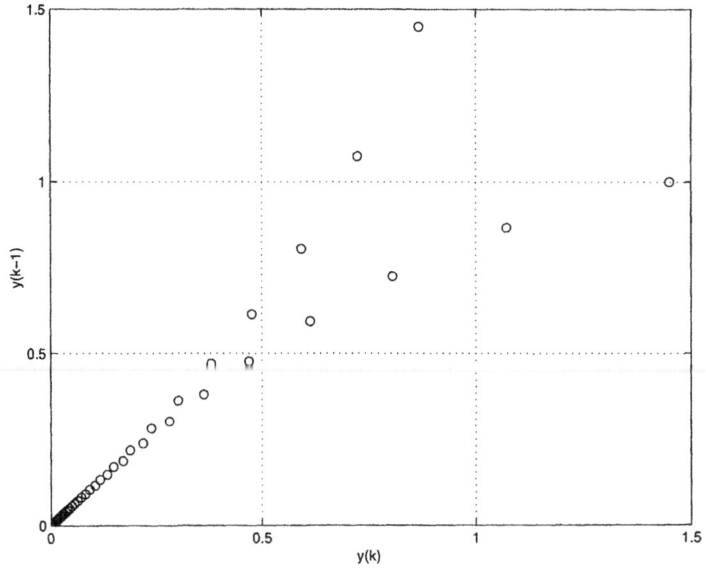

(b) Convergence of $\boldsymbol{Y}$ in the phase plane

Figure H.3 Convergence of the process $\boldsymbol{Y}(k) = \boldsymbol{A}\boldsymbol{Y}(k-1)$

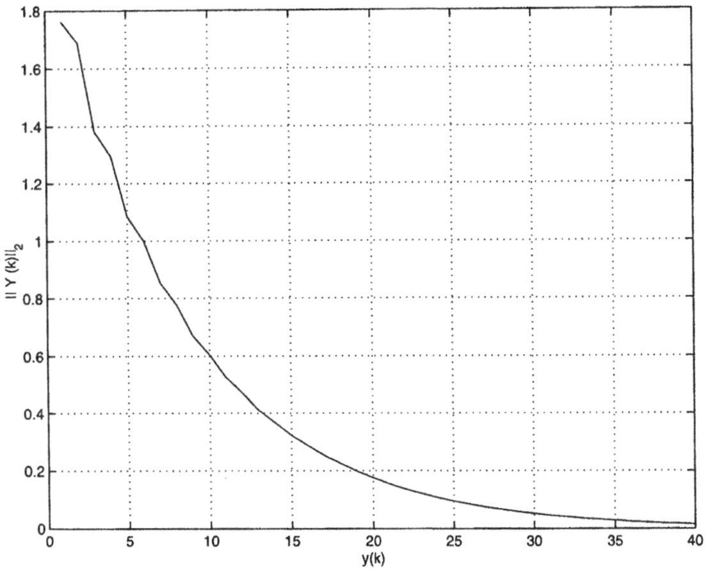

(c) Convergence of $\boldsymbol{Y}$ in the $\|\cdot\|_2$ norm

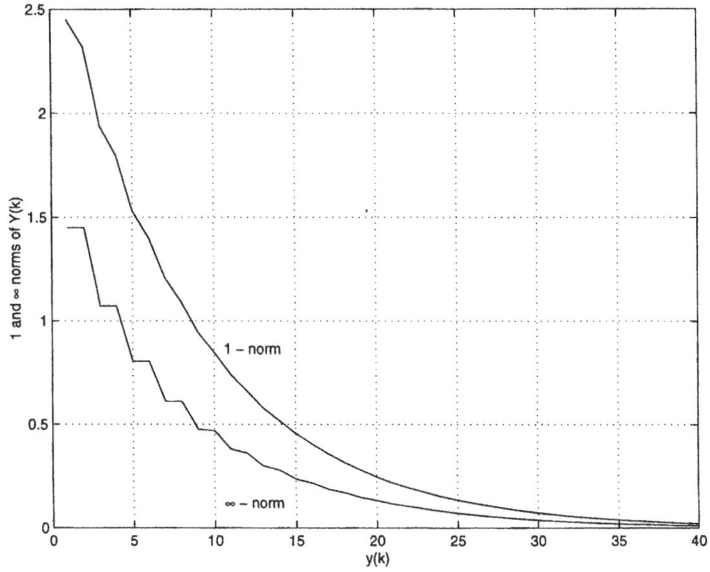

(d) Convergence of $\boldsymbol{Y}$ in the $\|\cdot\|_1$ and $\|\cdot\|_\infty$ norm

Figure H.3 *Cont.*

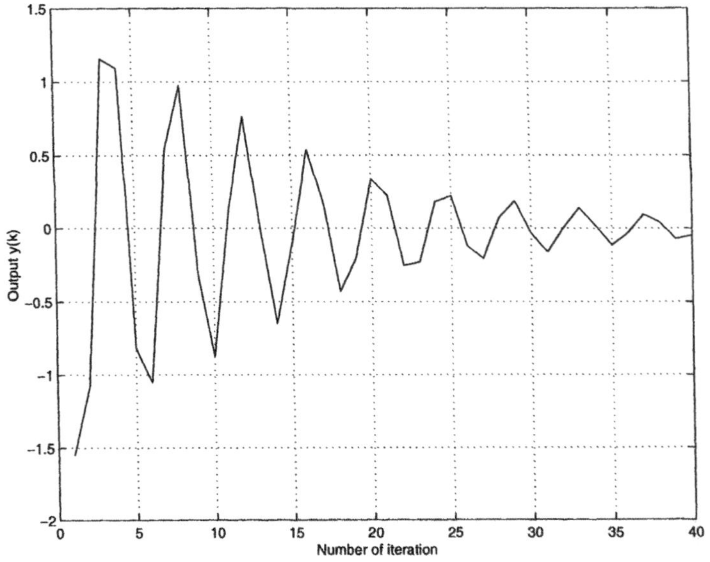

(a) The output $y(k)$

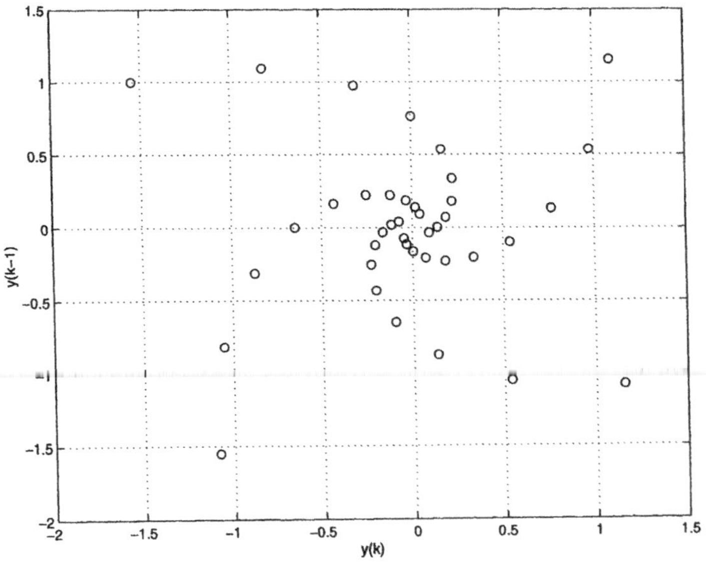

(b) Convergence of $\boldsymbol{Y}$ in the phase plane

Figure H.4 Convergence of the process $\boldsymbol{Y}(k) = \boldsymbol{A}\boldsymbol{Y}(k-1)$

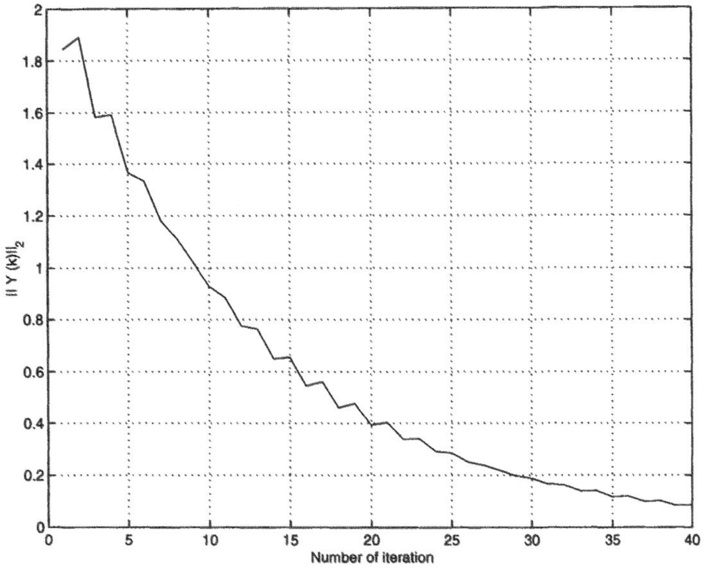

(c) Convergence of $\boldsymbol{Y}$ in the $\|\cdot\|_2$ norm

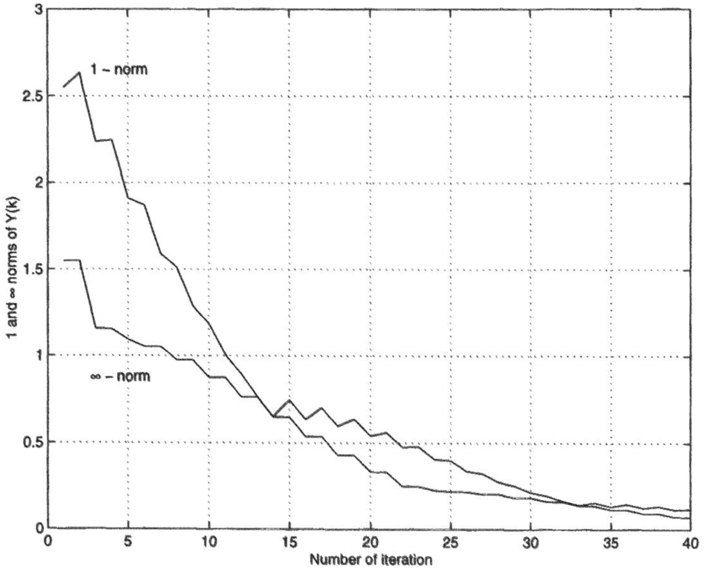

(d) Convergence of $\boldsymbol{Y}$ in the $\|\cdot\|_1$ and $\|\cdot\|_\infty$ norm

Figure H.4 *Cont.*

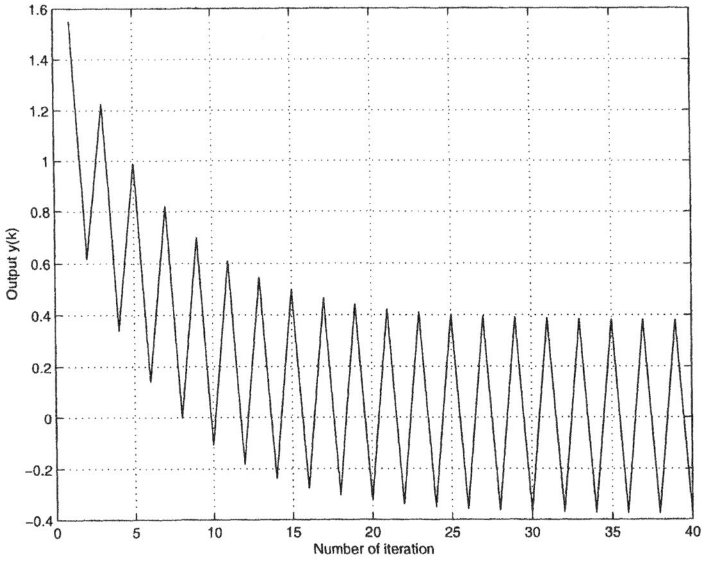

(a) The output $y(k)$

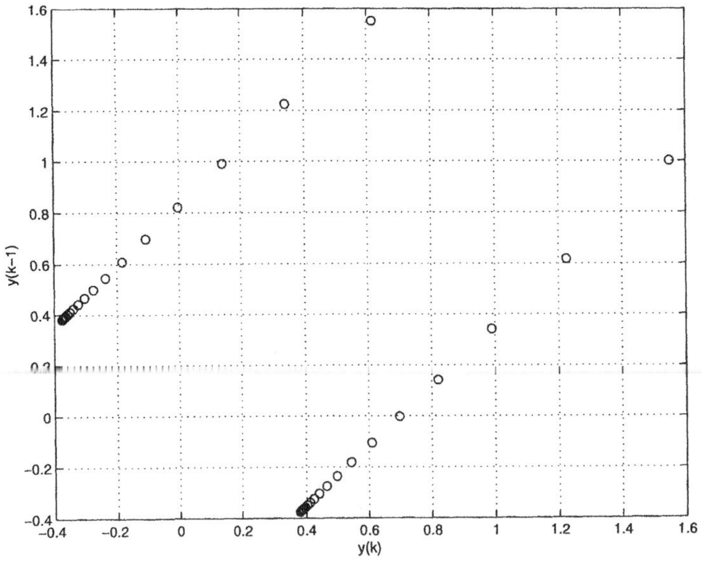

(b) Convergence of $\boldsymbol{Y}$ in the phase plane

Figure H.5 Convergence of the process $\boldsymbol{Y}(k) = \boldsymbol{A}\boldsymbol{Y}(k-1)$

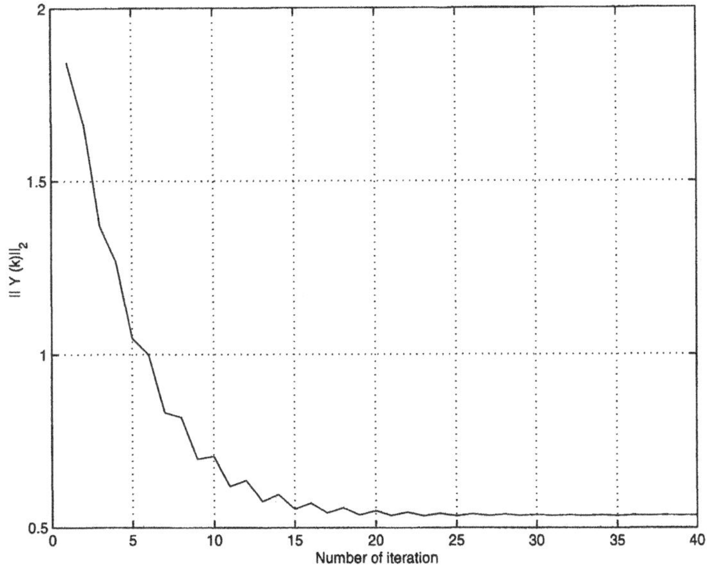

(c) Convergence of $\boldsymbol{Y}$ in the $\|\cdot\|_2$ norm

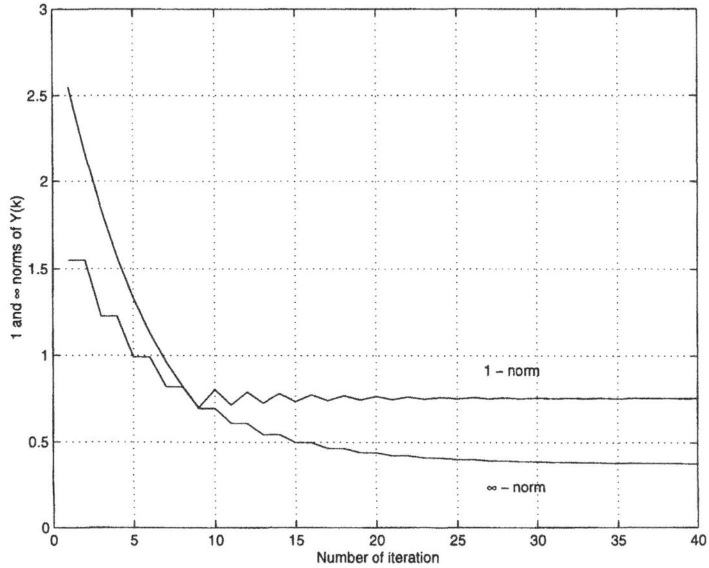

(d) Convergence of $\boldsymbol{Y}$ in the $\|\cdot\|_1$ and $\|\cdot\|_\infty$ norm

Figure H.5 *Cont.*

(a) For a system described by coefficients $a_1 = 0.15$, $a_2 = 0.85$, the relaxation is shown in Figure H.2. The output converges oscillatorily towards a point, as shown in Figure H.2(a). The geometric convergence of points $\boldsymbol{Y}(k)$ in the phase plane is oscillatory towards a point, forming a line in the plane (Figure H.2(b)). The 2 norm, shown in Figure H.2(c) and the 1 and ∞ norm, shown in Figure H.2(d) all exhibit a relaxive behaviour.

(b) For a system described by coefficients $a_1 = 0.15$, $a_2 = 0.65$, the corresponding relaxation diagrams are shown in Figure H.3. Here, $\|\boldsymbol{a}\|_1 < 1$, and the elements of $\boldsymbol{a}$ are positive. Relaxation in the 1 and 2 norm is aperiodic, whereas the output and phase plane convergence are oscillatory.

(c) For a system described by coefficients $a_1 = 0.15$, $a_2 = -0.85$, the relaxation diagrams are shown in Figure H.4. Here, $\|\boldsymbol{a}\|_1 = 1$, but not all the elements of $\boldsymbol{a}$ are positive. There is a clear difference between the relaxation behaviour of the output and the phase space diagram (Figure H.4(a) and H.4(b)) and the convergence in the norm (Figure H.4(c) and H.4(d)).

(d) For a system described by coefficients $a_1 = -0.15$, $a_2 = 0.85$, the convergence is shown in Figure H.5. Here, $\|\boldsymbol{a}\|_1 = 1$, but not all elements of $\boldsymbol{a}$ are nonnegative. The process (H.1) converges in all the norms (Figure H.5(c) and H.5(d)), but not in either the output or the geometric sense in the phase plane (Figure H.5(a) and H.5(b)), where it achieves its limit cycle, for there are two distinct points, with the same norm, to which the process converges.

Hence the following observation.

Observation H.2.1. *For the system (H.1), with $\|\boldsymbol{a}\|_1 = 1$, convergence in the norm does not imply convergence in the geometric sense.*

These situations depict the need for the use of alternative techniques for the analysis of relaxation and learning algorithms for nonlinear systems, such as the use of CMT and FPI as introduced in Chapter 7 and Appendix G (Mandic 2000b,c).

Appendix I

The Main Notions in Stability Theory

The roots of stability theory in physics and engineering can be traced down to Alexander Mikhailovitch Lyapunov (1857–1918). He developed the so-called 'Lyapunov's second method' in his PhD thesis '*The General Problem of the Stability of Motion*' in 1892[1]. The following are definitions of notions from the stability theory. They are similar to those given in Sastry and Bodson (1989), but rearranged and updated due to the specific requirements of this work. Let us consider the following differential equation:

$$\dot{x} = f(t, x), \qquad x(t_0) = x_0. \tag{I.1}$$

Definition I.1.2 (autonomous systems). The system defined by (I.1) is called autonomous or time-invariant, if f does not depend on t.

Definition I.1.3 (linearity). The system is said to be linear if $f(t, x) = A(t)x$ for some $A : \mathbb{R}^+ \to \mathbb{R}^{n \times n}$.

The properties of systems defined on a closed ball B_h with radius h centred at $\mathbf{0} \in \mathbb{R}^n$, are considered

(i) *locally*, if true for all $x_0 \in B_h$,

(ii) *globally*, if true for all $x_0 \in \mathbb{R}^n$,

(iii) *in any closed ball*, if true for all $x_0 \in B_h$, with h arbitrary,

(iv) *uniformly*, if true for all $t_0 > 0$.

Definition I.1.4 (Lipschitz function). The function f is said to be Lipschitz in x if, for some $h > 0$, there exists $l \geqslant 0$ such that

$$\|f(t, x_1) - f(t, x_2)\| \leqslant L\|x_1 - x_2\| \tag{I.2}$$

for all $x_1, x_2 \in B_h$, $t \geqslant 0$. The constant L is called the *Lipschitz* constant.

[1] Lyapunov was interested in the problem of equilibrium figures of a rotating liquid, which was also treated by Maclaurin, Jacobi and Laplace.

Definition I.1.5 (equilibrium). Point $\boldsymbol{x}$ is called an equilibrium point of (I.1), if $f(t, \boldsymbol{x}) = 0$ for all $t \geqslant 0$.

Definition I.1.6 (stability). Point $\boldsymbol{x} = 0$ is called a stable equilibrium point of (I.1), if, for all $t_0 \geqslant 0$ and $\varepsilon > 0$, there exists $\delta(t_0, \varepsilon)$ such that

$$\|\boldsymbol{x}_0\| < \delta(t_0, \varepsilon) \Rightarrow \|\boldsymbol{x}(t)\| < \varepsilon, \quad \text{for all } t \geqslant t_0, \tag{I.3}$$

where $\boldsymbol{x}(t)$ is the solution of (I.1) with initial conditions $\boldsymbol{x}_0$ and t_0.

Definition I.1.7 (uniform stability). Point $\boldsymbol{x} = \mathbf{0}$ is called a uniformly stable equilibrium point of (I.1) if, in the preceding definition, δ can be chosen independent of t_0.

In other words, the equilibrium point is not growing progressively less stable with time.

Definition I.1.8 (asymptotic stability). Point $\boldsymbol{x} = \mathbf{0}$ is called an asymptotically stable equilibrium point if

(a) $\boldsymbol{x} = \mathbf{0}$ is a stable equilibrium point,

(b) $\boldsymbol{x} = \mathbf{0}$ is attractive, i.e. for all $t_0 \geqslant 0$, there exists $\delta(t_0)$, such that

$$\|\boldsymbol{x}_0\| < \delta \Rightarrow \lim_{t \to \infty} \|\boldsymbol{x}(t)\| = 0. \tag{I.4}$$

Definition I.1.9 (uniform asymptotic stability). Point $\boldsymbol{x} = \mathbf{0}$ is called a uniformly asymptotically stable equilibrium point of (I.1) if

(a) $\boldsymbol{x} = \mathbf{0}$ uniformly stable,

(b) trajectory $\boldsymbol{x}(t)$ converges to $\mathbf{0}$ uniformly in t_0.

The previous definitions are local, since they concern neighbourhoods of the equilibrium point. The following definition treats global asymptotic stability (GAS).

Definition I.1.10 (global asymptotic stability). Point $\boldsymbol{x} = \mathbf{0}$ is called a globally asymptotically stable equilibrium point of (I.1), if it is asymptotically stable and $\lim_{t \to \infty} \|\boldsymbol{x}(t)\| = 0$, for all $\boldsymbol{x}_0 \in \mathbb{R}^n$.

Global uniform asymptotic stability is defined likewise.

Definition I.1.11 (exponential stability). Point $\boldsymbol{x} = \mathbf{0}$ is called an exponentially stable equilibrium point of (I.1) if there exist m, $\alpha > 0$, such that the solution $\boldsymbol{x}(t)$ satisfies

$$\|\boldsymbol{x}(t)\| \leqslant m e^{-\alpha(t-t_0)} \|\boldsymbol{x}_0\| \tag{I.5}$$

for all $\boldsymbol{x}_0 \in B_h$, $t \geqslant t_0 \geqslant 0$. The constant α is called the rate of convergence.

By GAS we mean that the system is stable for any $\boldsymbol{x}_0 \in \mathbb{R}^n$. Exponential stability in any closed ball is similar except that m and α may be functions of h. Exponential stability is uniform with respect to t_0.

Appendix J

Deseasonalising Time Series

Forecasting methods which do not account for seasonal characteristics are likely to perform poorly when applied to time series that contain such components. We present three different methods of deseasonalising time series – each method being appropriate for different forms of seasonal behaviour.

The first method assumes that the seasonal component of the time series is constant. The series is written as

$$x_k = m_k + s_k + y_k, \tag{J.1}$$

where m_k is the trend component, s_k is the seasonal component and y_k is the random noise component. It is desirable to remove the trend and seasonal components and then, if the remaining random noise component is linear and stationary, it can be fitted by an ARIMA model. Denote the length of the time series by N and the period of the seasonal component by d. In the air pollutant series, for instance, two seasonal effects would be expected; one with a period of a day (24 hours), the other with period of a year ($365 \times 24 = 8760$ hours).

A moving average is applied to the series to eliminate the seasonal component. If d is even let $d = 2q$, otherwise let $d = 2q + 1$. Let

$$\begin{aligned} \hat{m}_k &= (0.5x_{k-q} + x_{k-q+1} + \cdots + x_{k+q-1} + 0.5x_{k+q}) && \text{if } d \text{ is even,} \\ \hat{m}_k &= (x_{k-q} + x_{k-q+1} + \cdots + x_{k+q-1} + x_{k+q}) && \text{if } d \text{ is odd.} \end{aligned}$$

For each hour, the seasonal component can be estimated by the average deviation

$$w_c = \frac{1}{\sum_{k=1}^{N} I((k-c) \bmod d = 0)} \sum_{k=1}^{N} I((k-c) \bmod d = 0)(x_k - \hat{m}_k),$$

where $c = 1, \ldots, d$ and $I(\cdot)$ is the indicator function, $I(x) = 1$ if x is true, otherwise $I(x) = 0$. To ensure the seasonal components s_c sum to zero, they are calculated as

$$\hat{s}_c = w_c - \frac{1}{d} \sum_{i=1}^{d} w_i$$

for $c = 1, \dots, d$. The series $\hat{s}_c$ is replicated approximately N/d times, i.e. until it is of the same length as x_k, and the deseasonalised series is defined to be

$$d_k = x_k - \hat{s}_k.$$

The trend of the deseasonalised series d_k can now be estimated simply with a polynomial fit via least squares.

The second method of deseasonalising differs from the previous one in that it is no longer assumed that the seasonality component is constant. This is known as SARIMA modelling, or seasonal ARIMA (Box and Jenkins 1976). In terms of a season with period 24 hours, it is reasonable to expect that the behaviour of the series at a particular time is closely related to the behaviour of the series at the same time during the previous day, as well as being related to the behaviour of the series during the hour immediately preceding. Thus, a multiplicative model is used, involving a component which is a standard ARIMA model with parameters p, d and q, and a seasonal component, given by

$$z_k = \sum_{i=1}^{P} a_i z_{k-i*S} + \sum_{i=1}^{Q} b_i \epsilon_{k-i*S} + \epsilon_k, \tag{J.2}$$

where $z_k = \nabla^D x_k$. This model is written as

$$(p, d, q) \times (P, D, Q)_S,$$

where S is the period of seasonality.

Finally, when it is suspected that there is an approximately sinusoidal seasonal behaviour present in the time series – a likely scenario in air pollutant measurements given the interaction between climate and air pollutants and the known result that temperature approximately follows a sine wave with period one year – the sinusoidal component can be removed directly. This is done via the following filter

$$y_k = (1 - aB + B^2)x_k, \tag{J.3}$$

where $a = 2\cos(2\pi/b)$ and $b = 24 \times 365$, the number of measurements taken in one year; B is the *backshift* operator, defined by

$$B^j(x_k) = x_{k-j}.$$

The result (J.3) follows from the use of the *Z-transform* to model the sine wave and find its inverse, details of which can be found in many signal processing texts (see, for example, Vaidyanathan and Mitra 1987).

References

Agarwal A and Mammone RJ 1993 Long-term memory for neural networks. In *Artificial Neural Networks for Speech and Vision* (ed. Mammone RJ), pp. 171–193. Chapman & Hall.

Aguirre LA 2000 A nonlinear dynamical approach to system identification. *IEEE Circuits Systems Soc. Newslett.* **1**(2), 10–23.

Ahmad S, Tesauro G and He Y 1990 Asymptotic convergence of backpropagation: numerical experiments. In *Advances in Neural Information Processing Systems 2* (ed. Touretzky DS), pp. 606–613. Morgan Kaufmann.

Alippi C and Piuri V 1996 Experimental neural networks for prediction and identification. *IEEE Trans. Instrument. Measurement* **45**, 670–676.

Al-Ruwaihi KM 1997 CMOS analogue neurone circuit with programmable activation functions utilising MOS transistors with optimised process/device parameters. *IEE Proc. Circuits Devices Systems* **144**, 318–322.

An PE, Brown M and Harris CJ 1994 Aspects of instantaneous on-line learning rules. In *Proc. Int. Conf. on Control*, vol. I, pp. 646–651.

Anderson BDO, Bitmead RR, Johnson Jr CR, Kokotovic PV, Kosut RL, Mareels IMY, Praly L and Riedle BD 1986 *Stability of Adaptive Systems: Passivity and Averaging Analysis.* MIT Press.

Apostol TM 1997 *Modular Functions and Dirichlet Series in Number Theory*, 2nd edn. Springer.

Arena P, Fortuna L, Muscato G and Xibilia MG 1998a *Neural Networks in Multidimensional Domains: Fundamentals and New Trends in Modelling and Control.* Springer.

Arena P, Fortuna L and Xibilia MG 1998b On the capability of neural networks. In *Proc. IEEE Int. Conf. on Circuits and Systems*, vol. 4, pp. 1271–1274.

Aristotle 1975 *Posterior Analytics* (translated in English by J Barnes). Clarendon Press, Oxford.

Attali JG and Pages G 1997 Approximations of functions by a multilayer perceptron: a new approach. *Neural Networks* **10**, 1069–1081.

Bailer-Jones CA, MacKay DJC and Whiters PJ 1998 A recurrent neural network for modelling dynamical systems. *Network: Computation in Neural Systems* **9**, 531–547.

Baldi P and Atiya AF 1994 How delays affect neural dynamics and learning. *IEEE Trans. Neural Networks* **5**, 612–621.

Baltersee J and Chambers JA 1998 Non-linear adaptive prediction of speech signals using a pipelined recurrent neural network. *IEEE Trans. Signal Processing* **46**, 2207–2216.

Barnett S and Storey C 1970 *Matrix Methods in Stability Theory.* Nelson.

Barron AR 1993 Universal approximation bounds for superpositions of a sigmoidal function. *IEEE Trans. Information Theory* **39**, 930–945.

Basaglia A, Fornaciari W and Salice F 1995 Behaviour-driven minimal implementation of digital ANNs. In *Proc. Int. Conf. on Neural Networks*, vol. 4, pp. 1644–1649.

Basar T and Bernhard P 1995 *H^∞-Optimal Control and Relaxed Minimax Design Problems.* Birkhäuser.

Bauer P, Mansour M and Duran J 1993 Stability of polynomials with time-varying coefficients. *IEEE Trans. Circuits and Systems. I. Fundamental Theory and Applications* **40**, 423–426.

Beiu V 1998 On Kolmogorov's superpositions and Boolean functions. In *Proc. 5th Brazilian Symp. on Neural Networks*, pp. 55–60.

Bellman R 1961 *Adaptive Control Processes: A Guided Tour.* Oxford University Press.

Bengio Y 1995 *Neural Networks for Speech and Sequence Recognition.* International Thomson Publishing.

Bengio Y, Simard P and Frasconi P 1994 Learning long-term dependencies with gradient descent is difficult. *IEEE Trans. Neural Networks* **5**, 157–166.

Benvenuto N and Piazza F 1992 On the complex backpropagation algorithm. *IEEE Trans. Signal Processing* **40**, 967–969.

Bershad NJ 1986 Analysis of the normalised LMS algorithm with Gaussian inputs. *IEEE Trans. Acoustics Speech Signal Processing ASSP-34*:793–806.

Bershad NJ, McLaughlin S and Cowan CFN 1990 Performance comparison of RLS and LMS algorithms for tracking a first order Markov communications channel. In *Proc. Int. Symp. on Circuits and Systems*, vol. 1, pp. 266–270.

Bershad NJ, Shynk JJ and Feintuch PL 1993a Statistical analysis of the single-layer backpropagation algorithm. Part I. Mean weight behaviour. *IEEE Trans. Signal Processing* **41**, 573–582.

Bershad NJ, Shynk JJ and Feintuch PL 1993b Statistical analysis of the single-layer backpropagation algorithm. Part II. MSE and classification performance. *IEEE Trans. Signal Processing* **41**, 583–591.

Beule D, Herzel H, Uhlmann E, Kruger J and Becker F 1999 Detecting nonlinearities in time series of machining processes. *Proc. American Control Conf.*, pp. 694–698.

Bharucha-Reid AT 1976 Fixed point theorems in probabilistic analysis. *Bull. Am. Math. Soc.* **82**, 641–657.

Billings SA 1980 Identification of nonlinear systems—a survey. *IEE Proc.* D **127**(6), 272–285.

Billings SA and Voon WSF 1986 Correlation based model validity tests for non-linear models. *Int. J. Control* **44**, 235–244.

Billings SA, Jamaluddin HB and Chen S 1992 Properties of neural networks with applications to modelling non-linear dynamical systems. *Int. J. Control* **55**, 193–224.

Birkett AN and Goubran RA 1997 Nonlinear adaptive filtering with FIR synapses and adaptive activation functions. In *Proc. Int. Conf. on Acoustics, Speech and Signal Processing (ICASSP'97)*, pp. 3321–3324.

Blondel VD and Tsitsiklis JN 2000 A survey of computational complexity results in systems and control. *Automatica* **30**, 1249–1274.

Box GEP and Jenkins GM 1970 *Time Series Analysis: Forecasting and Control.* Holden-Day.

Box GEP and Jenkins GM 1976 *Time Series Analysis: Forecasting and Control*, 2nd edn. Holden-Day.

Brockwell PJ and Davis RA 1991 *Time Series: Theory and Methods.* Springer, New York.

Burr DJ 1993 Artificial neural networks: a decade of progress. In *Artificial Neural Networks for Speech and Vision* (ed. Mammone RJ). Chapman & Hall.

Casdagli MC and Weigend AS 1993 Exploring the continuum between deterministic and stochastic modeling. In *Time Series Prediction: Forecasting the Future and Understanding the Past* (ed. Weigend AS and Gershenfeld NA), pp. 347–366. Addison Wesley.

Chambers JA, Sherliker W and Mandic DP 2000 A normalised gradient algorithm for an adaptive recurrent perceptron. In *Proc. Int. Conf. on Acoustics, Speech and Signal Processing (ICASSP-2000)*, vol. I, pp. 396–399.

Chang P-R and Hu J-T 1997 Optimal nonlinear adaptive prediction and modeling of MPEG video in ATM networks using pipelined recurrent neural networks. *IEEE J. Selected Areas Commun.* **15**, 1087–1100.

Chen C and Aihara K 1997 Chaos and asymptotical stability in discrete-time neural networks. *Physica* D **104**(3–4):286–325.

Chen S and Billings SA 1989 Representations of non-linear systems: the NARMAX model. *Int. J. Control* **49**, 1013–1032.

Chen T 1997 The past, present, and future of neural networks for signal processing. *IEEE Signal Processing Mag.* **14**(6), 28–48.

Chon KH and Cohen RJ 1997 Linear and nonlinear ARMA model parameter estimation using an artificial neural network. *IEEE Trans. Biomedical Engng* **44**(3), 168–174.

Chon KH, Holstein-Rathlou N-H, Marsh DJ and Marmarelis VZ 1998 Comparative nonlinear modeling of renal autoregulation in rats: Volterra approach versus artificial neural networks. *IEEE Trans. Neural Networks* **9**, 430–435.

Chon KH, Hoyer D, Armoundas AA, Holstein-Rathlou N-H and Marsh DJ 1999 Robust nonlinear autoregressive moving average model parameter estimation using stochastic recurrent artificial neural networks. *Ann. Biomed. Engng* **27**, 538–547.

Ciarlet PG 1989 *Introduction to numerical linear algebra and optimization.* Cambridge University Press.

Cichocki A and Unbehauen R 1993 *Neural Networks for Optimization and Signal Processing.* Wiley.

Clarke TL 1990 Generalization of neural networks to the complex plane. In *Proc. Int. Joint Conf. on Neural Networks (IJCNN90)*, vol. II, pp. 435–440.

Connor J, Atlas LE and Martin DR 1992 Recurrent networks and NARMA modeling. In *Advances in Neural Information Processing Systems 4* (ed. Moody JE, Hanson SJ and Lippmann RP), pp. 301–308. Morgan Kaufmann.

Connor JT, Martin RD and Atlas LE 1994 Recurrent neural networks and robust time series prediction. *IEEE Trans. Neural Networks* **5**, 240–254.

Cotter NE 1990 The Stone–Weierstrass theorem and its applications to neural networks. *IEEE Trans. Neural Networks* **1**, 290–295.

Cybenko G 1989 Approximation by superpositions of a sigmoidal function. *Math. Control Signals Systems* **2**, 303–314.

deFigueiredo RJP 1997 Optimal neural network realisations of nonlinear FIR and IIR filters. In *Proc. IEEE Int. Symp. on Circuits and Systems*, pp. 709–712.

Dennis Jr JE and Schnabel RB 1983 *Numerical Methods for Unconstrained Optimization and Nonlinear Equations.* Prentice-Hall Series in Computational Mathematics, 1983.

DeRusso PM, Roy RJ, Close CM and Desrochers AA 1998 *State Variables for Engineers*, 2nd edn. Wiley.

Devaney RL 1989 *An Introduction to Chaotic Dynamical Systems.* Addison-Wesley.

Devaney RL 1999 The Mandelbrot set and the Farey tree, and the Fibonacci sequence. *Am. Math. Mon.* **106**, 289–302.

Diaconis P and Freedman D 1999 Iterated random functions. *SIAM Rev.* **41**, 719–741.

Dillon RM and Manikopoulos CN 1991 Neural net nonlinear prediction for speech data. *Electron. Lett.* **27**(10), 824–826.

Dogaru R, Murgan AT, Ortmann S and Glesner M 1996 Searching for robust chaos in discrete time neural networks using weight space exploration. In *Proc. IEEE Int. Conf. on Neural Networks*, vol. 2, pp. 688–693.

Douglas SC 1994 A family of normalized LMS algorithms. *IEEE Signal Processing Lett.* **1**(3), 49–51.

Douglas SC 1997 Adaptive filters employing partial updates. *IEEE Trans. Circuits and Systems. II. Analog and Digital Signal Processing* **44**, 209–216.

Douglas SC and Pan W 1995 Exact expectation analysis of the LMS adaptive filter. *IEEE Trans. Signal Processing* **43**, 2863–2871.

Douglas SC and Rupp M 1997 *A posteriori* updates for adaptive filters. In *Conf. Record 31st Asilomar Conf. on Signals, Systems and Computers*, vol. 2, pp. 1641–1645.

Draye J-PS, Pavisic DA, Cheron GA and Libert GA 1996 Dynamic recurrent neural networks: A dynamical analysis. *IEEE Trans. Systems Man. Cybernetics. Part B. Cybernetics* **26**, 692–706.

Dreyfus G and Idan Y 1998 The canonical form of nonlinear discrete-time models. *Neural Computation* **10**, 133–136.

Dror G and Tsodyks M 2000 Chaos in neural networks with dynamics synapses. *Neurocomputing* **32–33**:365–370.

Duch W and Jankowski N 1999 Survey of neural transfer functions. *Neural Computing Surv.* **2**, 163–212.

Dumitras A and Lazarescu V 1997 On viewing the transform performed by a hidden layer in a feedforward ANN as a complex Möbius mapping. In *Proc. Int. Conf. on Neural Networks*, vol. 2, pp. 1148–1151.

Eberhart RC 1990 Standardization of neural network terminology. *IEEE Trans. Neural Networks* **1**, 244–245.

Elliott DL 1993 A better activation function for artificial neural networks. Tech. Rep. TR 93-8. Institute for Systems Research, University of Maryland.

Elman JL 1990 Finding structure in time. *Cognitive Sci.* **14**, 179–211.

Elsken T 1999 Smaller nets may perform better: special transfer functions. *Neural Networks* **12**, 627–645.

Ermentrout B 1998 Neural systems as spatio-temporal pattern-forming systems. *Rep. Prog. Phys.* **61**, 353–430.

Fiesler E 1994 Neural network formalization. *Computer Standards Interfaces* **16**(3), 231–239.

Fiesler E and Beale R (eds) 1997 *Handbook of Neural Computation*. Institute of Physics Publishing and Oxford University Press.

Forti M and Tesi A 1994 Conditions for global stability of some classes of nonsymmetric neural networks. In *Proc. 33rd Conf. on Decision and Control*, pp. 2488–2493.

Foxall R, Krcmar I, Cawley G, Dorling S and Mandic DP 2001 On nonlinear processing of air pollution time series. In *Proc. Int. Conf. on Artificial Neural Nets and Genetic Algorithms, ICANNGA-2001*, pp. 477–480.

Frasconi P, Gori M and Soda G 1992 Local feedback multilayered networks. *Neural Computation* **4**, 120–130.

Fukushima K 1975 Cognitron: a self-organized multilayered neural network. *Biol. Cybernet.* **23**, 121–136.

Funahashi K 1989 On the approximate realisation of continuous mappings by neural networks. *Neural Networks* **2**, 183–192.

Funahashi K and Nakamura Y 1993 Approximation of dynamical systems by continuous time recurrent neural networks. *Neural Networks* **6**, 801–806.

Gemon S, Bienenstock E and Doursat R 1992 Neural networks and the bias/variance dilemma. *Neural Computation* **4**(1), 1–58.

Gent CR and Sheppard CP 1992 Predicting time series by a fully connected neural network trained by backpropagation. *Computing Control Engng J.*, pp. 109–112.

Georgiou GM and Koutsougeras C 1992 Complex domain backpropagation. *IEEE Trans. Circuits and Systems. II. Analog and Digital Signal Processing* **39**, 330–334.

Gershenfeld NA and Weigend AS 1993 The future of time series: learning and understanding. In *Time Series Prediction: Forecasting the Future and Understanding the Past* (ed. Weigend AS and Gershenfeld NA), pp. 1–70. Addison-Wesley.

Gholkar VA 1990 Mean square convergence analysis of LMS algorithm. *Electron. Lett.* **26**, 1705–1706.

Gill PE, Murray W and Wright MH 1981 *Practical Optimization*. Academic Press, London.

Girosi F and Poggio T 1989a Networks and the best approximation property. Tech. Rep. TR 1164. Artificial Intelligence Laboratory, Massachusetts Institute of Technology.

Girosi F and Poggio T 1989b Representation properties of networks: Kolmogorov's theorem is irrelevant. *Neural Computation* **1**, 465–469.

Golub GH and Van Loan CF 1996 *Matrix Computation*, 3rd edn. The Johns Hopkins University Press.

Gorban AN and Wunsch II DC 1998 The general approximation theorem. In *Proc. IEEE World Congress on Computational Intelligence*, vol. 2, pp. 1271–1274.

Grossberg S 1974 Classical and instrumental learning by neural networks. *Prog. Theoret. Biol.* **3**, 51–141.

Guarnieri S, Piazza F and Uncini A 1999 Multilayered feedforward networks with adaptive spline activation function. *IEEE Trans. Neural Networks* **10**, 672–683.

Guo L and Ljung L 1995 Exponential stability of general tracking algorithms. *IEEE Trans. Automatic Control* **40**, 1376–1387.

Guo L, Ljung L and Wang G-J 1997 Necessary and sufficient conditions for stability of LMS. *IEEE Trans. Automatic Control* **42**, 761–770.

Haber R and Unbehauen H 1990 Structure identification of nonlinear dynamic systems—a survey on input/output approaches. *Automatica* **26**, 651–677.

Hakim NZ, Kaufman JJ, Cerf G and Meadows HE 1991 Volterra characterization of neural networks. In *Conf. Record 25th Asilomar Conf. on Signals, Systems and Computers*, vol. 2, pp. 1128–1132.

Han JY, Sayeh MR and Zhang J 1989 Convergence and limit points of neural network and its applications to pattern recognition. *IEEE Trans. Systems Man. Cybernetics* **19**, 1217–1222.

Hassibi B, Sayed AH and Kailath T 1996 H^∞ optimality of the LMS algorithm. *IEEE Trans. Signal Processing* **44**, 267–280.

Hayes M 1997 *Statistical Signal Processing*. Wiley.

Haykin S 1994 *Neural Networks: A Comprehensive Foundation*. Prentice-Hall.

Haykin S 1996a *Adaptive Filter Theory*, 3rd edn. Prentice-Hall.

Haykin S 1996b Neural networks expand SP's horizons. *IEEE Signal Processing Mag.* **13**(2), 24–49.

Haykin S 1999a Lessons on adaptive systems for signal processing, communications, and control. *IEEE Signal Processing Mag.* **16**(5), 39–48.

Haykin S 1999b *Neural Networks: A Comprehensive Foundation*, 2nd edn. Prentice-Hall.

Haykin S (ed.) 2000 *Unsupervised Adaptive Filtering*, vol. 1: *Blind Separation*. Wiley.

Haykin S and Li L 1995 Nonlinear adaptive prediction of nonstationary signals. *IEEE Trans. Signal Processing* **43**, 526–535.

Haykin S and Principe J 1998 Using neural networks to dynamically model chaotic events such as sea clutter. *IEEE Signal Processing Mag.* **15**(3), 66–81.

Haykin S, Sayed AH, Zeidler JR, Yee P and Wei PC 1997 Adaptive tracking of linear time-variant systems by extended RLS algorithms. *IEEE Trans. Signal Processing* **45**, 1118–1128.

Hebb DO 1949 *The Organization of Behaviour*. Wiley.

Hecht-Nielsen R 1987 Kolmogorov's mapping neural networks existence theorem. In *Proc. Int. Joint Conf. on Neural Networks*, vol. 3, pp. 11–14.

Helmke U and Williamson RC 1995 Neural networks, rational functions, and realisation theory. *Math. Control Signals Systems* **8**, 27–49.

Hertz J, Krogh A and Palmer RG 1991 *Introduction to the Theory of Neural Computation* Addison-Wesley.

Hilbert D 1901–1902 Mathematical problems. *Bull. Am. Math. Soc.* **8**, 437–479.

Hlavacek I and Krizek M 1998 *A posteriori* error estimates for three-dimensional axisymmetric elliptic problems. In *Finite Element Methods* (ed. Krizek M, Neittaanmaki P and Stenberg R), pp. 147–154. Marcel Dekker.

Hoehfeld M and Fahlman SE 1992 Learning with limited numerical precision using the cascade-correlation algorithm. *IEEE Trans. Neural Networks* **3**, 602–611.

Hopfield JJ and Tank DW 1985 'Neural' computation of decisions in optimisation problems. *Biol. Cybernetics* **52**, 141–152.

Horn RA and Johnson CA 1985 *Matrix Analysis*. Cambridge University Press.

Hornik K 1990 Approximation capabilities of multilayer feedforward networks. *Neural Networks* **4**, 251–257.

Hornik K 1993 Some new results on neural network approximation. *Neural Networks* **8**, 1069–1072.

Hornik K, Stinchcombe M and White H 1989 Multilayer feedforward networks are universal approximators. *Neural Networks* **2**, 359–366.

Hornik K, Stinchcombe M and White H 1990 Universal approximation of an unknown mapping and its derivatives using multilayer feedforward network. *Neural Networks* **3**, 551–560.

Hunt KJ, Sbarbaro D, Zbikowski R and Gawthrop PJ 1992 Neural networks for control systems—a survey. *Automatica* **28**, 1083–1112.

Hwang K and Briggs FA 1986 *Computer Architectures and Parallel Processing*. McGraw-Hill.

Ibnkahla M, Bershad NJ, Sombrin J and Castanie F 1998 Neural network modelling and identification of nonlinear channels with memory: algorithms, applications, and analytic models. *IEEE Trans. Signal Processing* **46**, 1208–1220.

Iiguni Y and Sakai H 1992 A real-time learning algorithm for a multilayered neural network based on the extended Kalman filter. *IEEE Trans. Signal Processing* **40**, 959–966.

Jacobs RA 1988 Increased rates of convergence through learning rate adaptation. *Neural Networks* **1**, 295–307.

Jenkins WK, Hull AW, Strait JC, Schnaufer BA and Li X 1996 *Advanced Concepts in Adaptive Signal Processing*. Kluwer.

Jin L and Gupta MM 1996 Globally asymptotical stability of discrete-time analog neural networks. *IEEE Trans. Neural Networks* **7**, 1024–1031.

Jin L, Nikiforuk PN and Gupta MM 1994 Absolute stability conditions for discrete-time recurrent neural networks. *IEEE Trans. Neural Networks* **5**, 954–964.

Jones LK 1990 Constructive approximations for neural networks by sigmoidal functions. *Proc. IEEE* **78**, 1586–1589.

Jordan MI 1986 Serial order: a parallel distributed processing approach. Tech. Rep. TR-8604, UC San Diego, Institute for Cognitive Science.

Jury EI 1978 Stability of multidimensional scalar and matrix polynomials. *Proc. IEEE* **66**, 1018–1047.

Kailath T 1980 *Linear Systems*. Prentice-Hall.

Kailath T, Sayed AH and Hassibi B 2000 *Linear Estimation*. Prentice-Hall.

Kainen PC, Kurkova V and Vogt A 1999 Approximation by neural networks is not continuous. *Neurocomputing* **29**, 47–56.

Kang HW, Cho YS and Youn DH 1998 Adaptive precompensation of Wiener systems. *IEEE Trans. Signal Processing* **46**, 2825–2829.

Katsuura H and Sprecher DA 1994 Computational aspects of Kolmogorov's superposition theorem. *Neural Networks* **115**, 455–461.

Kay SM 1993 *Fundamentals of Statistical Signal Processing: Estimation Theory*. Prentice-Hall International.

Khalaf AAM and Nakayama K 1998 A cascade form predictor of neural and FIR filters and its minimum size estimation based on nonlinearity analysis of time series. *IEICE Trans. Fundamentals* E **81**A(3), 364–373.

Khalaf AAM and Nakayama K 1999 A hybrid nonlinear predictor: analysis of learning process and predictability for noisy time series. *IEICE Trans. Fundamentals* E **82**A(8), 1420–1427.

Khotanzad A and Lu J 1990 Non-parametric prediction of AR processes using neural networks. In *Proc. Int. Conf. on Acoustics, Speech and Signal Processing (ICASSP-90)*, pp. 2551–2555.

Kim MS and Guest CC 1990 Modification of backpropagation networks for complex-valued signal processing in frequency domain. In *Proc. Int. Joint Conf. on Neural Networks (IJCNN90)*, vol. III, pp. 27–31.

Kim S-S 1998 Time-delay recurrent neural network for temporal correlations and prediction. *Neurocomputing* **20**, 253–263.

Kirkpatrick S, Gelatt CD and Vecchi MP 1983 Optimization by simulated annealing. *Science* **220**, 671–680.

Kohonen T 1982 Self-organized formation of topologically correct feature maps. *Biol. Cybernetics* **43**, 59–69.

Kolmogorov AN 1941 Interpolation and extrapolation von stationären zufäfolgen. *Bull. Acad. Sci. (Nauk)* **5**, 3–14.

Kolmogorov AN 1957 On the representation of continuous functions of several variables by superposition of continuous functions of one variable and addition. *Dokl. Akad. Nauk SSSR* **114**, 953–956.

Kosmatopoulos EB, Polycarpou MM, Christodoulou MA and Ioannou PA 1995 High-order neural network structures for identification of dynamical systems. *IEEE Trans. Neural Networks* **6**, 422–431.

Krcmar IR and Mandic DP 2001 A fully adaptive NNGD algorithm. Accepted for ICASSP-2001.

Krcmar IR, Bozic MM and Mandic DP 2000 Global asymptotic stability of RNNs with bipolar activation functions. In *Proc. 5th IEEE Seminar on Neural Networks Applications in Electrical Engineering (NEUREL2000)*, pp. 33–36.

Kreinovich V, Nguyen HT and Yam Y 2000 Fuzzy systems are universal approximators for a smooth function and its derivatives. *Int. J. Intellig. Syst.* **15**, 565–574.

Kreith K and Chakerian D 1999 *Iterative Algebra and Dynamic Modeling.* Springer.

Kruschke JK and Movellan JR 1991 Benefits of gain: speeded learning and minimal hidden layers in back-propagation networks. *IEEE Trans. Syst. Man. Cybernetics* **21**(1), 273–280.

Kuan C-M and Hornik K 1991 Convergence of learning algorithms with constant learning rates. *IEEE Trans. Neural Networks* **2**, 484–489.

Kung S-Y and Hwang J-N 1998 Neural networks for intelligent multimedia processing. *Proc. IEEE* **86**, 1244–1272.

Kurkova V 1992 Kolmogorov's theorem and multilayer neural networks. *Neural Networks* **5**, 501–506.

Kushner HJ 1984 *Approximation and Weak Convergence Methods for Random Processes with Applications to Stochastic System Theory.* MIT Press.

LaSalle JP 1986 *The Stability and Control of Discrete Processes.* Springer.

Leontaritis IJ and Billings SA 1985 Input–output parametric models for nonlinear systems. *Int. J. Control* **41**, 303–344.

Leshno M, Lin VY, Pinkus A and Schocken S 1993 Multilayer feedforward neural networks with a nonpolynomial activation function can approximate any function. *Neural Networks* **6**, 861–867.

Leung H and Haykin S 1991 The complex backpropagation algorithm. *IEEE Trans. Signal Processing* **39**, 2101–2104.

Leung H and Haykin S 1993 Rational neural networks. *Neural Computation* **5**, 928–938.

Li LK 1992 Approximation theory and recurrent networks. In *Proc. Int. Joint Conf. on Neural Networks*, vol. II, pp. 266–271.

Li L and Haykin S 1993 A cascaded neural networks for real-time nonlinear adaptive filtering. In *Proc. IEEE Int. Conf. on Neural Networks (ICNN'93)*, pp. 857–862.

Liang X-B and Yamaguchi T 1997 Necessary and sufficient conditions for absolute exponential stability of a class of nonsymmetric neural networks. *IEICE Trans. Inform. Syst.* E **80**D(8), 802–807.

Lin T, Horne BG, Tino P and Lee Giles C 1996 Learning long-term dependencies in NARX recurrent neural networks. *IEEE Trans. Neural Networks* **7**, 1329–1338.

Lin T-N, Lee Giles C, Horne BG and Kung S-Y 1997 A delay damage model selection algorithm for NARX neural networks. *IEEE Trans. Signal Processing* **45**, 2719–2730.

Lippmann RP 1987 An introduction of computing with neural nets. *IEEE Acoust. Speech Signal Processing Mag.* **4**(2), 4–22.

Lipschutz S 1965 *Theory and Problems of Probability.* Schaum's Outline Series. McGraw-Hill.

Ljung L 1984 Analysis of stochastic gradient algorithms for linear regression problems. *IEEE Trans. Information Theory* **IT-30**(2):151–160.

Ljung L and Sjöberg J 1992 A system identification perspective of neural nets. In *Proc. II IEEE Workshop on Neural Networks for Signal Processing (NNSP92)*, pp. 423–435.

Ljung L and Soderstrom T 1983 *Theory and Practice of Recursive Identification.* MIT Press.

Lo JT-H 1994 Synthetic approach to optimal filtering. *IEEE Trans. Neural Networks* **5**, 803–811.

Lorentz GG 1976 The 13th problem of Hilbert. In *Mathematical Developments Arising from Hilbert Problems* (ed. Browder FE). American Mathematical Society.

Lorenz EN 1963 Deterministic nonperiodic flow. *J. Atmos. Sci.* **20**, 130–141.

Luenberger DG 1969 *Optimization by Vector Space Methods.* Wiley.

Luh PB, Zhao X and Wang Y 1998 Lagrangian relaxation neural networks for job shop scheduling. In *Proc. 1998 IEEE Int. Conf. on Robotics and Automation*, pp. 1799–1804.

Maass W and Sontag ED 2000 Neural systems as nonlinear filters. *Neural Computation* **12**, 1743–1772.

McCulloch WS and Pitts W 1943 A logical calculus of the ideas immanent in nervous activity. *Bull. Math. Biophys.* **5**, 115–133.

McDonnell JR and Waagen D 1994 Evolving recurrent perceptrons for time-series modelling. *IEEE Trans. Neural Networks* **5**, 24–38.

Maiorov V and Pinkus A 1999 Lower bounds for approximation by MLP neural networks. *Neurocomputing* **25**, 81–91.

Makhoul J 1975 Linear prediction: a tutorial overview. *Proc. IEEE* **63**, 561–580.

Mandic DP 2000a NNGD algorithm for neural adaptive filters. *Electron. Lett.* **36**, 845–846.

Mandic DP 2000b On fixed points of a general neural network via Möbius transformations. In *Proc. 5th Int. Conf. on Mathematics in Signal Processing.*

Mandic DP 2000c The use of Möbius transformations in neural networks and signal processing. In *Proc. Xth IEEE Workshop on Neural Networks for Signal Processing (NNSP2000)*, pp. 185–194.

Mandic DP and Chambers JA 1998a Advanced PRNN based nonlinear prediction/System identification. In *Digest of the IEE Colloquium on Non-Linear Signal and Image Processing*, pp. 11/1–11/6.

Mandic DP and Chambers JA 1998b From an *a priori* RNN to an *a posteriori* PRNN nonlinear predictor. In *Proc. VIII IEEE Workshop on Neural Networks for Signal Processing (NNSP98)*, pp. 174–183.

Mandic DP and Chambers JA 1998c *A posteriori* real time recurrent learning schemes for a recurrent neural network based non-linear predictor. *IEE Proc. Vision Image Signal Processing* **145**, 365–370.

Mandic DP and Chambers JA 1999a Exploiting inherent relationships in RNN architectures. *Neural Networks* **12**, 1341–1345.

Mandic DP and Chambers JA 1999b Global asymptotic stability of nonlinear relaxation equations realised through a recurrent perceptron. In *Proc. Int. Conf. on Acoustics, Speech and Signal Processing (ICASSP-99)*, vol. 2, pp. 1037–1040.

Mandic DP and Chambers JA 1999c A nonlinear adaptive predictor realised via recurrent neural networks with annealing. In *Digest of the IEE Colloquium Statistical Signal Processing*, pp. 2/1–2/6.

Mandic DP and Chambers JA 1999d *A posteriori* error learning in nonlinear adaptive filters. *IEE Proc. Vision, Image and Signal Processing* **146**(6), 293–296.

Mandic DP and Chambers JA 1999e Relationship between the slope of the activation function and the learning rate for the RNN. *Neural Computation* **11**, 1069–1077.

Mandic DP and Chambers JA 1999f Toward an optimal PRNN based nolinear predictor. *IEEE Trans. Neural Networks* **10**, 1435–1442.

Mandic DP and Chambers JA 2000a Advanced RNN based NARMA predictors. *J. VLSI Signal Processing Syst. Signal Image Video Technol. 26*, 105–117.

Mandic DP and Chambers JA 2000b A normalised real time recurrent learning algorithm. *Signal Processing* **80**, 1909–1916.

Mandic DP and Chambers JA 2000c On robust stability of time-variant discrete-time nonlinear systems with bounded parameter perturbations. *IEEE Trans. Circuits Systems. I. Fundamental Theory and Applications* **47**, 185–188.

Mandic DP and Chambers JA 2000d On stability of relaxive systems described by polynomials with time-variant coefficients. *IEEE Trans. Circuits Systems. I. Fundamental Theory and Applications* **47**, 1534–1537.

Mandic DP and Chambers JA 2000e Relations between the *a priori* and *a posteriori* errors in nonlinear adaptive neural filters. *Neural Computation* **12**, 1285–1292.

Mandic DP and Chambers JA 2000f Towards an optimal learning rate for backpropagation. *Neural Processing Lett.* **11**(1), 1–5.

Mandic DP and Krcmar IR 2000 On training with slope adaptation for feedforward neural networks. In *Proc. 5th IEEE Seminar on Neural Networks Applications in Electrical Engineering (NEUREL2000)*, pp. 42–45.

Mandic DP and Krcmar IR 2001 Stability of the NNGD algorithm for nonlinear system identification. *Electron. Lett.* **37**, 200–202.

Mandic DP, Baltersee J and Chambers JA 1998 Nonlinear prediction of speech with a pipelined recurrent neural network and advanced learning algorithms. In *Signal Analysis and Prediction* (ed. Prochazka A, Uhlir J, Rayner PJW and Kingsbury NG), pp. 291–309. Birkhäuser, Boston.

Mandic DP, Chambers JA and Bozic MM 2000 On global asymptotic stability of fully connected recurrent neural networks. In *Proc. Int. Conf. on Acoustics, Speech and Signal Processing (ICASSP-2000)*, vol. VI, pp. 3406–3409.

Mane R 1981 On the dimension of the compact invariant sets of certain nonlinear maps. In *Dynamical Systems and Turbulence* (ed. Rand D and Young LS). Warwick 1980 Lecture Notes in Mathematics, vol. 898. Springer.

Marmarelis VZ 1989 Signal transformation and coding in neural systems. *IEEE Trans. Biomed. Engng* **36**(1), 15–24.

Mathews VJ 1991 Adaptive polynomial filters. *IEEE Signal Processing Mag.* **8**(3), 10–26.

Medler DA 1998 A brief history of connectionism. *Neural Computing Surveys* **1**, 61–101.

Mathews JH and Howell RW 1997 *Complex Analysis: for Mathematics and Engineering*, 3rd edn. Jones and Bartlett.

Mathews VJ and Xie Z 1993 A stochastic gradient adaptive filter with gradient adaptive step size. *IEEE Trans. Signal Processing* **41**, 2075–2087.

Mitaim S and Kosko B 1996 What is the best shape for a fuzzy set in function approximation? In *Proc. 5th IEEE Int. Conf. on Fuzzy Systems (FUZZ-96)*, pp. 542–560.

Mitaim S and Kosko B 1997 Adaptive joint fuzzy sets for function approximation. In *Proc. Int. Conf. on Neural Networks (ICNN-97)*, pp. 537–542.

Mhaskar HN and Micchelli C 1992 Approximation by superposition of sigmoidal and radial basis functions. *Adv. Appl. Math.* **13**, 350–373.

Minsky M and Papert SA 1969 *Perceptrons: An Introduction to Computational Geometry.* MIT Press.

Mozer MC 1993 Neural net architectures for temporal sequence processing. In *Time Series Prediction: Forecasting the Future and Understanding the Past* (ed. Weigend AS and Gershenfeld NA). Addison-Wesley.

Murtagh P and Tsoi AC 1992 Implementation issues of sigmoid function and its derivative for VLSI digital neural networks. *IEE Proc.* E **139**(3).

Nakagawa M 1996 An autonomously controlled chaos neural network. In *Proc. IEEE Int. Conf. on Neural Networks*, vol. 2, pp. 862–867.

Narendra KS 1996 Neural networks for control: theory and practice. *Proc. IEEE* **84**, 1385–1406.

Narendra KS and Parthasarathy K 1990 Identification and control of dynamical systems using neural networks. *IEEE Trans. Neural Networks* **1**, 4–27.

Nerrand O, Roussel-Ragot P, Personnaz L and Dreyfus G 1991 Neural network training schemes for non-linear adaptive filtering and modelling. In *Proc. Int. Joint Conf. on Neural Networks (IJCNN-91)*, vol. I, pp. 61–66.

Nerrand O, Roussel-Ragot P, Personnaz L and Dreyfus G 1993 Neural networks and nonlinear adaptive filtering: unifying concepts and new algorithms. *Neural Computation* **5**, 165–199.

Nerrand O, Roussel-Ragot P, Urbani D, Personnaz L and Dreyfus G 1994 Training recurrent neural networks: Why and how? An illustration in dynamical process modelling. *IEEE Trans. Neural Networks* **5**, 178–184.

Niranjan M and Kadirkamanathan V 1991 A nonlinear model for time series prediction and signal interpolation. In *Proc. Int. Conf. on Acoustics, Speech and Signal Processing (ICASSP-91)*, pp. 1713–1716.

Nitzberg R 1985 Application of the normalized LMS algorithm to MSLC. *IEEE Trans. Aerospace Electron. Syst.* **AES-21**(1):79–91.

Oppenheim AV, Buck JR and Schafer RW 1999 *Discrete-Time Signal Processing.* Prentice-Hall.

Papoulis A 1984 *Probability, Random Variables, and Stochastic Processes.* McGraw-Hill.

Pearson RK 1995 Nonlinear input/output modelling. *J. Process Control* **5**(4), 197–211.

Peng HC, Sha LF, Gan Q and Wei Y 1998 Combining adaptive sigmoid packet and trace neural networks for fast invariance learning. *Electron. Lett.* **34**, 898–899.

Personnaz L and Dreyfus G 1998 Comment on 'Discrete-time recurrent neural network architectures: a unifying review'. *Neurocomputing* **20**, 325–331.

Piazza F, Uncini A and Zenobi M 1992 Artificial neural networks with adaptive polynomial activation function. In *Proc. Int. Joint Conf. on Neural Networks*, vol. II, pp. 343–349.

Piazza F, Uncini A and Zenobi M 1993 Neural networks with digital LUT activation functions. In *Proc. 1993 Int. Joint Conf. on Neural Networks*, pp. 1401–1404.

Pineda FJ 1987 Generalization of backpropagation to recurrent neural networks. *Phys. Rev. Lett.* **59**, 2229–2232.

Pineda FJ 1989 Recurrent backpropagation and the dynamical approach to adaptive neural computation. *Neural Computation* **1**, 161–172.

Poddar P and Unninkrishnan KP 1991 Non-linear prediction of speech signals using memory neuron networks. In *Proc. IEEE Workshop NNSP I*, pp. 395–404.

Poggio T and Girosi F 1990 Networks for approximation and learning. *Proc. IEEE* **78**, 1481–1497.

Premaratne K and Mansour M 1995 Robust stability of time-variant discrete-time systems with bounded parameter perturbations. *IEEE Trans. Circuits and Systems. I. Fundamental Theory and Applications* **42**, 40–45.

Priestley MB 1991 *Non-Linear and Non-Stationary Time Series Analysis.* Academic Press, London.

Principe JC, deVries B and deOliveira PG 1993 The Gamma filter—a new class of adaptive IIR filters with restricted feedback. *IEEE Trans. Signal Processing* **41**, 649–656.

Principe JC, Euliano NR and Lefebvre WC 2000 *Neural and Adaptive Systems.* Wiley.

Puskoris GA and Feldkamp LA 1994 Neurocontrol of nonlinear dynamical systems with Kalman filter trained recurrent networks. *IEEE Trans. Neural Networks* **5**, 279–297.

Qin SZ, Su HT and Mc-Avoy TJ 1992 Comparison of four neural net learning methods for dynamic system identification. *IEEE Trans. Neural Networks* **3**, 122–130.

Rao DH and Gupta MM 1993 Dynamic neural units and function approximation. In *Proc. IEEE Int. Conf. on Neural Networks*, vol. 2, pp. 743–748.

Reed R 1993 Pruning algorithms—a survey. *IEEE Trans. Neural Networks* **4**, 740–747.

Regalia PA 1994 *Adaptive IIR Filtering in Signal Processing and Control.* Marcel Dekker.

Ridella S, Rovetta S and Zunino R 1997 Circular backpropagation networks for classification. *IEEE Trans. Neural Networks* **8**, 84–97.

Robinson AJ and Fallside F 1987 The utility driven dynamic error propagation network. Tech. Rep. CUED/F–INFENG/TR.1, Cambridge University Engineering Department.

Rose K 1998 Deterministic annealing for clustering, compression, classification, regression, and related optimization problems. *Proc. IEEE* **86**, 2210–2239.

Rosenblatt F 1958 The perceptron: a probabilistic model for information storage and organization in the brain. *Psychol. Rev.* **65**, 386–408.

Rosenblatt F 1962 *Principles of Neuro-Dynamics.* Washington, DC, Spartan.

Roy S and Shynk JJ 1989 Analysis of the data-reusing LMS algorithm. In *Proc. 32nd Midwest Symp. on Circuits and Systems*, vol. 2, pp. 1127–1130.

Ruck DW, Rogers SK, Kabrisky M, Maybeck PS and Oxley ME 1992 Comparative analysis of backpropagation and the extended Kalman filter for training multilayer perceptrons. *IEEE Trans. Pattern Analysis and Machine Intelligence* **14**, 686–691.

Rumelhart DE, Hinton GE and Williams R 1986 Learning internal representation by error propagation. *Nature* **323**, 533–536.

Sastry S and Bodson M 1989 *Adaptive Control: Stability, Convergence, and Robustness.* Prentice-Hall International.

Schetzen M 1981 Nonlinear system modelling based on the Wiener theory. *Proc. IEEE* **69**, 1557–1574.

Schnaufer BA and Jenkins WK 1993 New data-reusing LMS algorithms for improved convergence. In *Conf. Record 27th Asilomar Conf. on Signals and Systems*, vol. 2, pp. 1584–1588.

Sethares WA 1992 Adaptive algorithms with nonlinear data and error functions. *IEEE Trans. Signal Processing* **40**, 2199–2206.

Shadafan RS and Niranjan M 1993 A dynamic neural network architecture by sequential partitioning of the input space. In *Proc. IEEE Int. Conf. on Neural Networks*, pp. 226–231.

Sheu M-H, Wang J-F, Chen J-S, Suen A-N, Jeang Y-L and Lee J-Y 1992 A data-reuse architecture for gray-scale morphologic operations. *IEEE Trans. Circuits and Systems. II. Analog and Digital Signal Processing* **39**, 753–756.

Shynk J 1989 Adaptive IIR filtering. *IEEE Acoust. Speech and Signal Processing (ASSP) Mag.* **6**(2), 4–21.

Shynk JJ and Roy S 1990 Convergence properties and stationary points of a perceptron learning algorithm. *Proc. IEEE* **78**, 1599–1604.

Siegelmann H and Sontag ED 1995 On the computational power of neural networks. *J. Computat. Syst. Sci.* **45**, 132–150.

Siegelmann HT, Horne BG and Giles CL 1997 Computational capabilities of recurrent NARX neural networks. *IEEE Trans. Systems Man. Cybernetics. Part B. Cybernetics* **27**(2), 208–215.

Sjöberg J, Zhang Q, Ljung L, Benveniste A, Delyon B, Glorennec P-Y, Hjalmarsson H and Juditsky A 1995 Nonlinear black-box modelling in system identification: a unified overview. *Automatica* **31**, 1691–1724.

Slock DTM 1993 On the convergence properties of the LMS and the normalized LMS algorithms. *IEEE Trans. Signal Processing* **41**, 2811–2825.

Solo V and Kong X 1994 *Adaptive Signal Processing Algorithms: Stability and Performance.* Prentice-Hall.

Song M and Manry MT 1993 Conventional modeling of the multilayer perceptron using polynomial basis functions. *IEEE Trans. Neural Networks* **4**, 164–166.

Soria-Olivas E, Calpe-Maravilla J, Guerrero-Martinez JF Martinez-Sober M and Espi-Lopez J 1998 An easy demonstration of the optimum value of the adaptation constant in the LMS algorithm. *IEEE Trans. Education* **41**(1), 81.

Sprecher DA 1965 On the structure of continuous functions of several variables. *Trans. Am. Math. Soc.* **115**, 340–355.

Sprecher DA 1993 A universal mapping for Kolmogorov's superposition theorem. *Neural Networks* **6**, 1089–1094.

Stark H and Woods JW 1986 *Probability, Random Processes and Estimation Theory for Engineers.* Prentice-Hall.

Strogatz SH 1994 *Nonlinear Dynamics and Chaos With Applications to Physics, Biology, Chemistry, and Engineering.* Perseus Books.

Sum J, Leung C-S, Young GH and Kan W-K 1999 On the Kalman filtering method in neural network training and pruning. *IEEE Trans. Neural Networks* **10**, 161–166.

Szu H and Hartley R 1987 Nonconvex optimization by fast simulated annealing. *Proc. IEEE* **75**, 1538–1540.

Takens F 1981 On the numerical determination of the dimension of an attractor. In *Dynamical Systems and Turbulence* (ed. Rand D and Young LS). Warwick 1980 Lecture Notes in Mathematics, vol. 898. Springer.

Tanaka K 1996 Stability analysis of neural networks via Lyapunov approach. In *Proc. Int. Conf. on Neural Networks*, vol. 6, pp. 3192–3197.

Tarrab M and Feuer A 1988 Convergence and performance analysis of the normalized LMS algorithm with uncorrelated Gaussian data. *IEEE Trans. Information Theory* **34**, 680–691.

Theiler J, Lindsay PS and Rubin DM 1993 Detecting nonlinearity in data with long coherence times with internal delays. In *Time Series Prediction: Forecasting the Future and Understanding the Past* (ed. Weigend AS and Gershenfeld NA). Addison-Wesley.

Thimm G and Fiesler E 1997a High order and multilayer perceptron initialisation. *IEEE Trans. Neural Networks* **8**, 349–359.

Thimm G and Fiesler E 1997b Optimal setting of the weights, learning rate, and gain. Tech. Rep. IDIAP-RR-97-04, Institut Dalle Molle D'Intelligence Artificielle Perceptive, Martigny Valais, Switzerland.

Thimm G, Moerland P and Fiesler E The interchangeability of learning rate and gain in backpropagation neural networks. *Neural Computation* **8**, 451–460, 1996.

Tikhonov AN, Leonov AS and Yagola AG 1998 *Nonlinear ill-posed problems.* Applied Mathematics and Mathematical Computation. Chapman & Hall, London.

Townshend B 1991 Nonlinear prediction of speech. In *Proc. Int. Conf. on Acoustics, Speech and Signal Processing (ICASSP-91)*, pp. 425–428.

Treichler JR, Johnson Jr CR and Larimore MG 1987 *Theory and Design of Adaptive Filters.* Wiley.

Tsoi CA and Back A 1997 Discrete time neural network architectures: a unifying review. *Neurocomputing* **15**(3), 183–223.

Vaidyanathan PP and Mitra SK 1987 A unified structural interpretation of some well-known stability-test procedures for linear systems. *Proc. IEEE* **75**, 478–497.

Vecci L, Campolucci P, Piazza F and Uncini A 1997 Approximation capabilities of adaptive spline neural neworks. In *Proc. Int. Conf. on Neural Networks*, vol. 1, pp. 260–265.

Vitushkin AG 1954 On Hilbet's thirteenth problem. *Dokl. Akad. Nauk SSSR* **95**, 701–704.

Waibel A, Hanazawa T, Hinton G, Shikano K and Lang K 1989 Phoneme recognition using time-delay neural networks. *IEEE Trans. Acoustics, Speech and Signal Processing* **37**, 328–339.

Wallace WA 1992 *Galileo's Logical Treatises (A Translation, with Notes and Commentary, of His Appropriated Latin Questions on Aristotle's Posteriori Analytics)*. Kluwer.

Wan EA 1993 Time series prediction by using a connectionist network with internal delay lines. In *Time Series Prediction: Forecasting the Future and Understanding the Past* (ed. Weigend AS and Gershenfeld NA). Addison-Wesley.

Wang K and Michel AN 1994 On Lyapunov stability of a family of nonlinear time-varying systems. In *Proc. 33rd Conf. on Decision and Control*, pp. 2131–2136.

Weigend AS and Gershenfeld NA (eds) 1994 *Time Series Prediction: Forecasting the Future and Understanding the Past*. Santa Fe Institute Studies in the Sciences of Complexity. Addison-Wesley.

Werbos P 1974 Beyond Regression: New Tools for Prediction and Analysis in the Behavioral Science. PhD thesis, Harvard University, Cambridge, MA.

Werbos P 1990 Backpropagation through time: what it does and how to do it. *Proc. IEEE* **78**, 1550–1560.

Widrow B and Hoff ME 1960 Adaptive switching circuits. In *Proc. WESCON Convention*, vol. IV, pp. 96–104.

Widrow B and Lehr MA 1990 30 years of adaptive neural networks: perceptron, madaline, and backpropagation. *Proc. IEEE* **78**, 1415–1442.

Widrow B and Stearns SD 1985 *Adaptive Signal Processing*. Prentice-Hall.

Wiener N 1949 *The Extrapolation, Interpolation and Smoothing of Stationary Time Series with Engineering Applications*. Wiley.

Williams RJ 1992 Training recurrent networks using the extended Kalman filter. In *Proc. Int. Joint Conf. on Neural Networks (IJCNN'92)*, vol. IV, pp. 241–246.

Williams R and Zipser D 1989a A learning algorithm for continually running fully recurrent neural networks. *Neural Computation* **1**, 270–280.

Williams RJ and Zipser D 1989b Experimental analysis of the real-time recurrent learning algorithm. *Connection Sci.* **1**(1), 87–111.

Williams RJ and Zipser D 1995 Gradient-based algorithms for recurrent networks and their computational complexity. In *Backpropagation: Theory, Architectures, and Applications* (ed. Chauvin Y and Rumelhart DE). Lawrence Erlbaum Associates.

Williamson RC and Helmke U 1995 Existence and uniqueness results for neural network approximations. *IEEE Trans. Neural Networks* **6**, 2–13.

Wilkinson JH 1965 *The Algebraic Eigenvalue Problem*. Oxford University Press.

Wu L and Niranjan M 1994 On the design of nonlinear speech predictors with recurrent nets. In *Proc. IEEE Int. Conf. on Acoustics, Speech and Signal Processing (ICASSP-94)*, pp. 529–532.

Wu S-I 1995 Mirroring our thought processes. *IEEE Potentials* **14**, 36–41.

Yang J, Ahmadi M, Jullien GA and Miller WC 1998 Model validation and determination for neural network activation function modelling. In *Proc. 1998 Midwest Symp. on Circuits and Systems*, pp. 548–551.

Yee P and Haykin S 1999 A dynamic regularized radial basis function network for nonlinear, nostationary time series prediction. *IEEE Trans. Signal Processing* **47**, 2503–2521.

Yule GU 1927 On a method of investigating periodicities in disturbed series,with special reference to Wölfer's sunspot numbers. *Phil. Trans. R. Soc. Lond.* A **226**, 267–298.

Zakeri S 1996 On critical points of proper holomorphic maps on the unit disc. *Bull. Lond. Math. Soc.* **30**, 62–66.

Zeidler E 1986 *Nonlinear Functional Analysis and its Applications*, vol. 1: *Fixed-Point Theorems*. Springer.

Zhang J, Walter GG, Miao Y and Lee WNW 1995 Wavelet neural networks for function learning. *IEEE Trans. Signal Processing* **43**, 1485–1497.

Zhang S and Constantinides AG 1992 Lagrange programming neural networks. *IEEE Trans. Circuits and Systems. II. Analog and Digital Signal Processing* **37**, 441–452.

Zhang Z and Sarhadi M 1993 A modified neuron activation function which enables single layer perceptrons to solve some linearly inseparable problems. In *Proc. 1993 Int. Conf. on Neural Networks*, pp. 2723–2726.

Zhang Z and Sarhadi M 1993 Performance aspects of a novel neuron activation function in multilayer feedforward networks. In *Proc. 1993 Int. Joint Conf. on Neural Networks*, pp. 2727–2730.

Zurada JM and Shen W 1990 Sufficient condition for convergence of a relaxation algorithm in actual single-layer neural networks. *IEEE Trans. Neural Networks* **1**, 300–303.

Index

Printed and bound by CPI Group (UK) Ltd, Croydon, CR0 4YY

29/03/2023

03206572-0002